Unconventional Anticancer Metallodrugs and Strategies to Improve their Pharmacological Profile

Unconventional Anticancer Metallodrugs and Strategies to Improve their Pharmacological Profile

Special Issue Editor

Maria Contel

MDPI • Basel • Beijing • Wuhan • Barcelona • Belgrade

Special Issue Editor
Maria Contel
The City University of New York
USA

Editorial Office
MDPI
St. Alban-Anlage 66
4052 Basel, Switzerland

This is a reprint of articles from the Special Issue published online in the open access journal *Inorganics* (ISSN 2304-6740) from 2018 to 2019 (available at: https://www.mdpi.com/journal/inorganics/special_issues/Unconventional_Anticancer_Metallodrugs)

For citation purposes, cite each article independently as indicated on the article page online and as indicated below:

LastName, A.A.; LastName, B.B.; LastName, C.C. Article Title. *Journal Name* **Year**, *Article Number*, Page Range.

ISBN 978-3-03921-315-3 (Pbk)
ISBN 978-3-03921-316-0 (PDF)

Contents

About the Special Issue Editor

Maria Contel, Professor of Chemistry, received her PhD from the Public University of Navarra in Spain in 1996. She was a postdoctoral fellow at the Australian National University (1997–1999) and at the University of Utrecht (1999–2000). In 2001 she was awarded a Ramón y Cajal Senior Researcher Fellowship, the most prestigious fellowship for young investigators in Spain, at the University of Zaragoza. She joined Brooklyn College as an Assistant Professor (inorganic chemistry) in September 2006. She became a faculty member of the Graduate Center in the doctoral programs of Chemistry (2006), Biology (2014) and Biochemistry (2018). She was promoted to Associate Professor in 2011, was made a Tow Professor in 2015 and in August 2016 she became a full Professor. She has been a member of the University of Hawaii Cancer Center since 2014 and is currently the Chairperson of the Chemistry Department at Brooklyn College. Her research interests lie in the field of inorganic/organometallic chemistry and, more specifically, medicinal chemistry (anticancer and antimicrobial chemotherapeutics) and homogeneous catalysis (green chemistry).

 *inorganics*

Editorial

Unconventional Anticancer Metallodrugs and Strategies to Improve Their Pharmacological Profile

María Contel [1,2,3,4]

1   Department of Chemistry, Brooklyn College, The City University of New York, Brooklyn, NY 11210, USA;
    mariacontel@brooklyn.cuny.edu; Tel.: +1-718-951-5000 (ext. 2833)
2   Biology PhD Program, The Graduate Center, The City University of New York, New York, NY 10016, USA
3   Biochemistry PhD Program, The Graduate Center, The City University of New York,
    New York, NY 10016, USA
4   Chemistry PhD Program, The Graduate Center, The City University of New York, New York, NY 10016, USA

Received: 3 July 2019; Accepted: 6 July 2019; Published: 10 July 2019

For the past 41 years, metal-based drugs have been widely used for the treatment of cancer. Cisplatin and follow-up drugs carboplatin (Paraplatin™) and oxaliplatin (Eloxatin™) have been the gold standard for metallodrugs as antineoplastic agents in clinical settings. Although effective, these drugs, either alone or in combination therapy, have faced a number of clinical challenges resulting from their limited spectrum of activity, high toxicity producing significant side effects, resistance, poor water solubility, low bioavailability, and short circulating time. In the past two decades, various unconventional non-platinum metal-based agents have emerged as potential alternatives for cancer treatment. These compounds are highly effective and selective in cancers resistant to cisplatin and other chemotherapeutic agents. Research in this area has recently intensified with a relevant number of patents and clinical trials, in addition to reports in scientific journals including some excellent reviews and books published in 2018–2019 [1–6]. Some recent highlights include ongoing clinical trials with gold auranofin for the treatment of small and non-small lung cancer and high-grade ovarian, fallopian tube, and peritoneal cancer [7,8], as well as upcoming clinical trials with copper derivatives in metastatic pancreas cancer [9], and phase II clinical trials with a ruthenium-based photodynamic compound (TLD-1433) for non-muscle invasive bladder cancer [10]. In parallel to the synthesis of coordination and organometallic compounds comprising different metals and unconventional platinum-based derivatives, researchers have also worked on optimizing the mechanistic and pharmacological features of promising drug candidates [1,2]. This Special Issue is devoted to some of the latest advances in anticancer metallodrugs with a focus on unconventional anticancer agents, as well as novel activation, targeting, and delivery strategies aimed at improving their pharmacological profile. Twelve medicinal inorganic chemistry groups from different countries have provided contributions to this Special Issue.

Three groups contributed with superb and well-organized reviews. In the area of unconventional anticancer agents, Romero-Canelón et al. completed an overview on key ruthenium drug candidates and the knowledge acquired during the past two decades with the aim of discussing ideas to optimize their chemical design by incorporating new concepts [11]. Tinoco et al. contributed a review on the significant roles that copper and iron play in the molecular pathways involved in cell proliferation and metastasis, and the evaluation of selected chelators for these metals showing promise as anticancer drugs [12]. Efforts to optimize the pharmacological profile (cellular delivery, efficacy, and tumor responsiveness) of these chelators as well as a description of analytical tools used to quantify the metal levels and to track the metals intracellularly are described [12]. Lastly, Mieszawska et al. contributed a timely comprehensive review on the use of nano-based systems and biomacromolecules as carriers to facilitate the in vivo application of metal-based drugs (solubility, bioavailability, and delivery to tumor tissues). This review focuses on complexes comprising platinum, ruthenium, copper, and iron [13].

This Special Issue also contains nine original research articles. Seven of these articles focus on unconventional metal-based agents with promising anticancer activity and/or their interactions with relevant cancer biomolecular targets. Komeda et al. report on dinuclear platinum(II) complexes containing ammonia and a bridge ligand between the platinum(II) centers consisting of a tetrazolate moiety with lipophilic substituents in the C5 position [14]. The authors describe the interactions of these complexes with β-cyclodextrin and its positive influence on the in vitro and in vivo activity of the dinuclear platinum(II) complexes in colorectal cancer cells and tumors [14]. Gómez-et al. describe the synthesis and cytotoxicity of new *trans*-platinum complexes containing iodido and amine ligands, and their chemical behavior in solution and reactivity towards biomolecules [15]. They found a beneficial effect (increased reactivity towards model nucleobase 5′-GMP) when exposed to UVA irradiation. Density functional theory (DFT) calculations for these compound, and comparisons of reactivity and biological activity with other iodide platinum(II) derivatives are also included in this article [15].

This Special Issue also collects reports on the synthesis and anticancer properties of compounds containing metals other than platinum [16–20]. Navarro et al. report cell viability assays on selected human cancer cell lines of cationic ruthenium(II) compounds based on *p*-cymene, triphenylphosphine, and biologically active clotrimazole and ketoconazole as ligands [16]. Preliminary studies on the cell cycle and mechanism of cell death as well as the promising anti-migration activity of a selected compound with clotrimazole on a triple negative breast tumor cancer cell line were reported [16]. Da Costa Ferreira et al. contributed to this issue with a report on new copper(II) and zinc(II) complexes containing new oxindolimine ligands [17]. The cytotoxicity of these compounds against hepatocellular carcinoma and neuroblastoma cancer cell lines, as well as their reactivity toward Calf Thymus DNA and human serum albumin, was investigated. The main conclusion is a confirmation of DNA as an important target for these compounds and an indication that oxidative damage is not the leading mechanism of cell death [17]. Three other leading medicinal inorganic chemistry groups contributed original articles on gold compounds [18–20]. Gimeno et al. describe the excellent cytotoxicity observed in several cancer cell lines by neutral gold(I) compounds containing biologically relevant thiolates and a new phosphine ligand bearing a thiophene molecule [18]. Farrell and Beaton report novel cationic gold(III) compounds containing the 1-methylcytosine ligand and chelating diamines for greater specificity toward biomolecules, with the ultimate goal of avoiding undesirable nonselective interactions and providing a better understanding of the speciation [19]. They describe the interactions of these compounds with models for the HIV nucleocapsid protein NCp7. More specifically, the authors report the affinity of the gold(III) complexes with the "essential" tryptophan of the C-terminal zinc finger motif of NCp7 by fluorescence and ^{1}H NMR spectroscopy, and included results on the specifics of this interaction by circular dichroism spectroscopy and electrospray-ionization mass spectrometry. A nearly immediate interaction with the apopeptide and indications of reactions via a charge transfer mechanism is described for the first time [19]. Casini et al. present findings on the synthesis and characterization of a series of cationic and neutral gold(III) compounds featuring a pyridine-benzimidazole scaffold [20]. The potent and selective inhibition of the membrane water and glycerol channels aquaporins (aquaglyceroporin, AQP3) in human red blood cells (hRBC) and a higher activity of the neutral compounds on melanoma A375 cells with marked membrane level expression of AQP3 are described. The potential of these compounds in the development of chemical probes to study the function of this protein isoform in biological systems is also highlighted [20].

This Special Issue contains a relevant research article by Meier-Menches et al. on the development and validation of liquid-chromatography-based methods to assess the lipophilicity of cytotoxic platinum(IV) complexes [21], which is of interest to the medicinal inorganic chemistry community due to: (1) the current availability of high-performance liquid chromatography (HPLC) instruments in research laboratories, and (2) the potential of obtaining chromatographic lipophilicity parameters (φ_0 that can be interconverted to Log P and Log Kw) for other metal-based compounds [21].

Lastly, Salassa et al. provide a contribution on functionalized upconverting nanoparticles (UCNPs) with bone-targeting phosphonate ligands for imaging purposes [22]. The authors report the synthesis and characterization of a new series of phosphonate-functionalized NaGdF4:YbmEr UCNPs that show affinity for hydroxyapatite, which is the inorganic constituent of bones, and discuss their potential as bone targeting multimodal (MRI/PET) imaging agents. In vivo biodistribution studies of ^{18}F-labbeled functionalized UCPNs in rats revealed the favored accumulation of nanoparticles in bones over time [22].

I truly hope that the readers find the open access format articles in this Special Issue timely and relevant, and that the Issue contributes to increasing awareness about the real potential of optimized metal-based drugs as competitive anticancer agents. The inorganic medicinal community has demonstrated that the assumption that all anticancer metallodrugs behave as cisplatin and related platinum-based compounds in terms of spectrum of activity and selectivity is no longer valid.

Finally, I want to thank all the authors for their excellent and diverse contributions to this Special Issue as well as the participating reviewers for their high quality suggestions and evaluations of the articles submitted. Lastly, this Special Issue would not have been possible without the constant dedication, support, and patience of the members of the editorial staff of *Inorganics* from the beginning to the end of the process. I am very grateful to them.

References

1. Casini, A.; Vessières, A.; Meier-Menches, M. (Eds.) *Metal-Based Anticancer Agents*, 1st ed.; Metallobiology Series No 14; Royal Society of Chemistry: Cambridge, UK, 2019.
2. Sigel, A.; Sigel, H.; Freisinger, E.; Sigel, R.K.O. (Eds.) *Metallodrugs: Development and Action of Anticancer Agents*; Metal Ions in Life Sciences Series No 18; Walter de Gruyter GmbH: Berlin, Germany, 2018.
3. Engliner, B.; Pirker, C.; Heffeter, P.; Terenzi, A.; Kowol, C.R.; Keppler, B.K.; Berger, W. Metal Drugs and the Anticancer Immune Response. *Chem. Rev.* **2019**, *119*, 1519–1624. [CrossRef] [PubMed]
4. Monro, S.; Colon, K.L.; Yin, H.; Roque, J.; Konda, P.; Gujar, S.; Thummel, R.P.; Lilge, L.; Cameron, C.G.; McFarland, S.A. Transition Metal Complexes and Photodynamic Therapy from a Tumor-Centered Approach: Challenges, Opportunities, and Highlights from the Development of TLD1433. *Chem. Rev.* **2019**, *119*, 797–828. [CrossRef] [PubMed]
5. Kenny, R.G.; Marmion, C. Toward Multi-Targeted Platinum and Ruthenium Drugs—A New Paradigm in Cancer Drug Treatment Regimens? *Chem. Rev.* **2019**, *119*, 1058–1137. [CrossRef] [PubMed]
6. Wang, X.; Wang, X.; Jin, S.; Muhammad, N.; Guo, Z. Stimuli-Responsive Therapeutic Metallodrugs. *Chem. Rev.* **2019**, *119*, 1138–1192. [CrossRef] [PubMed]
7. Sirolimus and Auranofin in Treating Patients with Advanced or Recurrent Non-Small Cell Lung Cancer or Small Cell Lung Cancer. Available online: https://clinicaltrials.gov/ct2/show/NCT01737502 (accessed on 1 July 2109).
8. Auranofin in Treating Patients with Recurrent Epithelial Ovarian, Primary Peritoneal, or Fallopian Tube Cancer'. Available online: https://clinicaltrials.gov/ct2/show/NCT01747798 (accessed on 1 July 2109).
9. Disulfiram-Copper Gluconate in Met Pancreas Cancer w Rising CA19-9 on Abraxane-Gemzar, FOLFIRINOX or Gemcitabine. Available online: https://clinicaltrials.gov/ct2/show/NCT03714555 (accessed on 1 July 2109).
10. Health Canada Grants ITA Approval to Commence Phase II Clinical Study. Available online: https://theralase.com/pressrelease/health-canada-grants-ita-approval-to-commence-phase-ii-clinical-study/ (accessed on 1 July 2109).
11. Coverdale, J.P.C.; Laroiya-McCarron, T.; Romero-Canelón, I. Designing Ruthenium Anticancer Drug: What Have We Learnt from the Key Drug Candidates? *Inorganics* **2019**, *7*, 31. [CrossRef]
12. Gaur, K.; Vázquez-Salgado, A.M.; Duran-Camacho, G.; Domínguez-Martínez, I.; Benjamin-Rivera, J.A.; Fernández-Vega, L.; Carmona Sarabia, L.; Cruz Gracia, A.; Pérez-Deliz, F.; Méndez Román, J.A.; et al. Iron and Copper Intracellular Chelation as an Anticancer Drug Strategy. *Inorganics* **2018**, *6*, 126. [CrossRef]
13. Poursharifi, M.; Wlodarczyk, M.T.; Mieszawska, A.J. Nano-Based Systems and Biomecremules as Carriers for Metallodrugs in Anticancer Therapy. *Inorganics* **2019**, *7*, 2. [CrossRef]

14. Komeda, S.; Uemura, M.; Yoneyama, H.; Harusawa, S.; Hiramoto, K. In Vitro Cytotoxicity and In Vivo Antitumor Efficacy of Tetrazolato-Bridged Dinuclear Platinum(II) Complexes with a Bulky Substituent at Tetrazole C5. *Inorganics* **2019**, *7*, 5. [CrossRef]

15. Cubo, L.; Parro, T.; Carnero, A.; Salassa, L.; Matesanz, A.I.; Quiroga, A.G. Synthesis, Reactivity Studies, and Cytotoxicity of Two trans-Iodidoplatinum(II) Complexes. Does Photoactivation Work? *Inorganics* **2018**, *6*, 127. [CrossRef]

16. Colina-Vega, L.; Oliveira, K.M.; Cunha, B.N.; Cominetti, M.R.; Navarro, M.; Azevedo Batista, A. Anti-Proliferative and Anti-Migration Activity of Arene-Ruthenium(II) Comeplex with Azole Therapeutic Agents. *Inorganics* **2018**, *6*, 132. [CrossRef]

17. Caviccioli, M.; Monteiro Lino Zaballa, A.; de Paula, Q.A.; Bach Prieto, M.; Columbano Oliveira, C.; Civitareale, P.; Ciriolo, M.R.; Da Costa Ferreira, A.M. Oxidative Assets Toward Biomolecules and Cytotoxicity of New Oxindolimine-Copper(II) and Zinc(II) Complexes. *Inorganics* **2019**, *7*, 12. [CrossRef]

18. Goitia, H.; Villacampa, M.D.; Laguna, A.; Gimeno, M.C. Cytotoxic Gold(I) Complexes with Amidophosphine Ligands Containing Thiophene Moieties. *Inorganics* **2019**, *7*, 13. [CrossRef]

19. Beaton, J.; Farrell, N.P. Investigation of 1-Methylcytosine as a Ligand in Gold(III) Complexes: Synthesis and Protein Interactions. *Inorganics* **2019**, *7*, 1. [CrossRef]

20. Aikman, B.; Wenzel, M.N.; Mósca, A.F.; de Almeida, A.; Klooster, W.T.; Coles, S.J.; Soveral, G.; Casini, A. Gold(III) Pyridine-Benzimidazole Complexes as Aquaglyceroporin Inhibitors and Antiproliferative Agents. *Inorganics* **2018**, *6*, 123. [CrossRef]

21. Klose, M.H.M.; Theiner, S.; Varbanov, H.P.; Hoefer, D.; Oichler, V.; Galanski, M.; Meier-Menches, S.M.; Keppler, B.K. Development and Validation of Liquid Chromatography-Based Methods to Assess the Lipophilicity of Cytotoxic Platinum(IV) Complexes. *Inorganics* **2018**, *6*, 130. [CrossRef]

22. Alonso-de Castro, S.; Ruggiero, E.; Lekuona Fernández, A.; Cossío, U.; Baz, Z.; Otaegui, D.; Gómez-Vallejo, V.; Padro, D.; Llop, J.; Salassa, L. Functionalizing NaGdF4:Yb,Er Upconverting Nanoparticles with Bone-Targeting Phosphonate Ligands: Imaging and In Vivo Biodistribution. *Inorganics* **2019**, *7*, 60. [CrossRef]

 *inorganics*

Article

Gold(III) Pyridine-Benzimidazole Complexes as Aquaglyceroporin Inhibitors and Antiproliferative Agents

Brech Aikman [1,†], Margot N. Wenzel [1,†], Andreia F. Mósca [2,†], Andreia de Almeida [1,3], Wim T. Klooster [4], Simon J. Coles [4], Graça Soveral [2,*] and Angela Casini [1,*]

[1] School of Chemistry, Cardiff University, Main Building, Park Place, Cardiff CF10 3AT, UK; AikmanB@cardiff.ac.uk (B.A.); WenzelM3@cardiff.ac.uk (M.N.W.); dealmeidaa@cardiff.ac.uk (A.d.A.)

[2] Research Institute for Medicines (iMed.ULisboa), Faculty of Pharmacy, Universidade de Lisboa, 1649-003 Lisboa, Portugal; andreiafbm@ff.ulisboa.pt

[3] Tumour MicroEnvironment Group, Division of Cancer and Genetics, School of Medicine, Cardiff University, Tenovus Building, Cardiff CF14 4XN, UK

[4] School of Chemistry, University of Southampton, Southampton SO17 1BJ, UK; W.T.Klooster@soton.ac.uk (W.T.K.); S.J.Coles@soton.ac.uk (S.J.C.)

[*] Correspondence: gsoveral@ff.ulisboa.pt (G.S.); casinia@cardiff.ac.uk (A.C.); Tel.: +351-217946461 (G.S.); +44-29-2087-6364 (A.C.)

[†] These authors contributed equally to this work.

Received: 11 October 2018; Accepted: 15 November 2018; Published: 20 November 2018

Abstract: Gold compounds have been proven to be novel and versatile tools for biological applications, including as anticancer agents. Recently, we explored the potential of Au(III) complexes with bi-dentate N-donor ligands as inhibitors of the membrane water and glycerol channels aquaporins (AQPs), involved in different physiological and pathophysiological pathways. Here, eight new Au(III) complexes featuring a pyridine-benzimidazole scaffold have been synthesized and characterized via different methods. The stability of all the compounds in aqueous solution and their reactivity with glutathione have been investigated by UV–visible spectroscopy. The Au(III) compounds, tested for their AQPs inhibition properties in human Red Blood Cells (hRBC), are potent and selective inhibitors of AQP3. Furthermore, the compounds' antiproliferative effects have been studied in a small panel of human cancer cells expressing AQP3. The complexes show only very moderate anticancer effects in vitro and are mostly active against the melanoma A375 cells, with marked expression of AQP3 at the level of the nuclear membrane. In general, the AQP3 inhibition properties of these complexes hold promises to develop them as chemical probes to study the function of this protein isoform in biological systems.

Keywords: Gold(III) complexes; pyridine benzimidazole; aquaporins; cancer; stopped-flow spectroscopy; antiproliferative activity

---

1. Introduction

The severe side effects associated with chemotherapy necessitate the development of improved anticancer therapies. Specifically, the discovery of compounds that can disrupt cancerous cellular machinery by novel mechanisms of action is nowadays the focus of intense research. For example, metal-based compounds acting via the interaction with proteins and secondary DNA structures, as well as by alteration of the intracellular redox balance, have become prominent experimental therapeutic agents. Among them, gold complexes have attracted attention in the last years and numerous families of Au(I) and Au(III) compounds have been synthesized and studied for their anticancer properties in vitro and in vivo [1,2]. Overall, the investigation of the cytotoxic activity and related mode of action of cytotoxic gold-based complexes has enabled the identification of their preferential "protein targets", as it is increasingly evident that DNA is not the unique or major target for such compounds [3]. In this context, coordination cytotoxic Au(III) compounds have been identified as selective inhibitors of the membrane water channels aquaporins (AQPs) [1,4].

Among the 13 mammalian AQPs described so far, three sub-groups can be recognized based on permeability features: *orthodox aquaporins* (AQP0, AQP1, AQP2, AQP4, AQP5, AQP6 and AQP8), which are primarily water selective and facilitate water movement across cell membranes in response to osmotic gradients [5]; *aquaglyceroporins* (AQP3, AQP7, AQP9 and AQP10), facilitating the permeation of small uncharged solutes such as glycerol [6]; and *unorthodox* aquaporins (AQP11, AQP12), found in intracellular membranes and with reported permeability to water and glycerol [7–9]. Specifically, the aquaglyceroporins regulate the glycerol content in the epidermis, fat and other tissues and appear to be involved in skin hydration, cell proliferation, fat metabolism, and carcinogenesis. Several studies showed that AQPs are closely associated with cancer proliferation and invasion, and are expressed in at least 20 human cancers [10]. Moreover, AQPs expression is related to tumour types, grades, proliferation, migration and angiogenesis, rendering these transport proteins attractive as both diagnostic and therapeutic targets in cancer [10]. To validate the various roles of AQPs in health and disease, and to develop AQP-targeted therapies, the use of selective inhibitors in addition to genetic approaches, holds great promise. However, so far, no reported organic small-molecule AQPs inhibitor possesses sufficient isoform selectivity to be a good candidate for clinical development [11].

A few years ago, we reported on the potent and selective inhibition of human AQP3 by a series of Au(III) complexes with bidentate N^N ligands [12,13], which could potently and selectively inhibit glycerol permeation through hAQP3 in human red blood cells (hRBC). The most effective inhibitor of the series, Auphen ([Au(phen)Cl$_2$]Cl, phen = 1,10-phenanthroline) had an IC$_{50}$ of 0.8 ± 0.08 μM [12]. In a further study, Auphen's capacity of inhibiting cell proliferation was examined in various cell lines, including cancerous ones, with different levels of AQP3 expression, and showed a direct correlation between AQP3 expression levels and the inhibition of cell growth by the Au(III) compound [14]. AQP3 inhibition was also demonstrated in the cell lines where proliferation was mostly affected by treatment with the gold complex [14]. Structure–activity relationships to optimize the design of AQP3 inhibitors were then established investigating other Au(III) compounds with different N^N ligand scaffolds [13].

Pursuing the design of more potent and selective AQP3 inhibitors, we have recently observed that the cationic complex [Au(pbzMe)Cl$_2$]PF$_6$ (**C1**, pbzMe = 1-methyl-2-(pyridin-2-yl)-benzimidazole) is even more efficient than Auphen in inhibiting glycerol permeation via AQP3 [1], and ca. three orders of magnitude more effective than the neutral related complex [Au(pbzH)Cl$_2$] (**C10**, pbzH = 2-(pyridin-2-yl)-benzimidazole) [15]. Combined molecular dynamics (MD) and density functional theory (DFT) studies were able to show that **C1**, upon binding to Cys40 in AQP3, is able to induce protein conformational changes, leading to the shrinkage of the channel, and thus, preventing glycerol and water permeation [15].

Following these promising results, we have synthesized a new series of Au(III) complexes based on the 2-(2-pyridyl)benzimidazole (pbzH) N-donor ligand, which is also known to inhibit hepatic enzymes, [16] and exhibits anticancer activities *per se* [17]. In general, metal complexes based on 2-(2′-pyridyl)benzimidazole scaffolds have attracted attention in various established and potential application areas, including medicinal inorganic chemistry [18–20]. Thus, we report here on the synthesis and characterization of eight new cationic Au(III) derivatives with functionalization at the non-coordinated benzimidazole nitrogen. In addition, two neutral complexes featuring extended aromatic scaffolds (namely pyrene and anthracene), endowed with luminescence properties, have been obtained. The compounds have been tested for their AQPs inhibition properties in human Red Blood Cells (hRBC) using a stopped-flow method, and their effects compared to **C1** [Au(pbzMe)Cl$_2$]PF$_6$ and **C10** [Au(pbzH)Cl$_2$]. Furthermore, the compounds' antiproliferative effects have been studied in a small panel of human cancer cells with different levels of AQP3 expression.

2. Results

2.1. Synthesis and Characterization of Au(III) Complexes

The library of functionalised pyridylbenzimidazole ligands **L1–L9** has been obtained by nucleophilic substitution on the non-coordinated nitrogen atom of the commercially available pyridylbenzimidazole by reaction with a halogenated substituent (R–X) in the presence of a base (Scheme S1, Supplementary Materials) [21]. Several types of functional groups have been envisaged to study the influence of both the steric hindrance and the electronic effect on the biological properties of the final gold complexes. In parallel, two additional ligands (**L11–L12**) featuring luminescent properties [22] have also been synthesized with the idea to monitor their fate in cancer cells by fluorescence microscopy (Scheme S1, supplementary materials). The use of ligands **L1–L9**, which possess a functionalised amine, gives rise to Au(III) cationic complexes (corresponding **C1–C9**) by reaction between an equimolar amount of **L1–L9** and NaAuCl$_4$, in the presence of an excess of KPF$_6$ (Scheme 1, top). The pure cationic gold complexes can then easily be isolated following precipitation, washing and filtration. On the other hand, reaction between ligands **L10–L12** and NaAuCl$_4$ in the presence of a base leads to the formation of the neutral complexes **C10–C12** (Scheme 1, bottom) [15]. The identity and the purity of the complexes **C1–C12** was confirmed by NMR, IR and UV–Visible spectroscopies, as well as by mass spectrometry and in some cases by elemental analysis (See Experimental and Supplementary Materials for details, Figures S1–S24). The obtained results confirmed the purity of the compounds, which were all obtained in good yields. While all ligands were found to be soluble in most organic solvents, and thus, their NMR analysis was performed in CDCl$_3$, the complexes were insoluble in most cases, except in rare examples in acetone or acetonitrile. To ensure a similar analysis for all complexes, their NMR spectra were recorded in DMSO-d_6. The ^{1}H NMR spectra of the ligands were easily attributable and the most downfield shifted signals corresponded to the benzimidazole ring. In most cases, the signals of the protons of the pyridyl were found overlapping each other, in addition to the signals of the phenyl rings for the R substituents in the case of **L4–L9**. The ^{1}H NMR spectra of the complexes were similar to those of the corresponding ligands in terms of number of resonance signals; however, some signals (H$_a$, H$_c$ and H$_h$, see Scheme 2 for the numbering scheme) were clearly more affected by the presence of the Au(III)Cl$_2$ fragment [$\Delta\delta$ ($\delta_{coord} - \delta_{free}$) of 0.24 and 0.47 ppm]. The NMR analysis of both the ligands **L11–L12** and their corresponding complexes **C11** and **C12** was very challenging due to the electronic similarity and thus proximity on the spectra. However, the number and nature of the signals were compatible with the structures, and further analytical methods allowed us to confirm the purity of the compounds (ESI-MS, UV–Visible spectroscopy and IR). IR analysis of the complexes showed in all cases the presence of C=C bending, C–H stretching and C–N stretching bands, and confirmed the presence of specific chemical groups on the main scaffold: alkyl chains in the case of **C1–C3**, ester group in the case of **C5** and C–F bonds for **C6–C9**.

Crystals suitable for X-ray diffraction were obtained for complex **C6** by slow diffusion of pentane in a concentrated solution of the complex in a mixture of acetonitrile and dichloromethane at room temperature (see Supplementary Materials for details). The structure confirmed a bidentate coordination mode of the ligand **L6** onto the gold centre via the nitrogen of the pyridine and the benzimidazole rings, giving rise to square planar complexes. The 4-trifluoromethylbenzyl functional group added on the benzimidazole moiety always points out of the plane, as already described with similar ligands and copper complexes [18]. It is worth mentioning that the slow process of crystallisation (10–15 days) may favour the partial decomposition of the complexes, specifically the de-coordination of the gold centre and the exchange of counter anions. In fact, the structure of **C6** revealed the presence of AuCl$_4$ counterions in the lattice (see Supplementary Materials). Furthermore, we attempted to crystallize compound **C7** in the same conditions, but the resulting X-ray structure confirmed the de-coordination of the gold centre from one of the nitrogens of the pyridine ligand, thus leading to a neutral gold complex with three coordinated chlorido ligands (see supplementary materials).

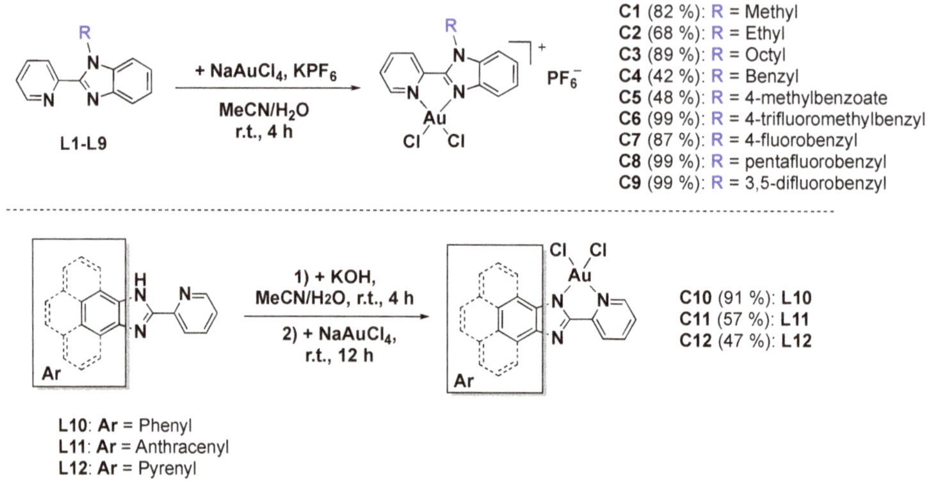
+ NaAuCl4, KPF6
MeCN/H2O
r.t., 4 h
L1-L9
PF6
C1 (82 %): R = Methyl
C2 (68 %): R = Ethyl
C3 (89 %): R = Octyl
C4 (42 %): R = Benzyl
C5 (48 %): R = 4-methylbenzoate
C6 (99 %): R = 4-trifluoromethylbenzyl
C7 (87 %): R = 4-fluorobenzyl
C8 (99 %): R = pentafluorobenzyl
C9 (99 %): R = 3,5-difluorobenzyl
1) + KOH,
MeCN/H2O, r.t., 4 h
2) + NaAuCl4,
r.t., 12 h
C10 (91 %): L10
C11 (57 %): L11
C12 (47 %): L12
L10: Ar = Phenyl
L11: Ar = Anthracenyl
L12: Ar = Pyrenyl

Scheme 1. Synthetic pathways to the series of cationic (top) and neutral (bottom) Au(III) complexes **C1–C9** and **C10–C12**, respectively.

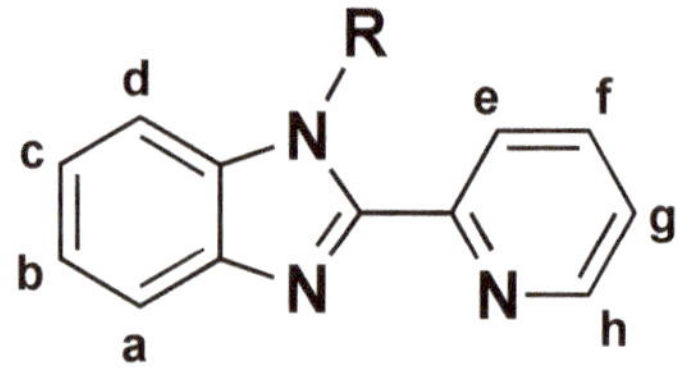

Scheme 2. ^{1}H labelling in the selected ligand scaffold.

The complexes **C1–C12** and their corresponding ligands **L1–L12** have also been investigated for their photophysical properties (see Figures S25–S48). Both ligands **L1–L10** and complexes **C1–C9** exhibit a strong absorption band centred around 310–315 nm which can be attributed to $\pi \rightarrow \pi^*$ transitions and/or ligand-to-metal charge transfers (LMCT) in the case of the Au(III) complexes. The absorption spectra of ligands **L11–L12** and corresponding complexes **C11–C12**, with the extended conjugated systems, show several bands attributed to the same transitions between 330 and 390 nm. Ligands **L1–L10** and complexes **C1–C9** all have single fluorescence emission bands centred around 375–380 nm, representing a Stokes shift of about 65 nm. Complexes **C11–C12** and their corresponding ligands possess extended aromatic and conjugated systems; thus, a shift in the emission bands is observed: ligand **L11** and complex **C11** emit at 415 nm whereas **L12** and **C12** exhibit an emission band around 450 nm (Figures S45–S48). The quantum yield of fluorescence (Φ_F) has also been assessed for all the reported compounds (see Figures S25–S48). While the ligands with the alkane substituents (**L1–L3**) have relatively high quantum yields (50–60%), the ligands with the functionalised benzyl groups (**L4–L9**) have decreased quantum yields between 27 and 42%. The ligands with the extended aromatic systems **L11** and **L12** have quantum yields of 74 and 61%, respectively (Figures S45 and S47). In general, upon coordination of the ligands to the gold(III) ion, almost all quantum yields of luminescence are decreased due to the "heavy metal effect", with the exception of complex **C12** (Φ_F = 71%).

2.2. UV–Visible Stability Studies

The stability of the gold complexes was investigated using UV–Visible spectroscopy before further biological testing. Thus, the absorbance of the compounds' solutions in PBS buffer (pH 7.4) was measured between 300 and 800 nm at regular time intervals during 24 h at room temperature, allowing the monitoring of possible compound's transformations such as hydrolysis, reduction and/or precipitation. In parallel, as Au(III) complexes tend to be reduced in physiological conditions to Au(I) and even Au(0), the reactivity of the compounds with the intracellular reducing agent glutathione (GSH) was monitored in the same conditions.

All the Au(III) compounds exhibit intense transitions in the 300–400 nm range, characteristic of the Au(III) chromophore, that may be straightforwardly assigned as LMCT bands. Complexes **C1**, **C2**, **C5**, **C6** and **C9** were found to be mostly stable over the first 6 h in PBS buffer (pH 7.4) with no significant change in the UV–Visible spectra (Figures S49, S50, S53, S54 and S57). The observed small spectral changes developing with time might be related to the occurrence of partial hydrolysis processes. Instead, the spectra of complexes **C3**, **C4**, **C7**, **C8** and **C10** were found to undergo major changes over the first few hours (Figures S51–S52, S55–S56 and S58), leading to the disappearance of the classical LMCT bands, suggesting disruption of the Au(III) complex. Complexes **C11**–**C12** were moderately stable in solution (Figures S59 and S60) featuring hypocromic shifts in their spectra over the first 6 h of incubation. In the presence of GSH, all the compounds immediately reacted to form adducts leading to the loss of the N-donor ligands from the gold centre, as demonstrated by the disappearance of the LMCT bands (Figures S49–S58). Only compounds **C11** and **C12** were scarcely reactive and maintained their spectral features over time after the addition of GSH (Figures S59 and S60).

2.3. Inhibition of Aquaporins

Based on the previosuly discussed stability studies, only the most stable gold complexes were tested for their AQP1 and AQP3 inhibition properties in hRBC by stopped-flow spectroscopy, according to previously reported procedures [12]. A representative IC$_{50}$ curve for the inhibition of glycerol permeation via AQP3 by two gold compounds is reported in Figure 1. The obtained results are summarized in Table 1 and show that all the new cationic complexes **C2**, **C4**, **C5**, **C6**, **C9** and **C11** are able to selectively inhibit glycerol permeation via AQP3, with IC$_{50}$ values in the sub-micromolar level, comparable to **C1** and Auphen. The antiproliferative activities of the corresponding ligands are presented in the supplementary (Table S1). The most effective compound was **C6**. In line with previous results, the neutral compound **C10** was ca. one order of magnitude less effective as an AQP3 inhibtior [15]. Of note, all the compounds were inactive as inhibitors of the orthodox water channel AQP1 in the same cellular model (data not shown).

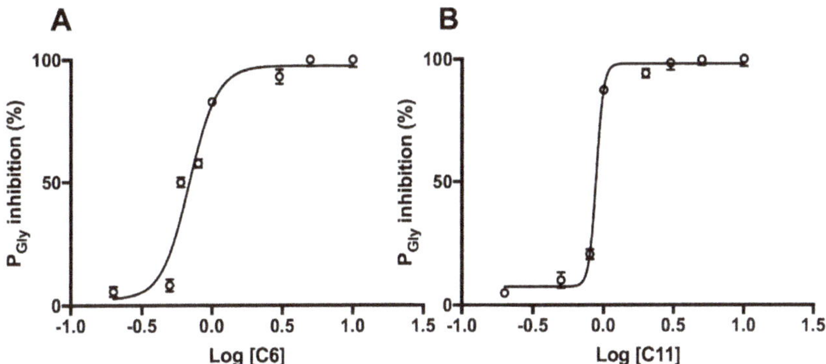

Figure 1. Representative IC$_{50}$ curve for the inhibition of glycerol permeation via AQP3 by the Au(III) complexes **C6** (**A**) and **C11** (**B**) in hRBC after 30 min incubation.

Table 1. AQP3 inhibitory effects (IC$_{50}$ values) measured in hRBC after 30 min incubation; and antiproliferative activities (EC$_{50}$ values) of Au(III) compounds in human SKOV-3, A375, MCF-7, and A549 cells after 72 h incubation, measured by the MTT assay.

Compound	AQP3 Inhibition	EC$_{50}$ [μM] [1]			
	IC$_{50}$ [μM] [1]	SKOV-3	A375	MCF7	A549
Auphen	0.80 ± 0.08	7.00 ± 2.00	1.7 ± 0.3	3.00 ± 0.05	1.07 ± 0.09
C1	1.018 ± 0.137	>80	>80	>80	>80
C2	0.881 ± 0.015	>80	>80	>80	>80
C3	1.825 ± 0.017 (*n* = 2)	41 ± 13	23 ± 1	40 ± 4	57 ± 2
C4	0.85 ± 0.21	56 ± 12	69 ± 3	63 ± 1	81 ± 9
C5	0.80 ± 0.10	>50	>50	25 (*n* = 1)	>50)
C6	0.69 ± 0.06	>50	34 (*n* = 1)	38 (*n* = 1)	47 (*n* = 2)
C7	n.d.	n.d.	n.d.	n.d.	n.d.
C8	n.d.	n.d.	n.d.	n.d.	n.d.
C9	0.72 ± 0.05	>50	>50	>50	>50
C10	>50	17 ± 7	5 ± 2	12 ± 1	>50
C11	0.82 ± 0.13 (*n* = 2)	33 ± 5	12 ± 2	29 ± 8	>50
C12	n.d.	41 ±13	13 ± 2	17 ± 3	45 ± 3

[1] Values represented as mean (±SEM) of at least three independent experiments (*n*), unless otherwise stated. n.d. = non determined.

2.4. Expression of AQP3 in Cancer Cells and Antiproliferative Activities

Afterward, the compounds—with the exception of the highly unstable **C7** and **C8**—were evaluated for their antiproliferative activities against a panel of human cancer cells in vitro. The cells were selected and studied for their level of expression of human AQP3. Information on the expression of AQP3 in cancer lines is limited and most of the data currently available refers to mRNA expression levels [23]. Specifically, cells were studied for the expression levels of human AQP3 by flow cytometry (Figure S61) and their AQP3 cellular distribution by immunocytofluorescence, respectively (Figure 1). As expected from the literature and mRNA levels [23], the breast cancer cell line MCF-7 highly express AQP3. Ovarian cancer cells SKOV-3 also showed marked AQP3 expression, followed by the lung cancer cells A549 and the skin malignant melanoma cells A375. Regarding cellular localization, in the case of the melanoma cells A375, AQP3 was found to be also localized in the nuclear membrane (Figure 2) as evidenced by fluorescence microscopy.

The antiproliferative activities of the complexes are summarized in Table 1. Overall, the compounds show very moderate anticancer effects in all cell lines, with compounds **C1** and **C2** being the least toxic (EC$_{50}$ > 80 μM). Compounds **C3** and **C4** were moderately cytotoxic, but their scarce stability in the aqueous environment may affect their antiproliferative effects. Compounds **C10**–**C12** were the most effective in the series, particularly against the melanoma A375 cells. It is worth mentioning that these three compounds are the only neutral ones, and may display different uptake and accumulation in cancer cells with respect to the cationic derivatives. Of note, the ligands **L1**–**L12** generally showed reduced antiproliferative effects with respect to the corresponding gold complexes (Table S1).

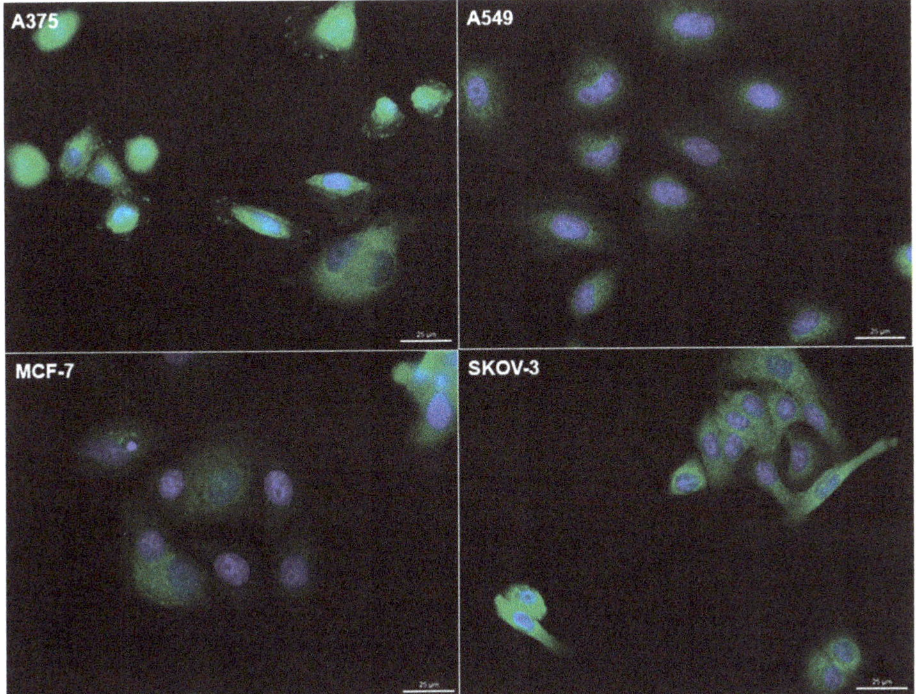

Figure 2. Human AQP3 expression and localization in human cancer cells by immunocytofluorescence. Human AQP3 expression (green) in A549, A375, SKOV-3 and MCF-7 cancer cells, with nuclei stained with DAPI (blue). Scale bars represent 25 μm.

3. Materials and Methods

3.1. General Information

All reagents and solvents used have been obtained from Haereus (Hanau, Germany), Sigma Aldrich (St. Louis, MO, United States), Fluorochem (Old Glossop, UK), Alfa Aesar (Ward Hill, MA, USA) or Acros (Loughborough, UK) and were used as received unless specified. The identity and purity ($\geq$95%) of the complexes were unambiguously established using high-resolution mass spectrometry and multinuclear NMR spectroscopy. ^{1}H, ^{13}C and ^{19}F NMR spectra were recorded on a Bruker Avance II300, II400 or II500 spectrometers (Bruker, Coventry, UK) at room temperature (r.t.) and referenced internally to residual solvent peaks [24]. The coupling constants are reported in Hertz. HR-ESI-MS spectra were obtained in acetonitrile/methanol on a Thermo Finnigan LCQ DecaXPPlus quadrupole ion-trap instrument (Thermo Fisher Scientific, Paisley, UK) operated in positive ion mode over a mass range of m/z 150–2000. IR spectra were measured on a Shimadzu IRAffinity-1S FT-IR (ATR) (Shimadzu, Milton Keynes, UK). Ligand functionalization reactions were monitored by thin-layer chromatography (Merck 60 F254 silica gel). Column chromatography was carried out manually using silica gel (Fluorochem; 40–63 μm, 60 Å) or on a Biotage Isolera automated flash purification system (ZIP cartridge, 5–10 g). The absorption and emission spectra of the ligands and corresponding complexes were recorded on Cary 5000 or 60 UV–Visible NIR (Agilent, Wokingham, UK), Cary Eclipse Fluorescence spectrophotometers (Agilent).

3.2. Compounds Synthesis

3.2.1. General Procedure for the Synthesis of Ligands **L1–L9**

Pyridylbenzimidazole ligand **L10** is commercially available. Functionalised pyridylbenzimidazole derivatives **L1** [21], **L4** [25], **L7** [22], **L8** [22], **L9** [22], **L11** [25], and **L12** [18] have been synthetized following protocols reported in the literature. Ligands **L2** and **L3** have been obtained adapting protocols used to produce **L1** and **L4**, and have also been already reported in the literature [26]. The purity of the compounds was confirmed by elemental analysis, and all of them showed purity greater than 98%. Compounds **L5** and **L6** have been obtained using the following general protocol: A solution of 1 eq. of pyridylbenzimidazole (488 mg, 2.50 mmol) and 1.5 eq. of K_2CO_3 (518 mg, 3.75 mmol) in 5 mL of DMF was stirred at room temperature for 30 min. Then, 1.2 eq. (3.00 mmol) of methyl-4-(bromomethyl)benzoate (in the case of **L5**; 687 mg) or 4-(trifluoromethyl)benzyl bromide (in the case of **L6**; 717 mg) were added to the mixture and left to stir at room temperature for approximately 4 h (the reaction was followed by TLC using a 1:1 ethyl acetate/hexane eluent system). The solution was then concentrated and water was added. The product was extracted several times by ethyl acetate. Organic layers were combined, washed with water and dried over $MgSO_4$. After filtration and evaporation of the solvent, the crude products **L5** and **L6** were obtained and purified by column chromatography either manually using a 1:1 mixture of ethyl acetate and hexane or on an automated flash purification system using a gradient of ethyl acetate in hexane.

L5. White powder, 83% isolated yield (712 mg, 2.07 mmol). R_f (1:1 AcOEt/hexane) ≈ 0.6. NMR ^{1}H (CDCl$_3$): 8.56 (ddd, J = 7.9, 4.0 and 2.7 Hz, 1H, H_a), 8.44 (dt, J = 8.0 and 1.0 Hz, 1H, H_d), 7.92–7.87 (m, 3H, 2xCH–Ph + H_e), 7.81 (td, J = 8.0 and 1.8 Hz, 1H, H_c), 7.33–7.25 (m, 5H, H_h + H_b + H_f + H_g + H_e), 7.20 (d, J = 8.5 Hz, 2H, 2xCH–Ph), 6.14 (s, 2H, CH_2), 3.77 (s, 3H, CH_3). NMR ^{13}C{^{1}H} (CDCl$_3$): 166.8 (CO_2Me), 150.4 (C^{IV}), 149.9 (C^{IV}), 148.7 (C_a), 142.8 (C^{IV}), 142.8 (C^{IV}), 137.0 (C_c), 136.8 (C^{IV}), 130.0 (2xCH–Ph), 129.3 (C^{IV}), 126.7 (2xCH–Ph), 124.7 (C_d), 124.0, 123.9, 123.0 (C_b, C_f, C_g), 120.3 (C_e), 110.5 (C_h), 52.2 (CH_3), 48.9 (CH_2). ESI-TOF-MS (positive mode) for $C_{21}H_{18}N_3O_2$ ([M + H]$^+$): calc. 344.1399, exp. 344.1404 (err. 1.5 ppm). UV–Vis (DMSO): λ_{max} (nm) (ε, cm^{-1}·mol^{-1}·dm^3) 313 (14,123). IR (ATR): ν (cm^{-1}) 1709 ($\nu_{C-H\ bending}$), 1428 ($\nu_{C-H\ stretching}$), 1276 ($\nu_{C-O\ stretching}$), 1110 ($\nu_{C-N\ stretching}$), 845 ($\nu_{C=C\ bending}$). Anal. Calcd for $C_{21}H_{17}N_3O_2$ (343.39): C, 73.45; H, 4.99; N, 12.24. Found: C, 73.48; H, 4.97; N, 12.20.

L6. White powder, 88% isolated yield (774 mg, 2.19 mmol). R_f (1:1 AcOEt/hexane) ≈ 0.6. NMR ^{1}H (CDCl$_3$): 8.55 (m, 1H, H_a), 8.44 (d, J = 8.0 Hz, 1H, H_d), 7.84 (d, J = 7.8 Hz, 1H, H_e), 7.79 (td, J = 7.8 and 1.8 Hz, 1H, H_c), 7.46 (d, 2H, 2xCH–Ph), 7.47–7.22 (m, 6H, 2xCH–Ph + H_h + H_b + H_f + H_g), 6.20 (s, 2H, CH_2). NMR ^{13}C{^{1}H} (CDCl$_3$): 150.4 (C^{IV}), 149.8 (C^{IV}), 148.7 (C_a), 142.8 (C^{IV}), 141.7 (C^{IV}), 137.1 (C_c), 136.7 (C^{IV}), 129.7 (q, C^{IV}–CF$_3$, $^2J_{C-F}$ = 32.0 Hz), 126.8 (2xCH–Ph), 125.7 (q, 2xCH–C–CF$_3$, $^3J_{C-F}$ = 3.8 Hz), 124.7 (C_d), 124.2 (q, C^{IV}F$_3$, $^1J_{C-F}$ = 270 Hz), 124.1, 123.9, 123.2 (C_b, C_f, C_g), 120.4 (C_e), 110.4 (C_h), 48.7 (CH_2). NMR ^{19}F{^{1}H} (CDCl$_3$): −62.5 (s, CF$_3$). ESI-TOF-MS (positive mode) for $C_{20}H_{15}N_3F_3$ ([M + H]$^+$): calc. 354.1218, exp. 354.1225 (err. 2.0 ppm). UV–Vis (DMSO): λ_{max} (nm) (ε, cm^{-1}·mol^{-1}·dm^3) 313 (23,130). IR (ATR): ν (cm^{-1}) 1444 ($\nu_{C-H\ stretching}$), 1328, 1157 ($\nu_{C-F\ stretching}$), 1107 ($\nu_{C-N\ stretching}$), 833 ($\nu_{C=C\ bending}$). Anal. Calcd for $C_{20}H_{14}N_3F_3$ (353.11): C, 67.98; H, 3.99; N, 11.89. Found: C, 67.95; H, 3.93; N, 11.82.

3.2.2. General Procedure for the Synthesis of the Cationic Complexes **C1–C9**

Complex **C1** was previously reported by us [15], and the same procedure was used to obtain complexes **C2–C9**. Thus, to a solution of 1 eq. of pyridylbenzimidazole ligand **L1–L9** in acetonitrile (2.5 mL) was added an aqueous solution (15 mL) of 1 eq. of NaAuCl$_4$ and 3 eq. of KPF$_6$, and the resulting mixture was stirred 3 h at room temperature. The resulting yellow to orange precipitate was filtered, washed with water, ethanol and diethyl ether and dried under vacuum.

C2. Yield: 68% (480 mg, 0.76 mmol). NMR ^{1}H (DMSO-d_6): 8.93 (m, 1H, H_a), 8.29 (m, 1H, H_d), 8.21 (m, 1H, H_c), 8.05 (m, 1H, H_e), 7.89 (m, 1H, H_h), 7.76 (m, 1H, H_b), 7.61 (m, 2H, H_f + H_g), 4.87 (m, 2H, CH_2),

1.50 (s, 3H, CH_3). NMR $^{13}C\{^1H\}$ (DMSO-d_6): 150.2 (C_a), 146.8 (C^{IV}), 144.3 (C^{IV}), 138.3 (C_c), 133.3 ($2xC^{IV}$), 128.0 (C_g), 127.6 (C_f), 126.0 (C_d), 125.8 (C_b), 115.8 (C_h), 112.9 (C_e), 41.4 (CH_2), 14.9 (CH_3). ESI-TOF-MS (positive mode) for $C_{14}H_{13}N_3Cl_2Au$ ([M^+]): calc. 490.0152, exp. 490.0134 (err. -3.7 ppm). UV–Vis (DMSO): λ_{max} (nm) (ε, $cm^{-1}\cdot mol^{-1}\cdot dm^3$) 313 (21,654). IR (ATR): ν (cm^{-1}) 1475, 1457 ($\nu_{C-H\ stretching}$), 1036 ($\nu_{C-N\ stretching}$), 841, 749 ($\nu_{C=C\ bending}$).

C3. Yield: 89% (561 mg, 0.78 mmol). NMR ^1H (DMSO-d_6): 8.92 (m, 1H, H_a), 8.29–8.19 (m, 2H, $H_d + H_c$), 8.08 (m, 1H, H_e), 7.90 (m, 1H, H_h), 7.78 (m, 1H, H_b), 7.66–7.58 (m, 2H, $H_f + H_g$), 4.85 (m, 2H, CH_2), 1.85 (m, 2H, CH_2), 1.17 (m, 10H, CH_2), 0.82 (m, 3H, CH_3). NMR $^{13}C\{^1H\}$ (DMSO-d_6): 150.3 (C_a), 146.9 (C^{IV}), 143.9 (C^{IV}), 138.8 (C_c), 136.9 (C^{IV}), 133.6 (C^{IV}), 127.5 (C_g), 127.1 (C_f), 126.7 (C_d), 126.6 (C_b), 115.7 (C_h), 113.8 (C_e), 46.3 (CH_2), 31.6 (CH_2), 29.5 (CH_2), 28.9 (CH_2), 28.8 (CH_2), 26.2 (CH_2), 22.5 (CH_2), 14.4 (CH_3). ESI-TOF-MS (positive mode) for $C_{20}H_{25}N_3Cl_2Au$ ([M^+]): calc. 574.1091, exp. 574.1100 (err. 1.6 ppm). UV–Vis (DMSO): λ_{max} (nm) (ε, $cm^{-1}\cdot mol^{-1}\cdot dm^3$) 312 (16,006). IR (ATR): ν (cm^{-1}) 2927 ($\nu_{C-H\ stretching}$), 1488, 1474 ($\nu_{C-H\ stretching}$), 1039 ($\nu_{C-N\ stretching}$), 831, 746 ($\nu_{C=C\ bending}$).

C4. Yield: 99% (403 mg, 1.61 mmol). NMR ^1H (DMSO-d_6): 8.83 (m, 1H, H_a), 8.30 (m, 1H, H_d), 8.17 (m, 1H, H_c), 7.87 (m, 1H, H_e), 7.78 (m, 1H, H_h), 7.70 (m, 1H, H_b), 7.55–7.49 (m, 2H, $H_f + H_g$), 7.29–7.25 (m, 5H, $5xCH–Ph$), 6.21 (s, 2H, CH_2). NMR $^{13}C\{^1H\}$ (DMSO-d_6): 149.9 (C_a), 147.2 (C^{IV}), 144.5 (C^{IV}), 138.1 (C_c), 135.5 (C^{IV}), 133.7 (C^{IV}), 128.6 (CH–Ph), 127.7 (CH–Ph), 126.9 (C_b), 126.5 (C_d), 125.7 (C_f), 125.7 (C_g), 116.1 (C_e), 113.0 (C_h), 48.7 (CH_2). ESI-TOF-MS (positive mode) for $C_{19}H_{15}N_3Cl_2Au$ ([M^+]): calc. 552.0309, exp. 552.0307 (err. -0.4 ppm). UV–Vis (DMSO): λ_{max} (nm) (ε, $cm^{-1}\cdot mol^{-1}\cdot dm^3$) 313 (21,527). IR (ATR): ν (cm^{-1}) 1497, 1475, 1454 ($\nu_{C-H\ stretching}$), 831, 746 ($\nu_{C=C\ bending}$).

C5. Yield 48% (265 mg, 0.351 mmol). NMR ^1H (DMSO-d_6): 8.80 (m, 1H, H_a), 8.31 (m, 1H, H_d), 8.17 (t, $J = 5.7$ Hz, 1H, H_c), 7.92–7.88 (m, 3H, $2xCH–Ph + H_e$), 7.77 (m, 1H, H_h), 7.69 (m, 1H, H_b), 7.58–7.50 (m, 2H, $H_f + H_g$), 7.41–7.39 (m, 2H, $2xCH–Ph$), 6.30 (s, 2H, CH_2), 3.80 (s, 3H, CH_3). NMR $^{13}C\{^1H\}$ (DMSO-d_6): 165.9 (CO_2Me), 149.9 (C_a), 145.0 (C^{IV}), 141.3 (C^{IV}), 138.3 (C_c), 134.5 (C^{IV}), 129.5 (CH–Ph), 129.1 (C^{IV}), 127.3 (CH–Ph), 126.6 (C_b), 125.9 (C_{f+g}), 125.7 (C_d), 116.5 (C_e), 112.9 (C_h), 52.2 (CH_3), 48.8 (CH_2). ESI-TOF-MS (positive mode) for $C_{21}H_{17}N_3O_2Cl_2Au$ ([M^+]): calc. 610.0363, exp. 610.0350 (err. -2.1 ppm). UV–Vis (DMSO): λ_{max} (nm) (ε, $cm^{-1}\cdot mol^{-1}\cdot dm^3$) 312 (23,448). IR (ATR): ν (cm^{-1}) 1706 ($\nu_{C-H\ bending}$), 1496, 1481 ($\nu_{C-H\ stretching}$), 1285, 1018 ($\nu_{C-O\ stretching}$), 1108 ($\nu_{C-N\ stretching}$), 845 ($\nu_{C=C\ bending}$).

C6. Yield: 99% (435 mg, 0.568 mmol). NMR ^1H (DMSO-d_6): 8.81 (d, $J = 4.3$ Hz, 1H, H_a), 8.32 (d, $J = 7.8$ Hz, 1H, H_d), 8.18 (td, $J = 7.8$ and 1.3 Hz, 1H, H_c), 7.92 (d, 1H, $J = 7.7$ Hz, H_e), 7.80 (d, 1H, $J = 7.8$ Hz, H_h), 7.72–7.68 (m, 3H, $H_b + 2xCH–Ph$), 7.59–7.53 (m, 2H, $H_f + H_g$), 7.52–7.49 (m, 2H, $2xCH–Ph$), 6.31 (s, 2H, CH_2). NMR $^{13}C\{^1H\}$ (DMSO-d_6): 150.2 (C_a), 147.3 (C^{IV}), 143.9 (C^{IV}), 140.3 (C^{IV}), 138.5 (C_c), 133.6 (C^{IV}), 132.9 (C^{IV}), 128.2 (q, $C^{IV}–CF_3$, $^2J_{C-F} = 31.5$ Hz), 127.9 ($2xCH–Ph$), 127.0 (C_b), 126.6 (C_f), 126.4 (C_g), 126.0 (C_d), 125.6 (q, $2xCH–C–CF_3$, $^3J_{C-F} = 3.9$ Hz), 123.8 (q, $C^{IV}F_3$, $^1J_{C-F} = 272$ Hz), 115.9 (C_e), 113.2 (C_h), 48.8 (CH_2). NMR $^{19}F\{^1H\}$ (DMSO-d_6): -61.0 (s, CF_3), -70.0 (d, PF_6, $^1J_{P-F} = 712.4$ Hz). ESI-TOF-MS (positive mode) for $C_{20}H_{14}N_3Cl_2F_3Au$ ([M^+]): calc. 620.0182, exp. 620.0212 (err. 4.8 ppm). UV–Vis (DMSO): λ_{max} (nm) (ε, $cm^{-1}\cdot mol^{-1}\cdot dm^3$) 312 (24,699). IR (ATR): ν (cm^{-1}) 1481 ($\nu_{C-H\ stretching}$), 1324, 1121, 1113 ($\nu_{C-F\ stretching}$), 1068 ($\nu_{C-N\ stretching}$), 834 ($\nu_{C=C\ bending}$). Anal. Calcd for $C_{20}H_{14}N_3Cl_2AuPF_9$ (766.18): C, 31.35; H, 1.84; N, 5.48. Found: C, 31.35; H, 1.88; N, 5.48.

C7. Yield: 87% (312 mg, 0.436 mmol). NMR ^1H (DMSO-d_6): 8.82 (d, $J = 4.2$ Hz, 1H, H_a), 8.31 (d, $J = 8.0$ Hz, 1H, H_d), 8.15 (t, $J = 7.6$ Hz, 1H, H_c), 7.86 (d, $J = 7.6$ Hz, H_e), 7.78 (d, 1H, $J = 7.1$ Hz, H_h), 7.88 (t, 1H, $J = 5.3$ Hz, H_b), 7.50 (m, 2H, $H_f + H_g$), 7.35–7.32 (m, 2H, $2xCH–Ph$), 7.15–7.11 (m, 2H, $2xCH–Ph$), 6.19 (s, 2H, CH_2). NMR $^{13}C\{^1H\}$ (DMSO-d_6): 162.9 (C^{IV}), 160.5 (C^{IV}), 150.0 (C_a), 147.4 (C^{IV}), 144.6 (C^{IV}), 138.3 (C_c), 133.9 (C^{IV}), 133.0 (d, $C^{IV}F$, $^1J_{C-F} = 203$ Hz), 129.4 (d, $2xCH–Ph$, $^3J_{C-F} = 8.4$ Hz), 126.7 (C_b), 126.0 (C_f), 126.0 (C_g), 125.9 (C_d), 116.3 (C_e), 115.6 (d, $2xCH–Ph$, $^2J_{C-F} = 21.5$ Hz), 113.1 (C_h), 48.2 (CH_2). NMR $^{19}F\{^1H\}$ (DMSO-d_6): -114.3 (s, Ph–F), -70.0 (d, PF_6, $^1J_{P-F} = 711.9$ Hz). ESI-TOF-MS (positive mode) for $C_{19}H_{14}N_3Cl_2FAu$ ([M^+]): calc. 570.0214, exp. 570.0236 (err. 3.9 ppm). UV–Vis (DMSO): λ_{max} (nm) (ε, $cm^{-1}\cdot mol^{-1}\cdot dm^3$) 312 (25,542). IR (ATR): ν (cm^{-1}) 1460 ($\nu_{C-H\ stretching}$),

1232 ($\nu_{\text{C-F stretching}}$), 835 ($\nu_{\text{C=C bending}}$). Anal. Calcd for $C_{18}H_{12}N_3O_2Cl_2AuPF_7$ (702.14): C, 31.86; H, 1.97; N, 5.87. Found: C, 31.82; H, 2.00; N, 5.88.

C8. Yield: 99% (394 mg, 0.500 mmol). NMR ^{1}H (DMSO-d_6): 8.76 (d, J = 4.4 Hz, 1H, H_a), 8.31 (d, J = 7.9 Hz, 1H, H_d), 8.14 (td, J = 7.8 and 1.6 Hz, 1H, H_c), 7.90–7.86 (m, 2H, H_h + H_e), 7.66 (m, 1H, H_b), 7.58–7.51 (m, 2H, H_f + H_g), 6.36 (s, 2H, CH$_2$). NMR ^{13}C{^{1}H} (DMSO-d_6): 149.3 (C_a), 148.5 (C^{IV}), 146.7 (C^{IV}), 146.0 (dd, C^{IV}F, $^1J_{\text{C-F}}$ = 246 Hz), 138.3 (C_c), 137.1 (C^{IV}), 135.2 (C^{IV}), 126.0 (C_b), 125.4 (C_g), 125.3 (C_f), 125.0 (C_d), 117.9 (C_e), 112.1 (C_h), 110.9 (t, C^{IV}–Ph, $^2J_{\text{C-F}}$ = 17 Hz), 39.9 (CH$_2$). NMR ^{19}F{^{1}H} (DMSO-d_6): -141.9 (Ph–F), -154.3 (Ph–F), -162.4 (Ph–F), -70.0 (d, PF$_6$, $^1J_{\text{P-F}}$ = 711.9 Hz). ESI-TOF-MS (positive mode) for $C_{19}H_{10}N_3Cl_2F_5Au$ ([M$^+$]): calc. 641.9837, exp. 641.9874 (err. 5.8 ppm). UV–Vis (DMSO): $\lambda_{\max}$ (nm) (ε, cm^{-1}·mol^{-1}·dm^3) 311 (23,441). IR (ATR): ν (cm^{-1}) 1474 ($\nu_{\text{C-H stretching}}$), 1337, 1128 ($\nu_{\text{C-F stretching}}$), 1029 ($\nu_{\text{C-N stretching}}$), 838 ($\nu_{\text{C=C bending}}$).

C9. Yield: 99% (443 mg, 0.603 mmol). NMR ^{1}H (DMSO-d_6): 8.82 (d, J = 4.2 Hz, 1H, H_a), 8.29 (d, J = 8.0 Hz, 1H, H_d), 8.20 (t, J = 7.6 Hz, 1H, H_c), 7.91 (d, 1H, J = 7.6 Hz, H_e), 7.78 (d, 1H, J = 7.1 Hz, H_h), 7.88 (t, 1H, J = 5.3 Hz, H_b), 7.57 (m, 2H, H_f + H_g), 7.19–7.09 (m, 4H, 4xCH–Ph), 6.20 (s, 2H, CH$_2$). NMR ^{13}C{^{1}H} (DMSO-d_6): 162.6 (d, C^{IV}F, $^1J_{\text{C-F}}$ = 247 Hz), 162.4 (d, C^{IV}F, $^1J_{\text{C-F}}$ = 247 Hz), 150.2 (C_a), 147.2 (C^{IV}), 143.6 (C^{IV}), 139.9 (t, C^{IV}Ph, $^3J_{\text{C-F}}$ = 9.6 Hz), 138.5 (C_c), 133.4 (C^{IV}), 132.4 (C^{IV}), 127.1 (C_b), 126.7, 126.5 (C_f, C_g), 126.1 (C_d), 115.7 (C_e), 113.2 (C_h), 110.6 (d, 2xCH–Ph, $^2J_{\text{C-F}}$ = 26.1 Hz), 103.5 (t, CH–Ph, $^2J_{\text{C-F}}$ = 25.8 Hz), 48.5 (CH$_2$). NMR ^{19}F{^{1}H} (DMSO-d_6): -109.1 (s, 2xPh–F), -70.0 (d, PF$_6$, $^1J_{\text{P-F}}$ = 712.6 Hz). ESI-TOF-MS (positive mode) for $C_{19}H_{13}N_3Cl_2F_2Au$ ([M$^+$]): calc. 588.0120, exp. 588.0150 (err. 5.1 ppm). UV–Vis (DMSO): $\lambda_{\max}$ (nm) (ε, cm^{-1}·mol^{-1}·dm^3) 313 (14,367). IR (ATR): ν (cm^{-1}) 1442 ($\nu_{\text{C-H stretching}}$), 1332, 1091 ($\nu_{\text{C-F stretching}}$), 1039 ($\nu_{\text{C-N stretching}}$), 833, 792 ($\nu_{\text{C=C bending}}$).

3.2.3. General Procedure for the Synthesis of the Neutral Complexes C10–C12

The synthesis of complex **C10** was already reported by us [15] and others [19], and was adapted to the synthesis of **C11** and **C12**. To a solution of 1 eq. of pyridylbenzimidazole ligand **L10–L12** in acetonitrile (1 mL) was added an aqueous solution of 1 eq. of KOH in water (6 mL). The reaction mixture was stirred at room temperature for 15 min. An aqueous solution of 1 eq. of NaAuCl$_4$ (6 mL) was then added, and the resulting mixture was stirred overnight at room temperature. The resulting dark brown precipitate was filtered; washed with water, ethanol and diethyl ether; and dried under vacuum.

C11. Yield: 57% (100 mg, 0.18 mmol). NMR ^{1}H (500 MHz, DMSO-d_6): 9.11–8.90 (m, 3H, 3xCH), 8.78 (m, 1H, CH), 8.42 (m, 2H, 2xCH), 8.21 (m, 1H, CH), 8.13 (m, 1H, CH), 7.86–7.74 (m, 3H, 3xCH), 7.67 (m, 1H, CH). NMR ^{13}C{^{1}H} (125.77 MHz, DMSO-d_6): 160.8 (C^{IV}), 150.3 (CH), 145.4 (C^{IV}), 145.0 (C^{IV}), 138.0 (CH), 134.9 (C^{IV}), 128.8 (CH), 127.6 (CH), 127.4 (CH), 125.5 (CH), 124.6 (CH), 124.5 (CH), 123.4 (CH), 122.7 (CH), 122.5 (CH), 121.0 (CH), 120.4 (CH). ESI-TOF-MS (positive mode) for $C_{20}H_{12}N_3ClAu$ ([M – Cl$^+$]): calc. 526.0385, exp. 526.0386 (err. 0.2 ppm). UV–Vis (DMSO): $\lambda_{\max}$ (nm) (ε, cm^{-1}·mol^{-1}·dm^3) 333 (20,891), 362 (16,528). IR (ATR): ν (cm^{-1}) 1428, 1452 ($\nu_{\text{C-H stretching}}$), 1034 ($\nu_{\text{C-N stretching}}$), 774 ($\nu_{\text{C=C bending}}$).

C12. Yield: 47% (87 mg, 0.15 mmol). NMR ^{1}H (500 MHz, MeCN-d_3): 8.63–8.47 (m, CH), 8.35 (m, CH), 8.27 (m, CH), 8.23 (m, CH), 8.14–7.98 (m, CH), 7.59–7.57 (m, CH). NMR ^{13}C{^{1}H} (125.77 MHz, DMSO-d_6): 161.2 (C^{IV}), 150.4 (CH), 149.6 (CH), 145.7 (C^{IV}), 145.5 (C^{IV}), 138.0 (CH), 135.6 (C^{IV}), 131.7 (C^{IV}), 128.3 (CH), 127.8 (CH), 127.2 (CH), 126.6 (CH), 125.0 (CH), 124.5 (C^{IV}), 123.4 (CH), 122.8 (C^{IV}), 122.6 (C^{IV}), 122.2 (C^{IV}), 120.0 (CH), 119.7 (CH), 119.5 (C^{IV}), 118.3 (CH). ESI-TOF-MS (positive mode) for $C_{22}H_{12}N_3ClAu$ ([M – Cl$^+$]): calc. 550.0385, exp. 550.0393 (err. 1.5 ppm). UV–Vis (DMSO): $\lambda_{\max}$ (nm) (ε, cm^{-1}·mol^{-1}·dm^3) 338 (28,904), 353 (28,264), 366 (19,892), 388 (20,031). IR (ATR): ν (cm^{-1}) 1442 ($\nu_{\text{C-H stretching}}$), 1064 ($\nu_{\text{C-N stretching}}$), 738 ($\nu_{\text{C=C bending}}$).

3.3. UV–Visible and Fluorescence Spectroscopy

The data are reported as the absorption maximum wavelength ($\lambda_{\max}$, in nm) and corresponding molar extinction coefficient at $\lambda_{\max}$ (ε, in L·mol^{-1}·cm^{-1}). Macro quartz cuvettes with a path length of 1 cm were used. The sample concentrations were chosen to obtain a maximum absorbance of around

0.8 to then measure the quantum yields. Relative quantum yields of fluorescence of the samples were obtained by comparing the areas under the corrected emission spectra with a standard absorbing in the same region than the sample. Measurements were performed in degassed DMSO (Sigma-Aldrich, spectroscopic grade $\geq$99.9%) at 298 K. Quinine sulfate (Φ_F = 0.546 in 0.5 M H_2SO_4, λ_{ex} = 366 nm) was used as the standard [26]. In all Φ_F calculations, the correction for the solvent refractive index (η) was applied: DMSO: η = 1.479; H_2O (or H_2SO_4): η = 1.333 [27]. The following equation was used to calculate the quantum yield of the sample ($\Phi_{F,x}$), in which $\Phi_{F,St}$ is the reported quantum yield of the standard, F is the integral photon flux, f is the absorption factor, and η is the refractive index of the solvent used. The x subscript denotes the sample, and St denotes the standard, and the fluorescence spectra between the sample and the standard were recorded at the same λ_{ex} [28]:

$$\Phi_{F,x} = \Phi_{F,St} \cdot \frac{F_x}{F_{St}} \cdot \frac{f_{St}}{f_x} \cdot \frac{\eta_x^2}{\eta_{St}^2}$$

3.4. X-ray Diffraction Analysis

Complexes **C6** and **C7** were allowed to form crystals by slow diffusion of pentane into a corresponding complex's concentrated solution in a mixture of acetonitrile and dicholoromethane at room temperature. The crystals were analysed at the UK National Crystallography Service in Southampton [29]. A suitable yellow block-shaped crystal from **C6** (0.050 $\times$ 0.040 $\times$ 0.030) mm^3 and from **C7** (0.080 $\times$ 0.050 $\times$ 0.020) mm^3, were selected and mounted on a MITIGEN holder in perfluoroether oil and mounted on a Rigaku 007HF diffractometer equipped with Varimax confocal mirrors and an AFC11 goniometer and HyPix 6000HE detector. The crystals were kept at T = 100.00(10) K during data collection. Using *Olex2* [30], the structures were solved with the *ShelXT* [31] structure solution program, using the Intrinsic Phasing solution method. The models were refined with version 2014/7 of *ShelXL* [32] using Least Squares minimisation, and the ORTEP views are presented in Figures S62 and S63. (CCDC 1868645 and 1868642 contain the supplementary crystallographic data for this paper. These data can be obtained free of charge via http://www.ccdc.cam.ac.uk/conts/retrieving.html (or from the CCDC, 12 Union Road, Cambridge CB2 1EZ, UK; Fax: +44 1223 336033; E-mail: deposit@ccdc.cam.ac.uk)).

3.5. UV–Visible Absorption Spectroscopy

The new Au(III) complexes were tested for their stability in aqueous media (1$\times$ Phosphate Buffer Saline (PBS, Corning), pH 7.4) in the presence or absence of 2 eq. GSH (vs. 1 eq. of Au(III) complex). The solutions were prepared from a stock solution of the Au(III) complex at a concentration of 10^{-2} M in DMSO and diluted to reach a concentration of 10^{-4} M in PBS. The absorption spectra were recorded over time at room temperature using a Cary 500 UV–Visible NIR spectrophotometer.

3.6. Aquaporins Inhibition

Venous blood samples were obtained from healthy human volunteers following a protocol approved by the Ethics Committee of the Faculty of Pharmacy of the University of Lisbon (Instituto Português de Sangue Protocol SN-22/05/2007). Informed written consent was obtained from all participants. Blood samples, collected in citrate anticoagulant (2.7% citric acid, 4.5% trisodium citrate and 2% glucose), were centrifuged at 750$\times$ g for 5 min at 4 °C. Plasma and buffy coat were discarded. Pelleted erythrocytes were washed three times in PBS (KCl 2.7 mM, KH_2PO_4 1.76 mM, Na_2HPO_4 10.1 mM, NaCl 137 mM, pH 7.4), diluted to 0.5% haematocrit and immediately used for experiments. hRBC mean volume in isotonic solution was determined using a CASY-1 Cell Counter (Schärfe System GmbH, Reutlingen, Germany) and was calculated as 82 fL. Stopped-flow experiments were performed on a HI-TECH Scientific PQ/SF-53 apparatus, with 2 ms dead time, temperature controlled and were interfaced with a microcomputer. Measurements of water permeability (Pf) and glycerol permeability (Pgly) were performed as described in References [12,33]. Briefly, 100 µL of the suspension of fresh erythrocytes (0.5%) was mixed with an equal volume of hyperosmotic PBS containing 200 mM sucrose (a non-permeable osmolyte that induces water outflow and subsequent cell shrinkage) and 200 mM glycerol (a permeable osmolyte that induces first fast cell shrinkage due to water outflow and then glycerol influx in response to its chemical gradient, followed by water influx with subsequent cell reswelling). The kinetics of cell volume change were measured from the time course of $90°$ scattered light intensity at 400 nm until a stable light scatter signal was attained. For each experimental condition, 5–7 replicates were analysed. Pf was estimated by Pf = k $(Vo/A)(1/Vw(osm_{out})\infty)$, where Vw is the molar volume of water, Vo/A is the initial cell volume to area ratio, $(osm_{out})\infty$ is the final medium osmolarity after the applied osmotic gradient, and k is the single exponential time constant fitted to the light scattering signal of erythrocyte shrinkage. Pgly was estimated by Pgly = k (Vo/A), where k is the single exponential time constant fitted to the light scattering signal of glycerol influx in erythrocytes.

In inhibition experiments, cells were incubated with different concentrations of complexes from freshly prepared stock solutions, for 30 min at r.t. before stopped-flow experiments. A time-dependent inhibition assay for all the tested compounds over several hours of incubation showed no further increase of inhibition after 30 min at r.t. The inhibitor concentration necessary to achieve 50% inhibition (IC_{50}) was calculated by nonlinear regression of the dose-response curves (Graph Pad Prism, Inc, San Diego, CA, USA) to the equation: $y = y_{min} + (y_{max} - y_{min})/(1 + 10((LogIC_{50} - Log[Inh])H))$, where y is the percentage inhibition obtained for each concentration of inhibitor [Inh], and H is the Hill slope. All solution osmolarities were determined from freezing point depression on a semi-micro osmometer (Knauer GmbH, Berlin, Germany) using standards of 100 and 400 mOsM.

3.7. Cell Lines and Culture Conditions

Human cell lines of lung adenocarcinoma (A549), breast carcinoma (MCF-7), ovarian adenocarcinoma (SKOV-3) and skin malignant melanoma (A375) were obtained from American Type Culture Collection (ATCC). The cells were cultured in Dulbecco's Modified Eagle Medium (DMEM, 4.5 g/L glucose, Corning, Thermo Fisher Scientific), supplemented with 10% foetal bovine serum (One-Shot FBS, EU-approved South American Origin, Thermo Fisher Scientific) and 1% penicillin/streptomycin (Gibco). All cell lines were cultured at 37 °C, in a humidified atmosphere of 5% CO_2 and passaged when reaching confluence.

3.8. Immunocytofluorescence

Round glass coverslips (Ø 13mm, VWR) sterilized by UV-light exposure were inserted in 24-well tissue culture-treated plates (Corning) prior to the addition of cells. Cells were seeded at a concentration of 125,000 cells/mL and incubated at 37 °C under humidified atmosphere with 5% CO_2, for 24 h. Glass coverslips were removed from the wells and washed three times with 1× PBS (Gibco). Afterwards, cells were fixed for 20 min with 4% formaldehyde (Alfa Aesar) and permeabilized for 5 min with 0.2% Triton X-100 (Alfa Aesar) at room temperature (r.t.). Glass coverslips were then incubated with 1:500 primary anti-AQP3 antibody (rabbit anti-human, PA1488 BosterBio, Pleasanton, CA, USA) in 1× PBS with 5% normal human serum (NHS, Invitrogen, UK), for 1 h at r.t., followed by incubation with 1:500 secondary goat anti-rabbit Alexa Fluor®488 (ab150077, Abcam, Cambridge, MA, US), in 1× PBS with 5% NHS for 1 h, at r.t. and in the dark. Before/after each step, cells were washed thrice with 1× PBS. After removing excess PBS, cells were mounted on glass microscope slides (VWR) using Mowiol® 4–88 (Sigma-Aldrich). Images were acquired on a Zeiss Axio Vert.A1 microscope and processed using Fiji (ImageJ) [34].

3.9. Flow Cytometry Analysis

For flow cytometry evaluation of AQP3 expression, samples of each cell line studied were prepared with 200,000 cells/sample. Cells were initially washed twice with 1× PBS (Gibco) and subsequently fixed with 4% formaldehyde (Alfa Aesar) for 30 min at room temperature (r.t.). Afterwards, cells were incubated with 1:500 dilution of anti-AQP3 antibody (rabbit anti-human, PA1488 Boster), in 1× PBS with 5% normal human serum (NHS, Invitrogen) and 0.1% Triton X-100 (Alfa Aesar) at r.t., for 1 h. Staining with primary antibody was followed by incubation with 1:500 secondary goat anti-rabbit Alexa Fluor®488 (ab150077, Abcam), in 1× PBS with 5% NHS and 0.1% Triton X-100 also for 1 h and at r.t., in the dark. Cells were kept on ice and away from direct light until analysed using a BD FACS Verse Flow Cytometer. Results were analysed using FlowJo 10.5.0. Firstly, samples were gated for live cells. Afterwards, stained and unstained samples were compared, in order to gate the positive population and the mean fluorescence, intensity (geometric mean) was taken from each positive sample peak. Data were normalized for the sample with the lowest AQP3 expression and results are shown as mean ± SEM of three independent experiments.

3.10. Antiproliferative Activities

For evaluation of cell growth inhibition, cells were seeded in 96-well plates (Corning) at a concentration of 15,000 cells/well, grown for 24 h in 200 μL complete medium. Solutions of the samples with the desired concentration (1 to 100 μM) were prepared by diluting a freshly prepared stock solution (10^{-2} M in DMSO) of the corresponding compound in aqueous DMEM medium, accordingly. Auphen (stock solution 10 mM in DMSO) was used as the reference compound. A negative control (medium only) was run for all the assays. After 24 h incubation, 200 μL of the compounds' dilutions were added to each well, and cells were incubated for an additional 72 h. Afterwards, medium was removed and 3-(4,5-dimethylthiazol-2-yl)-2,5-diphenyltetrazolium bromide (MTT, Fluorochem) in 10× PBS (Corning) was added to the cells, at a final concentration of 0.3 mg/mL, and incubated for 3–4 h. After, the MTT solution was discarded and replaced with DMSO, to allow the formed violet formazan crystals to dissolve. The optical density was quantified in quadruplicates for each experiment, at 550 nm using a multi-well plate reader (VICTOR X, Perking Elmer, UK). The percentage of surviving cells was calculated from the ratio of absorbance of treated to untreated cells. The EC_{50} values were calculated, using GraphPad Prism software, as the concentration showed 50% decrease in cell viability, compared to controls, using a nonlinear fit of concentration *vs.* response. Data is presented as mean ±SEM of at least three independent experiments.

4. Conclusions

In conclusion, we have synthesized a new series of Au(III) compounds, cationic and neutral, featuring a pyridine-benzimidazole scaffold, which has been characterized via different methods. Varying the ligands' substitution patterns influences the stability of the resulting Au(III) complexes in aqueous environment, as demonstrated by UV–visible spectroscopy. In general, all the compounds promptly react with the reducing agent GSH, except **C11–C12**. All the new gold-based complexes are potent inhibitors of human water and glycerol channel aqualgyceoroporin-3 (AQP3), while they are inactive as inhibitors of the water channel AQP1, as evidenced by stopped-flow spectroscopy in hRBC. While most of the compounds are scarcely active as antiproliferative agents against human cancer cells, the neutral complexes **C10–C12** showed promising anticancer activities, particularly in the melanoma A375 cancer cell line. Interestingly, while all the selected cell lines express AQP3, the melanoma cells display protein expression also at the level of the nuclear membrane, as shown by immunocytofluorescence. Thus, it may be hypothesized that **C10–C12** targets mainly AQP3 in the nuclear membrane after being taken up by the cancer cells. Overall, our results hold promise for the design of novel selective AQPs inhibitors to be used anticancer agents or as chemical probes to study the function of these interesting membrane channels.

Supplementary Materials: The following are available online at http://www.mdpi.com/2304-6740/6/4/123/s1, CIF and checkCIF files. Scheme S1: Synthetic pathways to ligands **L1–L12**. Figure S1: ^{1}H NMR (400.13 MHz, CDCl$_3$) spectrum of ligand **L5**. Figure S2: ^{13}C{^{1}H} NMR (100.61 MHz, CDCl$_3$) spectrum of ligand **L5**. Figure S3: ^{1}H NMR (400.14 MHz, CDCl$_3$) spectrum of ligand **L6**. Figure S4: ^{13}C{^{1}H} NMR (100.61 MHz, CDCl$_3$) spectrum of ligand **L6**. Figure S5: ^{1}H NMR (500.13 MHz, DMSO-d_6) spectrum of complex **C2**. Figure S6: ^{13}C{^{1}H} NMR

(125.77 MHz, DMSO-d_6) spectrum of complex **C2**. Figure S7: ^{1}H NMR (400.13 MHz, DMSO-d_6) spectrum of complex **C3**. Figure S8: ^{13}C{^{1}H} NMR (125.77 MHz, DMSO-d_6) spectrum of complex **C3**. Figure S9: ^{1}H NMR (400.13 MHz, DMSO-d_6) spectrum of complex **C4**. Figure S10: ^{13}C{^{1}H} NMR (125.77 MHz, DMSO-d_6) spectrum of complex **C4**. Figure S11: ^{1}H NMR (400.13 MHz, DMSO-d_6) spectrum of complex **C5**. Figure S12: ^{13}C{^{1}H} NMR (100.61 MHz, DMSO-d_6) spectrum of complex **C5**. Figure S13: ^{1}H NMR (400.13 MHz, DMSO-d_6) spectrum of complex **C6**. Figure S14: ^{13}C{^{1}H} NMR (100.61 MHz, DMSO-d_6) spectrum of complex **C6**. Figure S15: ^{1}H NMR (400.13 MHz, DMSO-d_6) spectrum of complex **C7**. Figure S16: ^{13}C{^{1}H} NMR (100.61 MHz, DMSO-d_6) spectrum of complex **C7**. Figure S17: ^{1}H NMR (400.13 MHz, DMSO-d_6) spectrum of complex **C8**. Figure S18: ^{13}C{^{1}H} NMR (100.61 MHz, DMSO-d_6) spectrum of complex **C8**. Figure S19: ^{1}H NMR (400.13 MHz, DMSO-d_6) spectrum of complex **C9**. Figure S20: ^{13}C{^{1}H} NMR (100.61 MHz, DMSO-d_6) spectrum of complex **C9**. Figure S21: ^{1}H NMR (500.17 MHz, DMSO-d_6) spectrum of complex **C11**. Figure S22: ^{13}C{^{1}H} NMR (125.77 MHz, DMSO-d_6) spectrum of complex **C11**. Figure S23: ^{1}H NMR (500.17 MHz, MeCN-d_3) spectrum of complex **C12**. Figure S24: ^{13}C{^{1}H} NMR (125.77 MHz, DMSO-d_6) spectrum of complex **C12**. Figure S25: Absorption and emission spectra of ligand **L1**. Figure S26: Absorption and emission spectra of complex **C1**. Figure S27: Absorption and emission spectra of ligand **L2**. Figure S28: Absorption and emission spectra of complex **C2**. Figure S29: Absorption and emission spectra of ligand **L3**. Figure S30: Absorption and emission spectra of complex **C3**. Figure S31: Absorption and emission spectra of ligand **L4**. Figure S32: Absorption and emission spectra of complex **C4**. Figure S33: Absorption and emission spectra of ligand **L5**. Figure S34: Absorption and emission spectra of complex **C5**. Figure S35: Absorption and emission spectra of ligand **L6**. Figure S36: Absorption and emission spectra of complex **C6**. Figure S37: Absorption and emission spectra of ligand **L7**. Figure S38: Absorption and emission spectra of complex **C7**. Figure S39: Absorption and emission spectra of ligand **L8**. Figure S40: Absorption and emission spectra of complex **C8**. Figure S41: Absorption and emission spectra of ligand **L9**. Figure S42: Absorption and emission spectra of complex **C9**. Figure S43: Absorption and emission spectra of ligand **L10**. Figure S44: Absorption and emission spectra of complex **C10**. Figure S45: Absorption and emission spectra of ligand **L11**. Figure S46: Absorption and emission spectra of complex **C11**. Figure S47: Absorption and emission spectra of ligand **L12**. Figure S48: Absorption and emission spectra of complex **C12**. Figure S49: UV–Visible spectra of the Au(III) complex **C1** (10^{-4} M) in PBS (pH 7.4) recorded over time (left); and of **C1** before and after addition of GSH (2 eq.) recorded over time at room temperature (right). Figure S50: UV–Visible spectra of the Au(III) complex **C2** (10^{-4} M) in PBS (pH 7.4) recorded over time (left); and of **C2** before and after addition of GSH (2 eq.) recorded over time at room temperature (right). Figure S51: UV–Visible spectra of the Au(III) complex **C3** (10^{-4} M) in PBS (pH 7.4) recorded over time (left); and of **C3** before and after addition of GSH (2 eq.) recorded over time at room temperature (right). Figure S52: UV–Visible spectra of the Au(III) complex **C4** (10^{-4} M) in PBS (pH 7.4) recorded over time (left); and of **C4** before and after addition of GSH (2 eq.) recorded over time at room temperature (right). Figure S53: UV–Visible spectra of the Au(III) complex **C5** (10^{-4} M) in PBS (pH 7.4) recorded over time (left); and of **C5** before and after addition of GSH (2 eq.) recorded over time at room temperature (right). Figure S54: UV–Visible spectra of the Au(III) complex **C6** (10^{-4} M) in PBS (pH 7.4) recorded over time (left); and of **C6** before and after addition of GSH (2 eq.) recorded over time at room temperature (right). Figure S55: UV–Visible spectra of the Au(III) complex **C7** (10^{-4} M) in PBS (pH 7.4) recorded over time (left); and of **C7** before and after addition of GSH (2 eq.) recorded over time at room temperature (right). Figure S56: UV–Visible spectra of the Au(III) complex C8 (10^{-4} M) in PBS (pH 7.4) recorded over time (left); and of **C8** before and after addition of GSH (2 eq.) recorded over time at room temperature (right). Figure S57: UV–Visible spectra of the Au(III) complex **C9** (10^{-4} M) in PBS (pH 7.4) recorded over time (left); and of **C9** before and after addition of GSH (2 eq.) recorded over time at room temperature (right). Figure S58: UV–Visible spectra of the Au(III) complex **C10** (10^{-4} M) in PBS (pH 7.4) recorded over time (left); and of **C10** before and after addition of GSH (2 eq.) recorded over time at room temperature (right). Figure S59: UV–Visible spectra of the Au(III) complex **C11** (10^{-4} M) in PBS (pH 7.4) recorded over time (left); and of **C11** before and after addition of GSH (2 eq.) recorded over time at room temperature (right). Figure S60: UV–Visible spectra of the Au(III) complex **C12** (10^{-4} M) in PBS (pH 7.4) recorded over time (left); and of **C12** before and after addition of GSH (2 eq.) recorded over time at room temperature (right). Figure S61: Normalized mean fluorescence intensity (MFI) of AQP3, detected using a secondary Alexa Fluor®488-labelled antibody. Results were normalized for the sample with the lowest expression (A375) and results are expressed as mean $\pm$SEM of three independent experiments. * $p < 0.03$. Figure S62: X-ray structure of crystals of **C6**. Thermal ellipsoids drawn at the 50% probability level. Figure S63: X-ray structure of crystals of **C7**. Thermal ellipsoids drawn at the 50% probability level. Table S1: Antiproliferative activities (EC$_{50}$ values) of ligands in human SKOV-3, A375, MCF-7, and A549 cells after 72 h incubation.

Author Contributions: Conceptualization, A.C. and G.S.; methodology, M.N.W., A.C., A.d.A., A.M., G.S.; formal analysis, A.M., A.d.A., B.A.; investigation, B.A., M.N.W., A.d.A., A.M.; resources, A.C. and G.S.; data curation, M.N.W., B.A., A.d.A., A.M.; X-ray data acquisition and analysis, W.T.K. and S.J.C; writing—original draft preparation, A.C., M.N.W.; writing—review and editing, all authors; visualization, B.A., A.d.A., A.M.; supervision, A.C., G.S.; funding acquisition, A.C., G.S.

Funding: This research received no external funding.

Acknowledgments: Cardiff University and Fundação para a Ciência e Tecnologia, Portugal (project PTDC/BTM-SAL/28977/2017 and PhD fellowship to A.M. SFRH/BD/52384/2013) are acknowledged for funding.

Conflicts of Interest: The authors declare no conflict of interest.

References

1. Casini, A.; Sun, R.W.-Y.; Ott, I. Medicinal Chemistry of Gold Anticancer Metallodrugs. In *Metallo-Drugs: Development and Action of Anticancer Agents*; Sigel, A., Sigel, H., Freisinger, E., Sigel, R.K.O., Eds.; De Gruyter: Berlin, Germany; Boston, MA, USA, 2018; pp. 199–217.
2. Shaw, C.F. Gold-based therapeutic agents. *Chem. Rev.* **1999**, *99*, 2589–2600. [CrossRef]
3. De Almeida, A.; Oliveira, B.L.; Correia, J.D.G.; Soveral, G.; Casini, A. Emerging protein targets for metal-based pharmaceutical agents: An update. *Coord. Chem. Rev.* **2013**, *257*, 2689–2704. [CrossRef]
4. de Almeida, A.; Soveral, G.; Casini, A. Gold compounds as aquaporin inhibitors: New opportunities for therapy and imaging. *Med. Chem. Commun.* **2014**, *5*, 1444–1453. [CrossRef]
5. Agre, P. Aquaporin Water Channels (Nobel Lecture). *Angew. Chem. Int. Ed.* **2004**, *43*, 4278–4290. [CrossRef] [PubMed]
6. Verkman, A.S. More than just water channels: Unexpected cellular roles of aquaporins. *J. Cell Sci.* **2005**, *118*, 3225–3232. [CrossRef] [PubMed]
7. Soveral, G.; Nielsen, S.; Casini, A. *Aquaporins in Health and Disease: New Molecular Targets for Drug Discovery*, 1st ed.; Soveral, G., Nielsen, S., Casini, A., Eds.; CRC Press: Boca Raton, FL, USA, 2016; ISBN 978-971-4987-0783-1.
8. Madeira, A.; Fernández-Veledo, S.; Camps, M.; Zorzano, A.; Moura, T.F.; Ceperuelo-Mallafré, V.; Vendrell, J.; Soveral, G. Human Aquaporin-11 is a water and glycerol channel and localizes in the vicinity of lipid droplets in human adipocytes. *Obesity* **2014**, *22*, 2010–2017. [CrossRef] [PubMed]
9. Ishibashi, K.; Tanaka, Y.; Morishita, Y. The role of mammalian superaquaporins inside the cell. *Biochim. Biophys. Acta Gen. Subj.* **2014**, *1840*, 1507–1512. [CrossRef] [PubMed]
10. Aikman, B.; De Almeida, A.; Meier-Menches, S.M.; Casini, A. Aquaporins in cancer development: Opportunities for bioinorganic chemistry to contribute novel chemical probes and therapeutic agents. *Metallomics* **2018**, *10*, 696–712. [CrossRef] [PubMed]
11. Soveral, G.; Casini, A. Aquaporin modulators: A patent review (2010–2015). *Expert Opin. Ther. Pat.* **2017**, *27*, 49–62. [CrossRef] [PubMed]
12. Martins, A.P.; Marrone, A.; Ciancetta, A.; Cobo, A.G.; Echevarría, M.; Moura, T.F.; Re, N.; Casini, A.; Soveral, G. Targeting aquaporin function: Potent inhibition of aquaglyceroporin-3 by a gold-based compound. *PLoS ONE* **2012**, *7*, e37435. [CrossRef] [PubMed]
13. Martins, A.P.; Ciancetta, A.; deAlmeida, A.; Marrone, A.; Re, N.; Soveral, G.; Casini, A. Aquaporin inhibition by gold(III) compounds: New insights. *ChemMedChem* **2013**, *8*, 1086–1092. [CrossRef] [PubMed]
14. Serna, A.; Galán-Cobo, A.; Rodrigues, C.; Sánchez-Gomar, I.; Toledo-Aral, J.J.; Moura, T.F.; Casini, A.; Soveral, G.; Echevarría, M. Functional Inhibition of Aquaporin-3 With a Gold-Based Compound Induces Blockage of Cell Proliferation. *J. Cell. Physiol.* **2014**, *229*, 1787–1801. [CrossRef] [PubMed]
15. de Almeida, A.; Mósca, A.F.; Wragg, D.; Wenzel, M.; Kavanagh, P.; Barone, G.; Leoni, S.; Soveral, G.; Casini, A. The mechanism of aquaporin inhibition by gold compounds elucidated by biophysical and computational methods. *Chem. Commun.* **2017**, *53*, 3830–3833. [CrossRef] [PubMed]
16. Murray, M.; Ryan, A.J.; Little, P.J. Inhibition of rat hepatic microsomal aminopyrine N-demethylase activity by benzimidazole derivatives. Quantitative structure-activity relationships. *J. Med. Chem.* **1982**, *25*, 887–892. [CrossRef] [PubMed]
17. Sontakke, V.A.; Ghosh, S.; Lawande, P.P.; Chopade, B.A.; Shinde, V.S. A simple, efficient synthesis of 2-aryl benzimidazoles using silica supported periodic acid catalyst and evaluation of anticancer activity. *ISRN Org. Chem.* **2013**, *2013*. [CrossRef] [PubMed]
18. Prosser, K.E.; Chang, S.W.; Saraci, F.; Le, P.H.; Walsby, C.J. Anticancer copper pyridine benzimidazole complexes: ROS generation, biomolecule interactions, and cytotoxicity. *J. Inorg. Biochem.* **2017**, *167*, 89–99. [CrossRef] [PubMed]

19. Serratrice, M.; Cinellu, M.A.; Maiore, L.; Pilo, M.; Zucca, A.; Gabbiani, C.; Guerri, A.; Landini, I.; Nobili, S.; Mini, E.; et al. Synthesis, structural characterization, solution behavior, and in vitro antiproliferative properties of a series of gold complexes with 2-(2′-pyridyl)benzimidazole as ligand: Comparisons of gold(III) versus gold(I) and mononuclear versus binuclear derivat. *Inorg. Chem.* **2012**, *51*, 3161–3171. [CrossRef] [PubMed]

20. Gümüş, F.; Pamuk, İ.; Özden, T.; Yıldız, S.; Diril, N.; Öksüzoğlu, E.; Gür, S.; Özkul, A. Synthesis, characterization and in vitro cytotoxic, mutagenic and antimicrobial activity of platinum(II) complexes with substituted benzimidazole ligands. *J. Inorg. Biochem.* **2003**, *94*, 255–262. [CrossRef]

21. Huang, W.-K.; Cheng, C.-W.; Chang, S.-M.; Lee, Y.-P.; Diau, E.W.-G. Synthesis and electron-transfer properties of benzimidazole-functionalized ruthenium complexes for highly efficient dye-sensitized solar cells. *Chem. Commun.* **2010**, *46*, 8992–8994. [CrossRef] [PubMed]

22. Mardanya, S.; Karmakar, S.; Das, S.; Baitalik, S. Anion and cation triggered modulation of optical properties of a pyridyl-imidazole receptor rigidly linked to pyrene and construction of INHIBIT, OR and XOR molecular logic gates: A combined experimental and DFT/TD-DFT investigation. *Sens. Actuators B Chem.* **2015**, *206*, 701–713. [CrossRef]

23. Petryszak, R.; Keays, M.; Tang, Y.A.; Fonseca, N.A.; Barrera, E.; Burdett, T.; Füllgrabe, A.; Fuentes, A.M.-P.; Jupp, S.; Koskinen, S.; et al. Expression Atlas update—An integrated database of gene and protein expression in humans, animals and plants. *Nucleic Acids Res.* **2016**, *44*, D746–D752. [CrossRef] [PubMed]

24. Fulmer, G.R.; Miller, A.J.M.; Sherden, N.H.; Gottlieb, H.E.; Nudelman, A.; Stoltz, B.M.; Bercaw, J.E.; Goldberg, K.I. NMR chemical shifts of trace impurities: Common laboratory solvents, organics, and gases in deuterated solvents relevant to the organometallic chemist. *Organometallics* **2010**, *29*, 2176–2179. [CrossRef]

25. Cao, Q.; Bailie, D.S.; Fu, R.; Muldoon, M.J. Cationic palladium(II) complexes as catalysts for the oxidation of terminal olefins to methyl ketones using hydrogen peroxide. *Green Chem.* **2015**, *17*, 2750–2757. [CrossRef]

26. Sunesh, C.D.; Mathai, G.; Choe, Y. Constructive Effects of Long Alkyl Chains on the Electroluminescent Properties of Cationic Iridium Complex-Based Light-Emitting Electrochemical Cells. *ACS Appl. Mater. Interfaces* **2014**, *6*, 17416–17425. [CrossRef] [PubMed]

27. Brouwer, A.M. Standards for photoluminescence quantum yield measurements in solution (IUPAC Technical Report). *Pure Appl. Chem.* **2011**, *83*, 2213–2228. [CrossRef]

28. Würth, C.; Grabolle, M.; Pauli, J.; Spieles, M.; Resch-Genger, U. Relative and absolute determination of fluorescence quantum yields of transparent samples. *Nat. Protoc.* **2013**, *8*, 1535–1550. [CrossRef] [PubMed]

29. Coles, S.J.; Gale, P.A. Changing and challenging times for service crystallography. *Chem. Sci.* **2012**, *3*, 683–689. [CrossRef]

30. Dolomanov, O.V.; Bourhis, L.J.; Gildea, R.J.; Howard, J.A.K.; Puschmann, H. *OLEX2*: A complete structure solution, refinement and analysis program. *J. Appl. Crystallogr.* **2009**, *42*, 339–341. [CrossRef]

31. Sheldrick, G.M. Crystal structure refinement with *SHELXL*. *Acta Crystallogr. Sect. C Struct. Chem.* **2015**, *71*, 3–8. [CrossRef] [PubMed]

32. Sheldrick, G.M. *SHELXT*—Integrated space-group and crystal-structure determination. *Acta Crystallogr. Sect. A Found. Crystallogr.* **2015**, *71*, 3–8. [CrossRef] [PubMed]

33. Campos, E.; Moura, T.F.; Oliva, A.; Leandro, P.; Soveral, G. Lack of Aquaporin 3 in bovine erythrocyte membranes correlates with low glycerol permeation. *Biochem. Biophys. Res. Commun.* **2011**, *408*, 477–481. [CrossRef] [PubMed]

34. Schindelin, J.; Arganda-Carreras, I.; Frise, E.; Kaynig, V.; Longair, M.; Pietzsch, T.; Preibisch, S.; Rueden, C.; Saalfeld, S.; Schmid, B.; et al. Fiji: An open-source platform for biological-image analysis. *Nat. Methods* **2012**, *9*, 676–682. [CrossRef] [PubMed]

Review

Iron and Copper Intracellular Chelation as an Anticancer Drug Strategy

Kavita Gaur [1,†], Alexandra M. Vázquez-Salgado [2,†], Geraldo Duran-Camacho [1,†], Irivette Dominguez-Martinez [1,†], Josué A. Benjamín-Rivera [1,†], Lauren Fernández-Vega [1,†], Lesly Carmona Sarabia [1,†], Angelys Cruz García [1], Felipe Pérez-Deliz [1], José A. Méndez Román [1], Melissa Vega-Cartagena [1], Sergio A. Loza-Rosas [1], Xaymara Rodriguez Acevedo [1] and Arthur D. Tinoco [1,*]

[1]   Department of Chemistry, University of Puerto Rico, Río Piedras Campus, Río Piedras, PR 00931, USA; kavitagaur05@gmail.com (K.G.); geraldo.duran1@upr.edu (G.D.-C.); irivette.dominguez@upr.edu (I.D.-M.); josuealberto@gmail.com (J.A.B.-R.); lauren.fernandez@upr.edu (L.F.-V.); leslyyohanac@gmail.com (L.C.S.); angelys.cruz1@upr.edu (A.C.G.); felipe.perez5@upr.edu (F.P.-D.); jose.mendez14@upr.edu (J.A.M.R.); melissa.vega@upr.edu (M.V.-C.); loza304@gmail.com (S.A.L.-R.); xaymara.rodriguez2@upr.edu (X.R.A.)

[2]   Department of Biology, University of Puerto Rico, Río Piedras Campus, Río Piedras, PR 00931, USA; alexandra.vazquez011@gmail.com

*   Correspondence: atinoco9278@gmail.com

†   These authors contributed equally to this work.

Received: 28 September 2018; Accepted: 29 November 2018; Published: 30 November 2018

Abstract: A very promising direction in the development of anticancer drugs is inhibiting the molecular pathways that keep cancer cells alive and able to metastasize. Copper and iron are two essential metals that play significant roles in the rapid proliferation of cancer cells and several chelators have been studied to suppress the bioavailability of these metals in the cells. This review discusses the major contributions that Cu and Fe play in the progression and spreading of cancer and evaluates select Cu and Fe chelators that demonstrate great promise as anticancer drugs. Efforts to improve the cellular delivery, efficacy, and tumor responsiveness of these chelators are also presented including a transmetallation strategy for dual targeting of Cu and Fe. To elucidate the effectiveness and specificity of Cu and Fe chelators for treating cancer, analytical tools are described for measuring Cu and Fe levels and for tracking the metals in cells, tissue, and the body.

Keywords: copper and iron chelators in cancer; transmetalation; metastasis; angiogenesis; metallomics

1. Introduction

In the U.S. cancer is second only to heart disease for the leading causes of death and is soon expected to become the number one cause [1]. The gold standard in the treatment of cancer remains the traditional approaches, surgery, radiation, and older drugs like cisplatin (and its second generation of compounds) but their success rates are extremely low and their applicability is limited [2]. Newer drugs, both broad and narrow spectrum, suffer from off target side effects and toxicity due to their lack of specificity for cancer cells and acquired cellular resistance. The omics-era of cancer research has brought a wealth of information regarding molecular details of cancer cells, the different types of cancer cells, the differences in the cause of the disease, and has also revealed new drug targets [3,4]. There is now much potential for personalized treatment especially in the field of immunotherapeutics [5–7], which seek to bolster the body's natural immune defense mechanism against aberrant cells.

A very promising direction in the development of anticancer drugs is inhibiting the molecular pathways that keep cancer cells alive and able to metastasize. Much attention has been directed toward copper (Cu) and iron (Fe) because they are at the root of cancer cell rapid proliferation and metastasis. Several Cu(I)/Cu(II)and Fe(II)/Fe(III) chelators, originally designed for other applications, are being repurposed as anticancer agents [8,9]. The chelators of interest operate by different proposed mechanisms including binding the metals extracellularly, depleting the metal ions from cancer cells or from intracellular sites that regulate cell function, or inhibiting molecular pathways without necessarily removing them. The common thread in all of these mechanisms is decreasing the bioavailability of functional forms of these metals. Using Cu and Fe chelators for anticancer treatment is extremely promising and several chelators are currently in anticancer clinical trials [10]. In this review article we discuss the roles of Cu and Fe in the progression and spreading of cancer. We also examine pertinent Cu and Fe chelators as potential anticancer drugs and describe efforts to improve their cellular uptake, efficacy, and exclusive cancer cell responsiveness. Transmetallation is introduced as a strategy to overcome the limitations inherent in the chelators binding other metals by manipulating coordination chemistry to target Cu and Fe chelation while releasing a cytotoxic metal into cancer cells. Finally, we briefly examine analytical tools that can be used to gauge the effectiveness of Cu and Fe chelators in inhibiting cancer growth and in targeting the cancer itself.

2. The Role of Cu in Cancer

2.1. Cu as an Important Biological Co-Factor

Cu is an important metal in many biological processes and is essential for cellular physiology. The concentration of Cu in blood is around 5 to 25 μM bound majorly with ceruloplasmin and other Cu binding proteins [11]. Its function stems from Cu's potent redox activity because it can easily change between Cu(II) and Cu(I) in the biological environment [11]. Grubman et al. describe that Cu, like calcium (Ca) and zinc (Zn), can play a modulatory role in cell signal transduction pathways [11]. Cu is an important biological co-factor of many essential enzymes, cuproenzymes, that take part in a variety of cellular processes in mammals [12]. With Cu as a co-factor, these cuproenzymes can be involved in different enzymatic functions, such as melanin synthesis (tyrosinase), energy metabolism (cytochrome *c* oxidase), tissue synthesis (lysil oxidase), antioxidative activity [Cu zinc superoxide dismutase (SOD1)], dopamine synthesis (dopamine-β-hydroxylase) and Fenton reactions, in which the potent redox activity of which can produce reactive oxygen species (ROS) [11]. Cu transportation into cells is regulated by the Cu transporter 1 (Ctr1) membrane protein when Cu(II) is reduced to Cu(I) by reductases. The metal is then transferred to Cu chaperones (one of which is ATOX1), and subsequently delivered to several organelles inside the cell. Cu can bind to Cu-dependent proteins in the Golgi apparatus, such as antioxidants to prevent ROS formation by free Cu ions. Some of the Cu forms a labile Cu(I) pool [13]. As a way of regulating these pathways and high levels of Cu, the Cu importer is internalized and degraded [14].

Changes in the homeostasis of Cu play an important role in different types of diseases, such as inflammation, neurodegeneration, and cancer. In cancer, Cu is involved in several aspects, including metastasis and angiogenesis [11]. Figure 1 shows the role of Cu in a normal cell and in a cancer cell. During angiogenesis, a hallmark of cancer progression, it has been demonstrated that reduced levels of Cu also decreases COX1 and ATP levels. At the same time, oxidative phosphorylation is reduced, which results in decreased growth of proliferative cancer cells [15]. Depletion of Cu was found to inhibit the epithelial-to-mesenchymal transition (EMT). EMT describes the process in which cells lose polarity and cell-cell adhesion. Hence, chelation of Cu can potentially reduce the migratory and invasive properties of cancer cells [11,16]. Different Cu chelators are used in the treatment of Wilson disease, including D-penicillamine (D-Pen), tetrathiomolybdate (TM), and trientine (Trien) (Figure 2), which can modulate cell invasiveness [17,18]. Recent studies demonstrate that small molecules can specifically target the Cu chaperone for superoxide dismutase (CCS) and antioxidant 1 Cu chaperone (ATOX1) and inhibit the cellular Cu transport [19]. ATOX1 on the other hand, plays a major role in Cu homeostasis and it has been seen that its deficiency may lead to the accumulation of high levels of intracellular Cu, and the cause was found impaired with the efflux of cellular Cu [20].

Barresi el al also demonstrated that ATOX1 deficiency could induce copper dyshomeostasis [21]. In the study, they used small Cu chelator and ionophore (binds and transports the metal ions into the cells) molecules to analyze the impact of silencing ATOX1 in colon carcinoma (Caco-2) cells. They observed that the Cu ionophore 5-chloro-8-hydroxyquinoline (ClHQ), was able to exhibit Cu-dependent toxicity in Caco-2 cells and the toxicity was enhanced by ATOX1 silencing. Conversely the Cu chelator *N*,*N*,*N′*,*N′*-tetrakis(2-pyridylmethyl)ethylenediamine (TPEN) generates toxicity which was reversed by the addition of Cu(II), consistent with its ability to remove metal ions. They also observed increased Caco-2 cells sensitivity to TPEN upon gene silencing of ATOX1 [21].

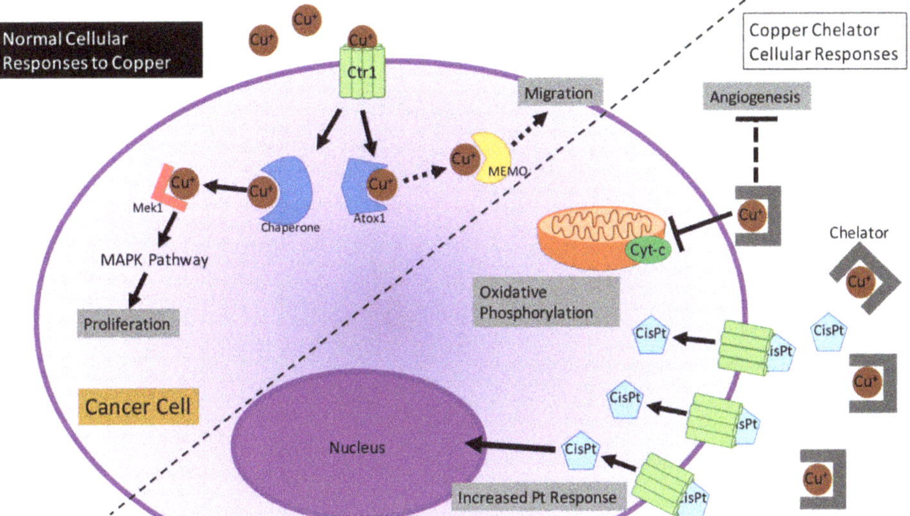

Figure 1. Normal cellular responses to Cu and altered cellular responses with Cu chelator treatment in cancer cells. Once Cu(II) is reduced to Cu(I), it can enter the cell through the high-affinity Cu transporter Ctr1. Cu(I) can be bound to Cu chaperones and transported in the cytoplasm to several organelles, not shown here [14,23,24]. Cu can be delivered to many Cu-binding proteins, including Mek1 and MEMO. Mek1, when activated by Cu binding, will activate the MAPK pathway, which leads to cell proliferation. Antioxidant 1 Cu chaperone (ATOX1) mediates the delivery of Cu to Cu-binding proteins located in the lamellipodia (not shown here), such as MEMO, which is separately known to regulate cancer cell migration and could have implications in metastasis [12,24]. Cu has been associated with angiogenesis [23]. Cu chelator tetrathiomolybdate (TM) treatment can delay the activation of angiogenesis on pancreatic islets. In pancreatic neuroendocrine tumor cells, treatment with TM reduced the activity of Cyt-c, which results in decreased oxidative phosphorylation [23]. CisPt can enter cancer cells through Ctr1. When Cu is present in high levels (a common property of many cancers) cells downregulate the expression of Ctr1 [12]. This results in less CisPt entering the cells. Here, we show the use of Cu chelators to overcome CisPt resistance and make cancer cells more sensitive to this chemotherapeutic agent [12,14,15,23–25].

2.2. Current Studies of Cu in Different Cancer-Altered Biological Processes

In the previous section, we introduced Cu as a metal involved in angiogenesis and cancer metastasis. Here, we dig deeper on the role of Cu in these cancer-altered biological processes, including cell growth and survival. During the last several decades, it has been discovered that Cu plays a role in the development and progression of cancer. The mechanisms by which it does are being explored. Cu has not yet been found to help in the normal-to-cancer cell transformation, but it plays a role in the development of new blood vessels called angiogenesis and the spreading of malignant tumors to secondary sites called metastasis. Ding et al. evaluated the functions of CD147, a signaling receptor in the signaling pathway of extracellular divalent Cu in hepatocellular carcinoma cells (HCC). Their data show that Cu(II)can bind to the extracellular membrane-proximal domain of CD147. This binding activates the PI3K/Akt signaling pathway, important for cell survival and growth, causing CD147 to self-dissociate. At the same time, they noticed the up-regulation of matrix metalloproteinases MMP-2 and MMP-14 in HCC cells. MMP-2 up-regulation induces cell invasion by stimulating neighboring fibroblasts in the presence of Cu(II). Without CD147, extracellular Cu(II) cannot affect the regulation of metalloproteinases. The action of CD147, however, cannot induce the production of MMPs. Ding et al. associated the action of extracellular Cu(II) to the function of CD147 as an interdependent process [22].

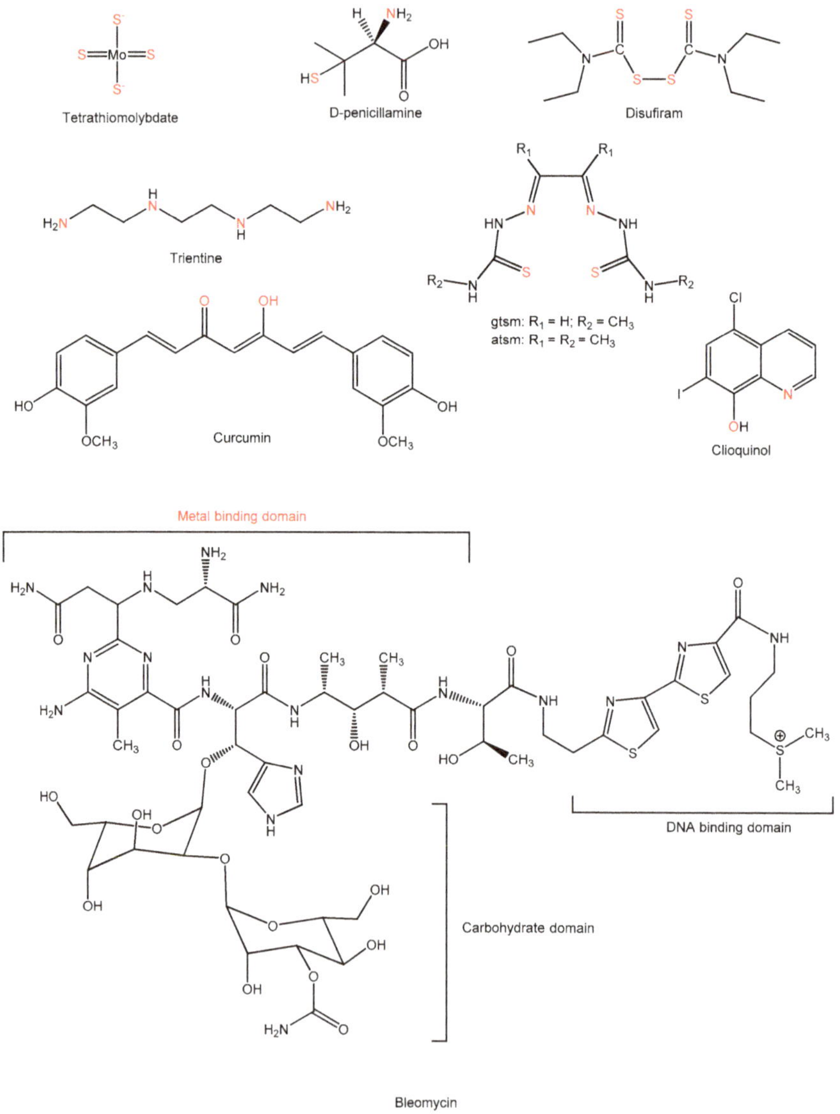

Figure 2. The structure of select Cu chelators for potential anticancer applications.

Animal studies have demonstrated that Cu-chelating drugs show antiangiogenic effects. The mechanisms by which Cu regulates angiogenesis are unknown. Ishida et al. assessed the effects of Cu chelation with tetrathiomolybdate (Figure 2) on the formation of angiogenic islets [23]. They noticed that the number of angiogenic islets decreased significantly after treatment with TM during premalignant lesions, but not during the formation of tumors and their subsequent growth suggesting that TM treatment can delay the angiogenic switch or the "activation" of angiogenesis [23]. Ishida et al. also demonstrated that elevated levels of Cu in drinking water increased the proliferation rate of cancer cells. They determined that intracellular Cu regulated the activity of cytochrome *c* oxidase (CcO) in cancer cells. CcO forms the last complex of the electron transport chain to produce ATP during oxidative phosphorylation. Depletion of Cu by TM decreases CcO activity in pancreatic neuroendocrine tumor cells (βTC3) thereby reducing oxidative phosphorylation in cancer cells. For ATP production under Cu-limited conditions, they proved that cells rely on glycolysis. These findings suggest new target strategies for cancer treatment where Cu chelators can be tested in combination with classical chemotherapy (platinum-based drugs) or with a glycolysis inhibitor, which would potentially eliminate overall production of ATP in cancer cells [23].

For cancer metastasis to occur, many signaling pathways and processes dependent on cell motility and migration must first occur. Cells move through the formation of protrusion "toes" at the leading edge, known as lamellipodia and filopodia [12]. Several Cu-dependent proteins have been implicated in cancer metastasis. ATOX1 is a Cu chaperone that picks up Cu from the Ctr1 and delivers it to ATPases found in the Golgi Apparatus [12]. It has not been determined if ATOX1 plays a direct role in Cu-dependent cancer metastasis. Blockhuys et al. show the presence of ATOX1 in the nucleus (acting as a transcription factor) and cytoplasm, but more interestingly, in the lamellipodia of MCF-7 and MDA-MB-231 breast cancer cell lines. ATOX1 silencing with siRNAs did not affect the formation of lamellipodium in MDA-MB-231 cells. Nevertheless, using a wound-healing assay, they determined that ATOX1 is required for breast cancer cell migration in vitro. ATOX1 may be mediating the delivery of Cu to Cu-binding proteins located in the lamellipodia, such as MEMO (Mediator of ErbB2-driven cell motility), which is separately known to regulate cancer cell migration. Another study demonstrated similar results [26], that ATOX1 was being delivered to the lamellipodia of vascular smooth muscle cells and that it was essential for platelet-derived growth factor-mediated migration.

McDonald et al. evaluated the behavior of MEMO as a Cu-dependent redox protein in the process of metastasis and migration [27]. Studying breast cancer, they found that MEMO acts in the migration and invasion of cancer cells in vitro promoting a more oxidized environment and that MEMO stimulates the production of ROS. In vivo studies in 6-week-old female non-obese diabetic/severe combined immunodeficient mice demonstrated that the activity of MEMO was present in the spontaneous metastasis of breast cancer to the lungs. The authors found that the abundance of MEMO can be correlated as an independent factor of early distant metastasis [27].

Recent studies examine the role of Cu in many biological processes that are commonly altered in cancer development and progression. Cu is a co-factor of enzymes that participates in several biochemical processes, such as oxidative phosphorylation. Turski et al. found that the MAPK (mitogen activated protein kinase) signaling pathway can be stimulated by the high-affinity Cu transporter Ctr1, which is overexpressed in genetically engineered mice models (GEMMs) of human cervical carcinoma when the Ras pathway is activated by external growth factors [23,24]. The Ras protein is a GTPase encoded by the Ras oncogene, which can subsequently activate several oncogenic signaling pathways, including the MAPK pathway. This pathway is involved in regulating cell proliferation, survival, motility, and metabolism, all of which can be altered in cancer cells. Even more, mutations that hyperactivate molecular components of the MAPK pathway show direct association with the development and progression of cancer. Tuski et al. found that reduction of intracellular Cu, using a Cu chelator, results in decreased activation of the MAPK pathway [24]. At a molecular level, they found that Cu binds to Mek1, a component of the said pathway, which phosphorylates Erk [28].

Their research with in vitro cell culture and fly and mice models demonstrated that Cu-dependent Ctr1 affects MAPK pathway activation.

In a related finding, increased levels of Ctr1 mRNA in colorectal cancer resulted in a parallel increase in transcript levels for Cu efflux pump ATP7A, copper metabolism Murr1 domain containing 1 (COMMD1), the cytochrome *c* oxidase assembly factors (synthesis of cytochrome *c* oxidase 1 (SCO1) and cytochrome c oxidase Cu chaperone 11 (COX11)), the cupric reductase enzyme six transmembrane epithelial antigen of the prostate (STEAP3), and the metal-regulatory transcription factors (MTF1, MTF2) and specificity protein 1 (SP1) [29]. This data suggests that due to the high proliferation rates, cancer cells require a greater amount of Cu than the normal cells. Considering that many cancer patients' serum and cancer cells possess increased levels of Cu, these results have strong implications on the potential role of Cu on proliferation and metastasis [30].

BRAF (V600E) is a kinase that phosphorylates Mek1, which results in the downstream signaling activation of the MAPK pathway. In a study, half of the melanoma patients showed mutations on the BRAF^{V600E} gene. Current drug treatments for BRAF^{V600E}-positive melanoma focus on administering inhibitors for BRAF and Mek1/2 but as with many other drugs, subsets of melanomas develop resistance. Due to the Cu-dependence of Mek1, downregulating Ctr1 (Cu transporter) has been proposed as a treatment option. Nevertheless, melanoma cells counteract this by producing Cu-independent Mek1. Brady et al. demonstrated that the use of TM, which was tested in vivo by Ishida et al., reduces the levels of Cu and activation of the MAPK pathway, resulting in the inhibition of BRAF^{V600E}-positive melanoma tumor cell growth of cells resistant to BRAF and Mek inhibitors. Other findings of this study showed toxicities in GEMMs for melanoma, but increased survival benefits after treatment with TM alone or in combination with a BRAF inhibitor, vemurafenib. This study highlights the potential of TM as a treatment for BRAF^{V600E}-positive melanoma patients alone or in combination with other melanoma treatments [25].

Altogether, these findings highlight Cu's role in cancer progression, metastasis, and angiogenesis and support the study of Cu chelators as potential cancer therapy alone or combined with other treatments already approved or in cancer clinical trials. We have shown the importance of studying Cu-dependent proteins, such as MEMO and signaling receptor CD147 and their involvement in migration and metastasis processes.

2.3. Chelation Strategies

Several strategies to target Cu for cancer therapy are the use of Cu chelators and Cu ionophores. By definition, chelators are the compounds that sequester the metal ions and can make them bio-unavailable, in contrast, ionophores usually elevate intracellular bioavailability of metal ions [31,32]. Features of both chelators (intracellular chelation) and ionophores (extracellular chelation) can be utilized in anticancer drug designing strategy [33]. Denoyer et al. assessed recent literature on the effects of Cu transporters in cells resistant against platinum complexes. Cisplatin (CisPt) and other platinum-based drugs can enter cells through Ctr1. When Cu is present in high levels (a common property of many cancers) cells downregulate the expression of Ctr1. This results in less CisPt entering the cells [14]. The use of Cu chelators to overcome this resistance has become more popular (Figure 1). Cu chelators were first synthesized for Wilson's disease patients that have Cu overload usually localized to the liver. Clinical data is variable between cancer types, but there seems to be a common behavior of Cu chelation being most useful in preventing cancer progression of micrometastasis, where angiogenesis has not reached later stages yet, as demonstrated by Ishida et al. [23]. This suggests that the role of Cu in cancer progression is mainly due to facilitating angiogenesis. Cu chelators could play important functions in inhibiting Cu dependent pathways for cancer progression. Cu(I) is a soft Lewis Acid that is typically bound by the S donor atoms and intermediate N donor atoms in ligands. It tends to form coordination compounds with coordination numbers 2, 3, and 4 (Figure 3) [34]. Cu(II) is an intermediate Lewis Acid that in coordination compounds is found bound to N donor atoms and hard O atoms. Cu(II) coordination compounds are very commonly of coordination numbers 4, 5, and 6 (Figure 3) [34]. Due to the reducing environment of cells, Cu dominantly exists intracellularly in the +1 oxidation state as opposed to the +2 form in blood. In the following, we discuss a selection of Cu chelators for anticancer application (Figure 2). Several of these ligands have the ability to bind Cu in both the +1 and +2 oxidation states and so are capable of binding Cu(II) extracellularly and Cu(I) intracellularly.

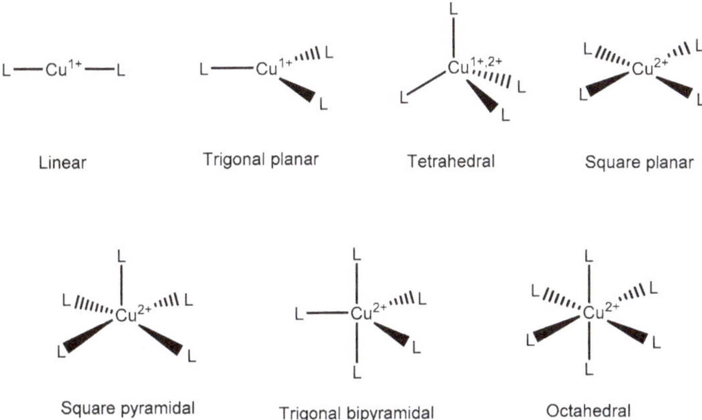

Figure 3. Typical geometries for Cu(I) and Cu(II) chelation.

2.3.1. D-Penicillamine

D-Pen is a D-isomer of dimethylcysteine which is used for the treatment of Wilson's disease [35]. Additionally, it is used for the treatment of rheumatoid arthritis, Cu poisoning, and lead poisoning [36,37]. D-pen was developed by Walshe in 1956 and sold under the trade name of Cuprimine and Depen. It is an α-amino acid metabolite of penicillin which was discovered by Walshe from the urine of patients with liver disease and treated with penicillin [38]. Later, he isolated the compound and identified its Cu chelating properties. D-Pen (ββ dimethyl cysteine) is a thiol that contains a sulfhydryl group which binds to Cu(I) and eliminates it via urine both in normal people and in patients with Wilson's disease [35]. Depending on the number of functional groups involved in the formation of the complex, D-pen can function as a monodentate, bidentate, or even tridentate ligand [39]. Walshe et al. proposed the possibilities of the bond formation between Cu(I) and D-pen, (i) a single Cu could bind to a single thiol group (C–S–Cu$^+$), (ii) a Cu atom might be involved in binding with the thiol groups of two D-pen molecules on both sides (–S–Cu–S–), or (iii) a ring compound might be formed by the linkage of one Cu atom with the thiol and amino groups of one D-pen [35]. The pKa of the thiol group of D-pen is 7.9 which makes it more reactive thus the degree of ionization for D-pen at physiological pH would be higher than that compared to cysteine (pKa 8.3) and Glutathione (GSH; pKa 8.8) [40]. Moreover, it has been seen that D-pen oxidizes at higher rates in the presence of Cu compared to other thiols (e.g., GSH, cysteine, Homocysteine, *N*-acetylcysteine) [41]. The recommended dose for penicillamine is 1 g/day [42]. The drug is highly efficient in reducing the excess Cu in the body and its safe long-term application has been investigated in patients. Conversely, it also shows some severe side effects such as neurological deterioration, induction in degenerative dermopathy, and cutaneous side effects [43,44]. According to a study on 179 patients with rheumatoid arthritis, after treatment with 5 mg D-Pen (control group) and with 100 mg (drug group) [45], convalescence was observed in 27% patients of the control group and 65% patients of the drug group. Some adverse reactions were also found in 34% of the control and 49% of the drug group. In the trial, the common adverse effects in the drug group were skin rashes, taste disturbances, gastrointestinal upset, and proteinuria.

Despite the side-effects, the ability of D-Pen to chelate Cu has also been investigated as an antiangiogenic property [46]. Brem et al. demonstrated that D-pen was able to remove excess serum Cu levels and reduced it to <50 µg/dL after two months of oral administration [47]. Alireza et al. demonstrated that penicillamine and low Cu diet reduces the serum Cu levels and serum Vascular Endothelial Growth Factor in the patients that underwent stereotactic radiosurgery for recurrent glioblastoma multiforme [48]. D-pen has been found to inhibit human endothelial cell proliferation in vitro and neovascularization in vivo [49]. Daizo et al. investigated a suppression in the growth of 9L gliosarcoma tumors implanted in rats when treated with D-Pen [47]. It has been studied that in the process of Cu chelation by D-pen Cu(II) reduces to its Cu(I) form which leads to the generation of hydrogen peroxide (H_2O_2) and other ROS [40,50]. It has been studied that D-Pen in combination with Cu promotes cell death in endothelial and lymphocyte cells due to the generation of ROS [30]. Starkebaum et al. concluded that D-Pen in the presence of Cu oxidizes to form H_2O_2 along with the superoxide anion ($O_2{}^-$) leading to the inhibition of lymphocyte mitogenesis [51]. They hypothesized that D-pen initially reduces Cu(II) to Cu(I) probably due to Cu chelation. The reduced Cu, Cu(I) then induces a reduction of oxygen and produces $O_2{}^-$, leading to the production of H_2O_2 [51]. Another report by Gupte et al. supported the possibility of the formation of ROS in human leukemia and breast cancer cells [52]. They reported that at low concentration D-Pen in the presence of Cu produces concentration-dependent H_2O_2 mediated cytotoxicity in cancer cells. Hence, D-pen could be employed as an anti-cancer agent owing to its dual behavior of exhibiting anti-angiogenic properties due to Cu chelation and cytotoxic properties due to the generation of ROS. However, some physicochemical properties of D-pen hinder its in-vivo anticancer activity because of limited intracellular delivery; for instance, high hydrophilicity, rapid elimination and metal catalyzed oxidation (due to which it forms an inactive D-pen disulfide or mixed disulfides) [53].

2.3.2. Tetrathiomolybdate

Tetrathiomolybdate (TM) is a highly potent Cu chelating agent. Structurally it has an atom of molybdenum with four sulfur substitutions and a bidentate ligand. TM possesses dual mechanisms of action: (a) when given with meals it forms a tripartite complex of TM, Cu, and food protein; (b) in the dosage without meals, it is absorbed into the blood and forms a tripartite complex with albumin and the freely available serum Cu [28,54]. Both forms of complexed Cu are unavailable for cellular uptake. Due to its efficiency to chelate Cu and excellent safety profile, it was first used in patients with Wilson's disease. Several clinical studies have demonstrated efficacy in treating Wilson's disease, especially as a first-line treatment for neurologically affected patients [55,56]. The role of Cu metabolism in angiogenesis has led to the testing of TM on angiogenesis and, consequently, on carcinogenesis. TM has been shown to cause the inhibition of nuclear factor kappa B (NF-κB), a master switch for transcription of many cytokines which results in anti-angiogenesis [57]. Several studies suggest that TM suppresses tumor growth and angiogenesis [58,59]. Brewer et al. carried out a Phase I clinical trial in 18 patients with metastatic cancer and reported the reduction of ceruloplasmin, a cuproenzyme in charge of carrying Cu in the blood, without any toxicity in five of six patients with stable disease [60]. Phase II clinical trials in patients with advanced kidney cancer concluded that TM could reduce Cu levels. TM was found to be well tolerated, although its anti-cancer activity was limited to the stabilization of disease for a median of 34.5 weeks [61]. Another phase II trial of TM was performed on the patients after surgery for malignant mesothelioma [62]. This trial concluded that TM has antiangiogenic effects in malignant pleural mesothelioma patients after surgery with minimal toxicity.

Some cancer cells overexpress many receptors and proteins in their membrane and other organelles. One of them is the hypoxia-inducible factor-1α (HIF-1α), which triggers adaptive responses during low oxygen conditions, including angiogenesis, invasion, metastasis, glycolysis, tumor survival, and proliferation [63]. Cu was shown to be a requirement for HIF-1α activation, activation of vascular endothelial growth factor (VEGF) expression in cells, and promotion of wound repair in mice. TM mediated a time and dose dependent reduction of HIF-1α protein levels in human endometrial and ovarian cancer cells after 48 h of treatment [63] due to Cu deprivation. Another study showed that TM could decrease the activity of complex IV, a Cu-dependent enzyme, in the mitochondria (cytochrome *c* oxidase). In human endometrial and ovarian cancer cells, TM was the most efficient chelator to exhibit this effect [63].

2.3.3. Trientine

Trien is sold under the name of Syprine, which is commonly used to treat Wilson's disease. It is a tetradentate ligand and chemically known as triethylene tetramine dihydrochloride and was developed as an alternative drug for penicillamine intolerant patients. Trien works primarily by promoting the urinary excretion of Cu from the body. Trien was found to produce neurological worsening at the beginning of treatment but appears much less common than that with penicillamine. Trien has been shown to induce apoptosis in murine fibrosarcoma cells in vitro and in vivo [64]. Moriguchi et al. examined the antiangiogenic effect of Trien against hepatocellular carcinoma by focusing on the relationship between Cu and interleukin-8 (IL-8), which is a potent angiogenic factor produced by hepatoma cells [65]. Moriguchi et al. concluded Trien efficiently chelates Cu and prevents it from functioning as a cofactor for angiogenesis, which resulted in reduced IL-8. Another study suggests that the Trien suppresses the development of tumors and angiogenesis in the murine hepatocellular carcinoma cells [66]. A recent in vivo study suggests that Trien significantly reduces the Cu in plasma and liver tissue, additionally with the inhibition of RFA-induced inflammatory gene expression and ROS-induced malondialdehyde production in liver [67]. Trien also causes some adverse side effects, for example, reversible sideroblastic anemia, lupus like reactions, and worsening of neurological manifestations [68].

2.3.4. Bleomycin

Bleomycin (BLM) is a group of glycopeptide antibiotics produced by *Streptomyces verticillus*. It is clinically used for the treatment of squamous cell carcinoma, malignant lymphoma, testicular cancer, cervical cancer, and other cancers [69]. Bleomycin causes DNA strand scission via the formation of an intermediate metal complex which requires a metal ion such as Cu(II) or Fe(III) [70,71]. BLM exhibits strong chelation with various metals and especially with Cu(II) [72]. Metal free BLM tends to bind to the Cu(II) in 1:1 stoichiometry with the formation constant value; log K = 12.63 [73]. Cu(II)-BLM complex obtains a square-pyramidal geometry [74]. Both forms—free ligand and complex form—are excellent anti-cancer agents. Due to the strong Cu(II) chelating ability BLM can reduce Cu storage in the tissue in Wilson's disease [73].

2.3.5. Curcumin

One of the most studied phytochemical compounds in Cu-based therapeutics for cancer is curcumin [75–78]. Curcumin is a bidentate ligand and binds more efficiently with redox-active metals such as Fe and Cu than the redox-inactive zinc [79]. Curcuminoid compounds have shown an excellent behavior in antitumor activity in vivo. Ahsan et al. found that curcumin in the presence of Cu(II), selectively exhibit calf thymus and supercoiled plasmid pBR322 DNA cleavage due to the formation of ROS [75]. They hypothesized that curcumin is capable of binding to DNA but in the presence of Cu(II), it shows cleavage due to the formation of hydroxyl radical ($\cdot$OH). Curcumin has shown to induce apoptosis by downregulating the transcription factor NF-κB in human multiple myeloma cells [80] According to a study curcumin exhibits enhanced cytotoxicity in the cancer cells in the presence of Cu(II) selectively over the other metal ions [81]. It has been found that curcumin in combination with Cu(II) exhibits ionophoric behavior and enhances the levels of intracellular Cu leading to the suppression of the NF-κB pathway and modification of mammalian target of rapamycin-raptor (mTOR) signaling in the cancer cells [81]. A study demonstrated that curcumin efficiently chelates the Cu(II) and prevents the tumor growth, angiogenesis, and induced apoptosis in A549 xenograft model [82].

2.3.6. Ionophores

Ionophores are a class of metal-binding chelators that are lipid soluble capable of transferring metal ions across biological membranes, between intra- and extracellular compartments, most commonly into the cells, but rarely out of cells [83–85]. In this section we discuss the cases of ionophores that carry metal ions into the cells. Ionophores usually transport metal ions inside the cells and can lead to the generation of toxicity in cancer cells. There are two different approaches that describes ionophoric behavior. First, moderate metal affinity allows some ionophores to bind to the metal ions from the higher concentration areas and deliver them in the area of lower concentration. Second, a suitable acid dissociation constant value (pKa) that affects deprotonation of the compound when it enters in the cellular compartments, this event induces the release of metal ions and can regulate cytotoxic behavior [86]. If the extracellular pH is higher than the pKa of the ionophore, it will form a complex with the metal ion and transport it to the environment where the pH is lower than the pKa of the compound [31]. Some Cu ionophores are selective towards cancer cells and they exhibit activity for a broad range of cancer types [14]. Bis-(thiosemicarbazone) ligands have been studied widely for their anticancer activities [87–90] and they can chelate Cu(II) after getting deprotonated at the N atoms and form neutral complexes, with the square planar N_2S_2 coordination geometry [90]. Cater et al. suggest that two bis-(thiosemicarbazone) Cu complexes glyoxalbis [N4-methylthiosemicarbazonato] Cu(II) [Cu(II)(gtsm)] and diacetylbis-[N4-methylthiosemicarbazonato] Cu(II) [Cu(II)(atsm)] (ligands; Figure 2), selectively kill prostate cancer cells in vitro and in vivo [90]. They found enhanced activity of both ligands when the extracellular Cu concentration increased and found them toxic when combined with Cu. They found that the Cu(II) dissociates from Cu(II)(gtsm) leading to the increase in intracellular bioavailable Cu(II) which was proven by its mechanistic action of inhibiting proteasomal chymotrypsin-like activity, an established feature associated with the Cu ionophores that increase intracellular bioavailable Cu(II) [90]. Cu in both oxidation states 1+ and 2+ can interact with electron donor groups; thiol and amino groups located on the active site of proteasome. This interaction leads to conformational changes which causes proteasome inhibition and apoptosis in cancer cells [86].

Two other compounds clioquinol and disulfiram have been investigated for their anticancer activities and are in clinical trials [31,91]. Clioquinol (5-chloro-7-iodo-8-hydroxyquinoline, CQ), and disulfiram (DSF) (Figure 2) both exhibit anticancer activities via Cu ionophoric [90] and follow the same mechanism of the inhibition of the proteasomal system chymotrypsin-like activity and induce apoptosis in cancer cells [14,83,86,92,93]. CQ is a derivative of 8-hydroxyquinoline which was initially developed as an effective amebicide for treating diarrhea [31]. CQ is a bidentate ligand which chelates metals via its N and O donor atoms and shows preference for binding to Cu(II) and Zn(II) [94]. It has been found that CQ at the micromolar concentration range induces apoptosis via a caspase-dependent pathway in eight different human cancer cell lines [31]. Studies revealed that without exhibiting any toxicity, CQ efficiently slows down the growth of xenografted tumor in mice models [84]. Caragounis et al. found that CQ elevates the intracellular Cu levels which results in the activation of PI3K (phosphoinositide 3-kinase), MAPK, and JNK (c-Jun N-terminal kinase) in amyloid precursor protein overexpressing CHO (Chinese-hamster ovary) cells [95]. In addition, when these cells are treated with the CQ and Cu(II) they exhibit a ~85–90% reduction of secreted amyloid β-peptides, Aβ-(1–40) and Aβ-(1–42) [95,96]. A sulfur-based chelator DSF is a member of dithiocarbamates and is used to treat alcoholism (FDA approved) [97]. DSF exhibits anticancer activity by the generation of ROS, inhibition of proteasome, and induction of apoptosis [94]. Denoyer et al. studied the effect of the treatment of transgenic adenocarcinoma of mouse prostate (TRAMP) cells and normal mouse prostate epithelial cells (PrECs) with the Cu-ionophores (CuII(gtsm), disulfiram and clioquinol) and found the production of toxic levels of ROS in TRAMP cells but not in the normal cells [32]. It has been studied that supplementation of DSF with Cu dramatically enhanced the inhibition of tumor growth in a prostate cancer mouse model [98]. DSF is currently under phase II clinical trial to study its impact in combination with Cu on metastatic breast cancer.

2.4. Efforts to Optimize Drug Delivery and Efficacy of Cu Chelator Agents

Different systems have been designed to deliver Cu chelating agents and improve their uptake from the body minimizing possible side effects. These systems take advantage of the tumor hallmarks. The enhanced vascular permeability of tumor tissue is a consequence of rapid tumor growth. Angiogenesis becomes deficient and produces blood vessels with large pores between the endothelial cells. In addition, the lymphatic drainage becomes inefficient due to the absence or malfunction of lymphatic vessels in tumors. These processes lead to the phenomenon of the enhanced permeation and retention (EPR) effect [99], which results in nanosized species penetrating the tumor vasculature and being retained for extended periods of time. Drug delivery strategies try to take advantage of this phenomenon by using nanoparticle carrier formulations of drugs. Optimal nanoparticle size range between 10 to 100 nm. The minimum size is to avoid secretion by the kidney and the larger size is to prevent phagocytic clearance by the reticuloendothelial system (RES) [100]. As drug-delivery agents, nano-carriers are capable of targeting cancer cells with enormous specificity and sensitivity especially if conjugated with ligands whose receptors are overexpressed in cancer cells versus healthy cells. If designed with a response system sensitive to the intracellular environment, they can release drugs in a regulated manner [100]. These structural fine tuning protect healthy cells from potentially toxic agents, prevent premature drug degradation, and control drug distribution over the body.

Efforts to improve the intracellular delivery of D-pen have been examined, taking into consideration the importance of the thiol group and the transport of high concentrations to cancer cells. This Cu chelating agent shows some disadvantages due to its high hydrophilicity. It is easy to oxidize to D-pen disulfide in vivo and it is rapidly removed from the blood [53]. One of the approaches for delivery of this drug has been the use of soluble macromolecules, such as peptides, protein, or polymers. Gupte and co-workers propose the synthesis of gelatin-D-pen conjugate with a reversible disulfide bond, resulting in the complete release of D-pen after 4 h in the presence of 1 mM glutathione at pH 7.4. Results showed cellular uptake and cytotoxicity in the HL-60 leukemia cell line [53]. Nevertheless, the system showed low efficacy. More recent methods have been reported, including the PGA-D-pen conjugate. Poly-L-glutamic acid (PGA) have the advantage of being a biocompatible and biodegradable polymer [101]. This arrangement increases the cellular uptake in vitro and the survival of mice in in vivo studies [101]. The chemical structure proposed for the gelatin-D-pen and PGA-D-pen conjugate is shown in Figure 4. Different strategies have been used to improve the cellular uptake of BLM. Norum et al. investigated the use of photodynamic therapy (PDT) and photochemical internalization (PCI) of BLM in CT26. CL25 mouse colon carcinoma cancer cells (Figure 5) [102]. PDT is a treatment well established for cancer in which a photosensitizer is used excited at a specific wavelength to produce reactive oxygen species to kill cancer cells. PCI is a recent therapy delivery technique that allows the release of macromolecules from the endosomes or lysosomes to the cytosol in the desire cancer cell. The in vivo studies implicated the use of athymic and thymic mice, in the presence and absence of T cells respectively. Delayed tumor growth in athymic mice was observed in both PCI and PDT of BLM but curative effects were not observed with the selected light dose used for PDT and PCI treatments. However, the results in thymic mice showed 90% and 70% of curative effects using PCI and PDT respectively, suggesting the importance of T cells and the immune system in general inducing cancer cells death [5,6,102].

Delivery systems have been developed to overcome the low solubility and poor bioavailability of curcumin [103]. Luo and co-workers designed curcumin-conjugated nanoparticles for delivery into A549 lung cancer cell line. Curcumin was conjugated with 1,4-(hydroxymethyl) phenylboronic acid (HPBA)-modified poly(ethylene glycol)-grafted poly(acrylic acid)polymer (PPH) in an effort to obtain a curcumin-coordinated ROS-responsive nanoparticle (PPHC) (Figure 6). Uniform spherical particles were obtained with a particle size around 163.8 nm that in the presence of H_2O_2 is able to release curcumin. Cellular studies demonstrated the efficient release of curcumin into A549 cells [103].

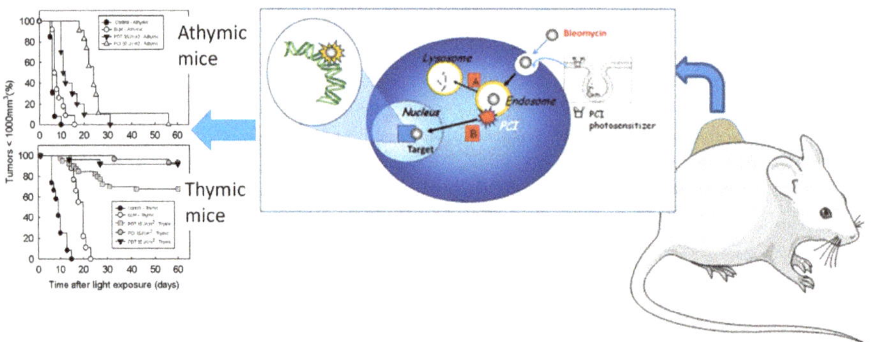

Figure 4. Proposed D-pen conjugates as drug delivery systems for D-pen. (**A**) Gelatin-D-pen conjugate. (**B**) Poly-L-glutamic acid (PGA)-D-pen conjugate. Reprinted with permission from *Bioconjugate Chemistry*, **19**, 1382–1388. Copyright 2008 American Chemical Society [53].

Figure 5. Proposed mechanism for photochemical internalization (PCI) stimulating the bleomycin release out of endosomes bringing a local therapeutic effect. Reprinted with permission from *Journal of Controlled Release*, 268, 120–127. Copyright 2017 Elsevier [102].

Due to the high affinity that sulfur exhibits for Cu(I) ions and the efficiency of nanomedicine, sulfur nanoparticles have been used as Cu chelators in melanoma and breast cancer cells [104]. It has been shown that Poly ethylene glycol (PEG) Sulfur nanoparticles (nano-S) are efficient as Cu chelators and selective to Cu enriched cancer cells. Cu depletion inhibited cell growth and the MEK/ERK proliferation pathway in A375 and MCF-7 cancer cells resulting in mitotic arrest [104].

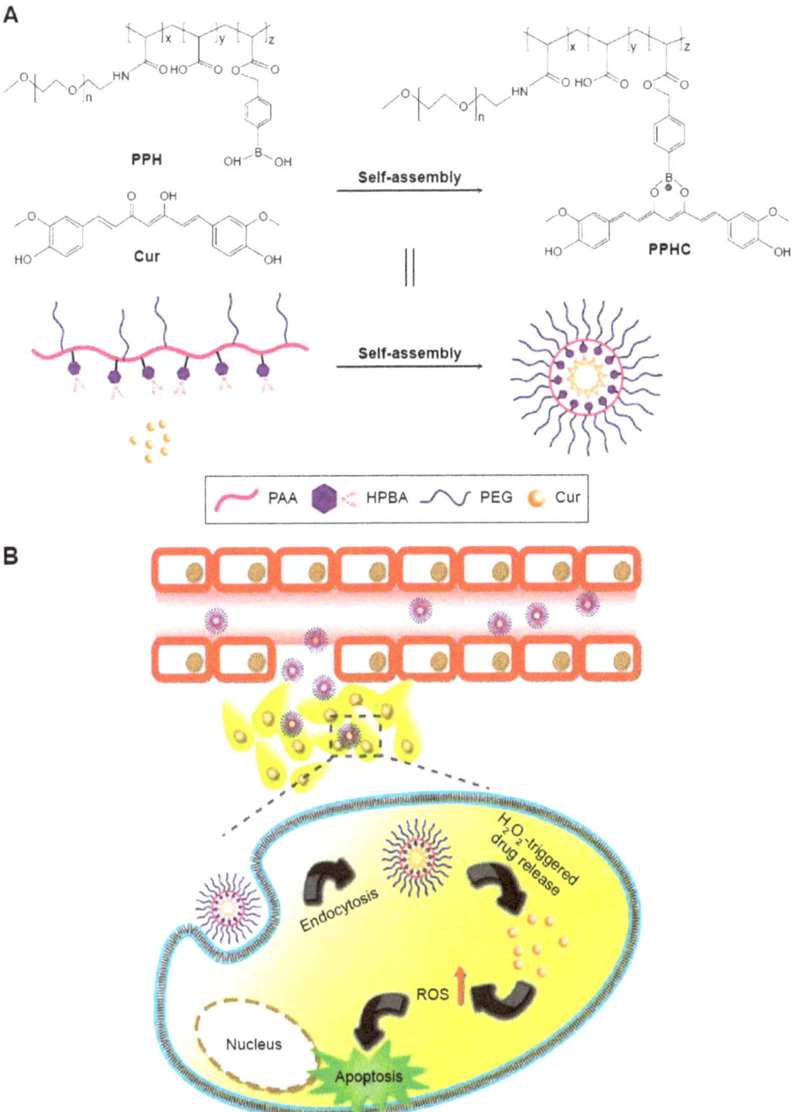

Figure 6. Schematic depiction of the preparation and intracellular delivery of reactive oxygen species (ROS)-sensitive responsive nanoparticle (PPHC) nanoparticles. (**A**) Preparation of ROS-responsive PPHC nanoparticles showing coordination between boronic acid and Cur. (**B**) Graphical representation of intracellular ROS-triggered drug delivery and cell apoptosis induced by amplified ROS signals. Abbreviations: PPH, 4-(hydroxymethyl) phenylboronic acid-modified PEG-grafted poly (acrylic acid) polymer; Cur, curcumin; PAA, poly(acrylic acid); HPBA, 4-(hydroxymethyl) phenylboronic acid; PEG, poly(ethylene glycol). Reprinted with permission from *International Journal of Nanomedicine*, **12**, 855. Copyright 2017 Dove Press [103].

In addition to efforts to improve the specificity of Cu chelators for cancer cells, other studies have been devoted to examining their synergism with other anticancer treatments. Cisplatin is used to treat different types of cancer diseases [105]. One of the issues with cisplatin is that it needs to be in high concentrations to be efficient and it could damage other cells and tissues at high dosages, while leaving the target drug resistant [105]. It has been shown that cisplatin-resistant mutants express the Cu transporter Ctr1 as a major mediator for cisplatin uptake in cells which gives platinum-based therapy higher efficacy [105]. The analysis of Cu exporters, ATP7A and ATP7B, confirmed that Ctr1 is the major determinant for cisplatin accumulation in yeast and mice cells [105]. But recent studies suggest that the transporters involved in the cellular accumulation of cisplatin are classified as; (i) Cu transporters: Ctr1, Ctr2, ATP7A, ATP7B; (ii) ABC transporter: ATOX1; (iii) organic cation transporter: OCT1; (iv) multidrug and toxin extrusion family members: MATE; (v) volume sensitive, the volume-regulated anion channel proteins: VRAC (LRRC8A/D-containing) [106]. Leucine-rich repeat-containing protein 8 (LRRC8) encoded by genes LRRC8A/D is the subunit of heteromer protein VRAC [107,108]. Cellular accumulation determination of Pt-based drugs in the cells having various LRRC8 genotypes of VRAC proteins revealed that 50–70% of long-term cisplatin uptake depends on LRRC8A and LRRC8D [109]. Studies with mice models of HPV16-induced cervical carcinoma demonstrated that the Cu chelator TM has a synergistic antitumor effect when combined with cisplatin [105]. The combination of both treatments shows an increase of cisplatin uptake into cancer cells compared to healthy cells which have a higher demand for Cu for proliferation and survival. Cancer cells overexpress Cu transporter Ctr1 which gives certain selectivity to the treatments mentioned before. The mechanism of Ctr1 is not clear but it is proposed that Cu starvation enhances cisplatin transport by changing the conformation of the Cu transporter allowing more cisplatin to enter cells and tissues. Also, TM serves as an anti-angiogenic agent and increases the efficiency of cisplatin delivery which indicates that combining both drugs will improve drug delivery and efficacy to treat different cancer diseases. A similar synergism was observed between cisplatin and the nano-S Cu chelator [104].

In conclusion, understanding Cu biochemistry especially in how Cu contributes to cancer development is useful in developing chelators as anticancer therapeutics. A number of strategies are being undertaken to improve the efficacy and uptake of Cu chelator agents into the body including formulations consisting of nanoparticles and polymers. Techniques requiring the use of light as in PDT and PCI have also demonstrated excellent value in these endeavors.

3. The Role of Fe in Cancer

3.1. Fe Transport and Regulation

Fe is the most abundant transition metal in the human body and is a vital nutrient required for several essential cell functions [110]. The forms of Fe present in the human body are heme, iron-sulfur clusters, and non-heme Fe [111,112]. Some of their biological functions are oxygen transport, energy metabolism, electron transport, cell cycle regulation, and DNA synthesis [112,113]. Fe is essential for replication, metabolism, and cell growth due to its requirement in the active site of the rate-limiting enzyme in DNA synthesis, ribonucleotide reductase (RR) [114]. The ability of Fe to convert between the ferrous form (Fe(II)) and the ferric form (Fe(III)) is the key factor in performing these biological functions [115].

Cells strictly regulate the absorption, storage, and distribution of Fe species because an imbalance in Fe homeostasis is detrimental [111]. Fe is ingested in the body in the heme and nonheme form (Fe(II)/Fe(III)) and also in metallic form fortified as in some cereal [112]. During digestion, the Fe is converted into the Fe(II) ion form and it is in this oxidation state that it is liberated from enterocytes into the blood circulation by ferroportin 1 (Fpn1), a transmembrane protein [112]. Serum transferrin (sTf) which is the major Fe transport protein can efficiently bind with Fe in its Fe(III) form [116]. For this reason, Fe needs to be oxidized for sTf to be uptaken and thus oxidation of Fe takes place at the site of Fe export. In enterocytes, the oxidation of Fe takes place by hephaestin, which belongs to a class of enzyme called multi-copper oxidase (MCOs) [112]. MCOs comprise a small family of copper-containing enzymes that oxidize substrates followed by the reduction of molecular oxygen to water [117]. Fe in +3 oxidation state is quickly uptaken by sTf in the blood, preventing loss of the metal due to its low solubility in its +3 state at pH 7.4. Fe in the blood is nearly 100% sTf bound. There is virtually little non-transferrin bound iron (NTBI), which consists of small anionic molecules such as citrate, which are likely labile in nature [118]. In cases of an overload of iron (hemochromatosis and thalassemia), Tf can be saturated and an increase in the NTBI occurs resulting in Fe(III) binding to additional anions such as phosphates and to serum albumin (the blood protein at highest concentration) and non-specific sites in Tf [118–120].

Under normal conditions, sTf exclusively regulates Fe transport into cells by binding to the transferrin receptor-1 (TfR1), a cell membrane-associated a homodimeric type II glycoprotein receptor that plays a major role in the regulation of cell growth [100,112,121]. Binding to TfR1 triggers a process known as endocytosis (Figure 7) and the subsequent endosome that forms is protected by becoming coated with the clathrin protein. The TfR1 homodimer is held together by disulfide linkages and constitutively endocytosed through the canonical clathrin-mediated pathway. Once in acidified endosomes, a receptor carrying iron loaded Tf undergoes a structural rearrangement that promotes the release of iron by Tf and then the iron-free Tf molecule is recycled back to the cell surface [121]. A combination of acidification and the reduction by STEAP3 results in the dissociation of the metal from sTf in the +2 oxidation state [122–126]. It is then released from the endosome into the cytoplasm via the divalent metal transporter 1 (DMT1). DMTs are transmembrane proteins that exist in isoforms, for instance DMT1 and DMT2 [127]. DMT1 specifically transports Fe(II) and other divalent but not all metal ions [128]. The Fe-free sTf is then returned to the membrane and released back into the blood to be recycled for further rounds of Fe cellular transport. Fe(II) ions are then trafficked to different parts of cells for a variety of functions [129] and also to the protein ferritin for storage. A small but important pool of labile Fe(II) remains in the cytoplasm for later insertion into biomolecules that depend on Fe for activity [130,131].

3.2. The Role of Fe in Cancer and Its Progression

Improperly sequestered Fe(II) or the excessive build-up of the labile Fe pool catalyzes the overproduction of ROS through Fenton chemistry, which can lead to cell death and disease [115]. Fenton in 1894, conducted an experiment that identified the role of Fe in the production of ·OH, a reaction called the Fenton reaction [132]. O_2^- and H_2O_2 damage the iron-sulfur clusters of dehydratases amongst other protein, releasing Fe ions and raising the levels of the labile Fe pool [133,134]. Any Fe(III) that is released from the proteins is quickly converted into the Fe(II) form because of the presence of O_2^- and the variety of reducing agents such as glutathione, NADH, and ascorbic acid in addition to redox active enzymes within the cell that create a reducing environment [134]. Fe(II) is then able to interact with oxygen leading to the production of H_2O_2 to initiate the Fenton reaction and yielding the ·OH radical [135]. This overall process is called the Haber–Weiss reaction [136]. In cells, O_2^- is chemically incapable of directly damaging DNA but serves as a reducer for Fe that is additively bound to DNA [134]. This reaction not only damages lipids and proteins, but also causes oxidative damage to DNA, including DNA base modifications and DNA strand breaks which can be mutagenic [111,114]. Mutagenesis by the excessive production of ROS could contribute to the initiation of cancer, in addition to being important in the promotion and progression phases [129]. Consequently, elevated levels of Fe have been identified as a risk factor for the development of cancer [137,138]. Fe-induced malignant tumors were first reported in 1959 by repeated intramuscular injection of Fe dextran complex in rats [139]. Many years later, several studies have reported observations of Fenton reactions in diverse types of cancer [114]. For these reasons, Fe is correlated with carcinogenesis and cancer progression.

The proliferation of tumor cells requires sustenance in the form of nutrients and oxygen [140]. Therefore tumor cells require more Fe which results in an increased expression of the transferrin 1 receptor (TfR1) for the endocytotic uptake of Fe(III) (Figure 7) [100,141]. Numerous studies showed that cancer cells overexpress TfR1 when compared to normal cells [131,142–144]. This is attributed to the increased need for Fe as a cofactor of RR, also overexpressed, involved in DNA synthesis of rapidly dividing cells [145]. Brookes et al. reported that the progression of colorectal cancer is associated with increased expression of Fe import proteins, such as DMT1 and TfR1, and decreased expression of the Fe export protein, Fpn1, which lead to an increased labile iron pool (LIP) [146]. Pinnix et al. demonstrated that Fpn1 abundance was reduced in aggressive some breast cancer cell lines when compared to healthy cells [147]. In addition, the metalloreductase, STEAP3, is overexpressed in cancer cells, which would increase the rate endosomal Fe(III) reduction to Fe(II) and subsequent release from the endosome [114]. In the process of metastasis, some studies have identified that tumor cells express Fe-containing matrix metalloproteinases that degrade the extracellular matrix and assist in the invasion of cancer cells [148,149]. As described previously, Cu contributes to the overexpression of matrix metalloproteinases, which shows an interplay between the excessive levels of Cu and Fe in cancer progression.

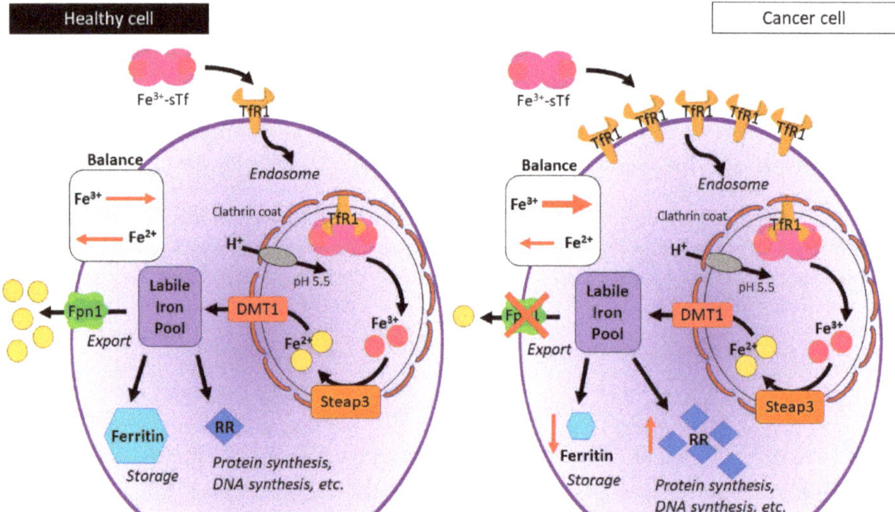

Figure 7. The endocytotic uptake of Fe(III) bound by sTf and trafficking and storage of Fe within the cell. In all cells, Fe(III)-bound sTf interacts with the TfR1, which then promotes the transport of Fe(III) across the cell membrane by endocytosis. A combination of pH decrease and Fe(III) reduction by STEAP3 within the endosome results in the release of Fe from sTf and transport as Fe(II) through the divalent metal transporter 1 (DMT1) into the cytosol. Fe(II) is then trafficked to several biomolecules that depend on it for activity such as ribonucleotide reductase (RR). Some of the is stored by ferritin. A portion of the Fe(II) remains part of a labile iron pool (LIP), readily accessible for use within the cell. In healthy cells there is a homeostasis between the amount of Fe that enters the cell through the transferrin receptor-1 (TfR1) and exported via ferroportin-1 (Fpn1). In cancer cells, there is a higher requirement for Fe and therefore these cells have an overexpression of TfR1 relative to healthy cells and a decreased expression of Fpn1. There is also less storage of the metal and an increased use of it. RR is overexpressed, which facilitates cancer proliferation.

In summary, many studies have demonstrated that an increase in the entry of Fe, a fall in its elimination, and an interruption in its storage inside the cell result in an accumulation of Fe that leads to an increase in the risk of cancer [114]. Fe inside the cell is available for DNA synthesis, cell proliferation, or the formation of ROS [135]. Due to the ability of Fe within the cell to catalyze ROS formation especially when exceeding homeostatic levels, it promotes many aspects of tumor development and progression [150].

3.3. Fe Chelators in Cancer Treatment

Several Fe chelators that were originally developed for the treatment of Fe overload in different diseases [151] were later found to reduce tumor growth by different possible processes that regulate the cell cycle, angiogenesis, or the suppression of metastases [8,152,153]. Fe in the +2 and +3 oxidation states forms coordination compounds that are typically six-coordinate. For this reason, Fe chelators for clinical use have been designed in bidentate, tridentate, or hexadentate modalities to satisfy this "preferred" coordination number (Figure 8). Fe(II) is an intermediate Lewis Acid whereas Fe(III) is a hard Lewis Acid and both Fe ions can bind to oxygen, nitrogen, and depending on the denticity even sulfur donor atoms [34]. The Fe(II) and Fe(III) chelators considered for anticancer application can be divided into the OO, ON, ONO, and XNS family of ligands (Figure 9). In recent years there has been an exhaustive collection of review articles on Fe chelators [141,154–156]. To avoid repetition, we herein highlight a few features about prominent Fe chelators that have the capacity to bind extracellular Fe(III) and intracellular Fe(II)/Fe(III). We explore how their role in suppressing cancer can provide insight into the next generation of anticancer chemotherapies.

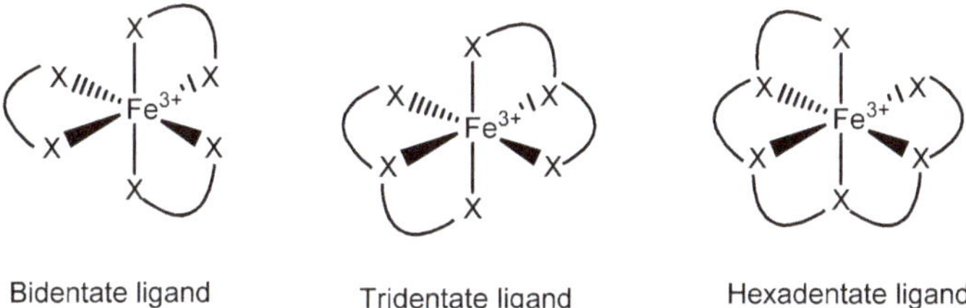

Figure 8. General Fe(II) and Fe(III) coordination by bidentate, tridentate, and hexadentate chelators to satisfy 6-coordinate structures.

3.3.1. OO and ON Ligands

Deferoxamine (DFO) (Figure 9) is a bacterial siderophore [157] that has the capacity to bind Fe in a hexadentate fashion forming a 1:1 metal:ligand complex. It is a member of the hydroxamate class of ligands. DFO is clinically used for the treatment of Fe overload and was the first Fe chelator to be tested for anticancer applications [158]. It inhibits melanoma and hepatoma cell growth both in vivo and in vitro by blocking their proliferation in the S phase of the cell cycle [159]. The mechanism of DFO in hepatocellular carcinoma cell lines (HCC) was investigated by monitoring Tf-^{59}Fe uptake in the cells [160]. Compared with the controls, the cells treated with DFO showed less uptake of ^{59}Fe, which indicates that the compound chelates the iron extracellularly. DFO was effective in the treatment of leukemia and neuroblastoma during preliminary clinical trials [159]. Nonetheless, DFO has a short plasma half-life, and is markedly hydrophilic, which makes it ineffective when administered orally. Deferiprone (DFP) (Figure 9) is a bidentate ligand that makes 3:1 ligand:Fe(III) complexes at physiological pH with Fe(III) [161]. DFP is approved in the United States for the treatment of thalassemia but also shows antiproliferation activity. It has been found that DFP acts as a pro-oxidant or protective anti-oxidant and can reduce Fe concentration in cells as much as DFO.

Studies have been conducted to explore modifications of Fechelators to incorporate functional groups with the hopes of inducing a synergistic antiproliferative effect. To that end Qiao et al. synthesized endoperoxide conjugates of derivatives of the Fe chelators hydroxamic acids, catechols, and 8-hydroxylquinolines (Figure 9) [162]. The endoperoxide is proposed to become activated in the presence of cellular Fe. Following treatment with the conjugates, the cell viability of five human cancer cells HL-60 (leukemia), A549 (lung), SW480 (colon), and SMMC7721 (liver), and the non-cancer hepatocellular cell (HL7702) was measured [162]. The 8-hydroxylquinoline conjugates exhibited the highest antiproliferative effect with IC$_{50}$ values that were generally less than 10 µM. These conjugates displayed a greater than three selectivity index, suggesting their selectivity towards attacking cancer cells. The mechanism of action of the conjugates was owed, in part, of inducing apoptosis and to the higher stability of the Fe compounds of these conjugates relative to the other chelators.

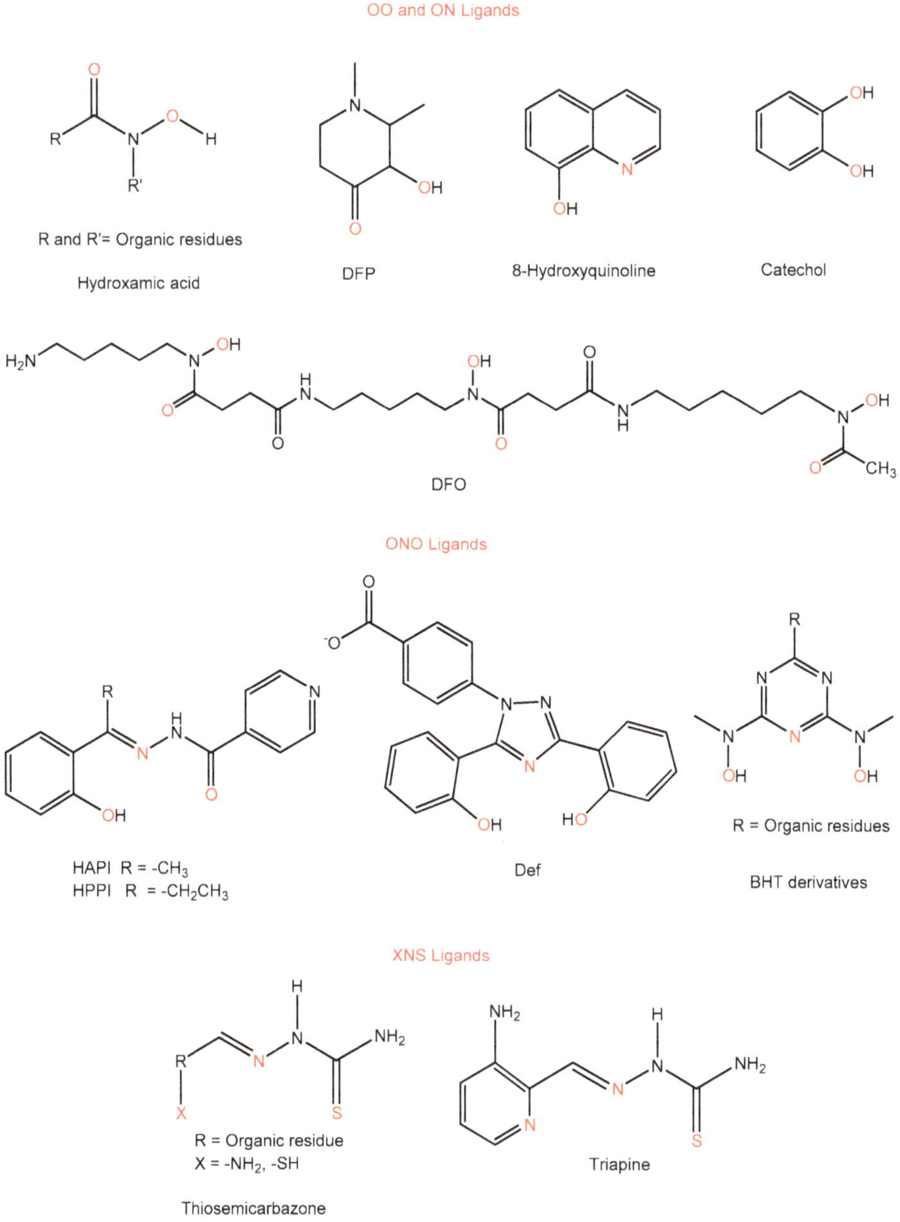

Figure 9. The structure of selected Fe chelators for potential anticancer application divided into the OO, ON, ONO, and XNS family of ligands. DFP: Deferiprone; DFO: Deferoxamine; BHT: 2,6-bis[hydroxy-(methyl)amino]-1,3,5-triazine; HAPI: (*E*)-*N'*-[1-(2-hydroxyphenyl)ethylidene]-isonicotinoylhydrazide; HPPI: (*E*)-*N'*-[1-(2-hydroxyphenyl)propylidene]isonicotinoylhydrazide.

3.3.2. ONO Ligands

Deferasirox (Def) (Figure 9) is a tridentate chelator that forms a 1:2 Fe(III):ligand complex at pH 7.4 [163]. It is commercially used as an oral drug for the treatment of Feoverload. Def is currently being tested against several cancer cell lines [164–167] because it has the capacity to induce DNA fragmentation, inhibit DNA synthesis, deplete Fe, trigger caspase-related apoptosis, amongst other activities.

The (*E*)-*N'*-[1-(2-hydroxyphenyl)ethylidene]isonicotinoylhydrazide (HAPI) and (*E*)-*N'*-[1-(2-hydroxyphenyl)propylidene]isonicotinoylhydrazide (HPPI) Fe semicarbazone chelators (Figure 9) also demonstrate antiproliferative properties. Vávrová et al. prepared derivatives of these chelators by modifying the hydrazide component of the molecules [168]. The derivatives were evaluated for their Fe-chelating abilities, anti/pro-oxidative properties, capacities to prevent Fe(III) uptake from transferrin, amongst other properties in MCF-7 (human breast cancer) and HL-60 (leukemia) cell lines. To compare intracellular Fe chelation, the calcein assay was used [169,170]. The HAPI analogues with a nitrogen heterocycle in the hydrazide part of the molecules showed similar or better Fe chelating activities than the standard chelators and greatly improved hydrolytic stability in plasma. It was also demonstrated that some of the derivatives with increased lipophilicity retained comparable antiproliferative activity as the unmodified HAPI and HPPI. The same compounds also showed high selectivity ratios, specifically the biphenyl-containing ligands.

The 2,6-bis[hydroxy-(methyl)amino]-1,3,5-triazine (BHT) family of tridentate chelators (Figure 9) [171] consists of the following properties: (i) strong binding of Fe(III) along with high Fe(III)/Fe(II) selectivity resulting in very low redox potential of the formed Fe(III) complex which precludes uncontrolled formation of reactive oxygen species in normal cells; (ii) rigidity of the ligand, resulting in size-selectivity towards Fe(III) binding; and (iii) balanced hydrophobicity of the ligand, allowing it to be soluble in aqueous media and membrane permeable. The in vitro antiproliferative behavior of BHT analogues against two human cancer cell lines, MDA-MB-231 (breast cancer), and MiaPaCa (pancreatic cancer) was assessed [171]. The results suggested that the substitution on the nitrogen or oxygen atoms of the hydroxyamino groups is essential for modulating the cytotoxic activity of the BHT analogues and improving selectivity toward cancer cells versus normal cells.

3.3.3. XNS Ligands

Like the semicarbazones, thiosemicarbazones (TSC) (Figure 9) are effective Fe(II) and Fe(III) chelators. TSC are versatile ligands that exhibit wide pharmacological properties for instance, antineoplastic, antibacterial, antiviral, and antifungal activity [33,172]. Their antineoplastic mechanism of action includes the activities for instance; (a) inhibition of cellular iron uptake from transferrin, (b) mobilization of iron from cells, (c) inhibition of RR, and (d) the formation of ROS [173,174]. One of the best studied TSC compounds is Triapine. Triapine (3-AP) was initially developed as a potent RR inhibitor, which uses radical-based catalysis to form deoxyribonucleotides. Its exact mechanism of inhibition is still being researched, but it is believed to be caused by ROS generation, specifically by the formation of the 3-AP Fe(II) complex, and/or depletion of cellular iron pools [8]. Triapine has undergone Phase I and II clinical trials for cancer treatment [153], because of the anti-proliferative activity against a number of cancers [172], including L1210 leukemia cells both in vitro and in vivo [175,176].

The use of iron chelators in the treatment of cancer does not require blockage or depletion of the metal from the cell but rather attenuating the toxicity or functionality of the metal to then detrimentally alter the intracellular processes of tumors. Koppenol et al. found that Fe complexes with some clinically used chelators such as desferrioxamine, deferiprone, and deferasirox exhibit a large negative electrode potential, which prevents the toxicity that is generated by Fe redox cycling from excessive labile Fe [177]. Some iron chelators interfere with the cell cycle and/or inhibit the reproduction of DNA, angiogenesis, cellular growth, amongst other molecular mechanisms that depend on Fe. Studies with iron chelators of clinical interest have revealed strategies that could prevent the onset of certain cancers and provide new molecular targets that are Fe dependent.

3.4. Efforts to Optimize Drug Delivery and Efficacy of Fe Chelator Agents

Fe chelators can prevent Fe acquisition or utilization by cancer cells. This results in a pronounced anti-proliferative effect due to the inactivation of the RR [178]. This discovery propitiated researchers to start developing novel Fe chelators, looking to exploit their strong Fe-binding capabilities for anti-proliferative cancer therapy. Many Fe-chelating compounds have been developed, and a few are even FDA-approved, such as DFO, DFP, and Def [179]. However, these compounds present limitations due to their acute toxicity and short plasma life [180]. For example, DFO is limited by its rapid excretion, metabolic breakdown, and low cellular uptake [159,180]. The demand exists for novel methods that can reduce the toxicity of these Fe chelators and enhance the bioavailability of these compounds in target tissues. We will discuss some innovative drug delivery methods, prochelation, and synergistic treatment approaches to improve the cellular specificity and/or efficacy of Fe chelators.

3.4.1. Nano-Approaches for Delivering Fe Chelators

Nanoparticles based on poly(lactic-*co*-glycolic acid) (PLGA) are being widely used as delivery agents because of their biocompatibility, ease of attachment of targeting molecules, and ability to fine tune the release of bound compounds [178]. Di-2-pyridylketone-4,4-dimethyl-3-thiosemicarbazone (Dp44mT) is an Fe chelator compound that has previously shown great anti-proliferative characteristics in several cancers including breast cancer and melanoma [178]. A PLGA-NP was successfully developed with high encapsulation efficiencies and efficient release of Dp44mT [179,180]. Cell viability assays indicate that Dp44mT in free form and PLGA-NP encapsulated were highly effective in killing U251 glioma cells, suggesting that encapsulation of this compound in PLGA NPs did not affect its activity [179]. These findings suggest that PLGA-NPs are suitable carriers for efficient encapsulation of Dp44mT, which increases the number of drug particles that actually reach the tumor. This results in a targeted delivery of this Fe chelator to malignant cells.

Abayaweera et. al. reported the synthesis, characterization, and efficacy of Fe/Fe3O4-nanoparticles co-labeled with a tumor-homing and membrane-disrupting oligopeptide and the Fe chelator Dp44mT [181]. This Fe chelator and the peptide sequence PLFAERL (D[KLAKLAKKLAKLAK])CGKRK were conjugated to the surface of the nanoparticles using dopamine anchors [181]. The peptide sequence has two important parts, one responsible for tumor targeting and the other is responsible for the disruption of mitochondrial cell walls and triggering apoptosis [182]. In this study, the activity of the nanocarrier was tested on the highly metastatic 4T1 murine breast cancer cell line and was compared with a non-cancerous murine skin fibroblast cell line (MSFs) [181]. It was observed that at an optimal ratio of 1 to 3.2 of the peptide sequence and the Fe chelator, the IC_{50} value was 2.2 times lower for the breast cancer cells than for the control cell line [181]. These were encouraging results but further studies in different compositions of nano-carriers are necessary to achieve higher efficacies and better specificity.

Polymeric micelles are self-assembly nanocarriers with core/shell structures formed by amphiphilic block copolymers [183]. The core can enclose the hydrophobic drug with a hydrophilic shell, increasing the solubility and stability [167,183]. PEG is commonly used as the shell because of its stealth property and hydrophilicity which allows micelle to avoid recognition by the mononuclear phagocyte system (MPS) [167]. MPS is a class of cells that occur in separated parts of the human body and that have in common the property of phagocytosis, whereby the cells engulf and destroy bacteria, viruses, and other foreign substances and ingest worn-out or abnormal body cells [184]. Theerasilp et al. presented the modification of the Fe-chelating drug Def (Figure 9), by conjugation of a pH-sensitive moiety encapsulated into PEG-b-PCL micelles as a drug delivery strategy for cancer chemotherapy [167]. Release studies of Def and its derivatives including methoxy (mDef) and imidazole-modified (iDef) deferasirox micelles were tested under simulated lysosomal condition (pH 6.0 and 5.5) representing different stages of endocytosis [167]. The results showed that in iDef-loaded micelles, a small change in pH environment (intracellular pH gradient) caused a significant effect of the drug release rate [167]. The cytotoxicity assays of the micelles loaded with Def, mDef and iDef demonstrated anti-proliferative properties suggesting that they have potential applications for cancer treatment, especially micelles loaded iDef [167]. These micelles showed cytotoxic effect against the PC-3 cell line in sub-micro molar level [167]. The IC_{50} of drug-loaded micelles was slightly higher than that of the free drug because of the prolonged release of the drug from micelles. The IC_{50} values of iDef-micelles in normal and cancer cell lines showed promising antiproliferative activity with minimal cytotoxicity to the normal cell lines [167].

3.4.2. Liposomes for Delivering Fe Chelators

Traditional drug delivery systems are engineered to yield a sustained release of bioactive compounds. Liposome drug delivery systems have played a significant role in delivering significant amounts of drug to improve therapeutic outcomes [185]. Recently, the liposome formulations are targeted to reduce toxicity and increase accumulation at the target site, which has been a limitation with conventional delivery systems for Fe chelators. Most clinical applications of liposomal drug delivery are focused on targeting tissue with or without expression of target recognition molecules on lipid membranes. This mode of drug delivery lends more safety and efficacy to administration of several classes of drugs like antiviral, antimicrobial, vaccines, and gene therapeutics. Recent developments in this field have demonstrated the specific binding properties of a drug-carrying liposome to a target cell such as a tumor cell and specific molecules in the body [179]. This ability can be exploited for the efficient delivery of Fe chelators.

O'Neill et al. examined the timed-release capabilities of lysolipid-based thermosensitive liposomes (LTSL) embedded in a chitosan-based thermo-responsive hydrogel matrix (denoted Lipogel). They tested their model by using DFO (Figure 9), the chelator of choice for removal of excess stored Fe. The group loaded their chitosan hydrogels with either free DFO or LTSL-encapsulated DFO and proceeded to test the release of DFO in PBS for 10 days. Their results demonstrate a sustained and prolonged release of DFO using LTSLs embedded in a hydrogel matrix [180]. With these experiments, they proved that the entrapment of drug-loaded liposomes in an injectable hydrogel permits local liposome retention, thus providing a prolonged release in target tissues [159,180]. They also showed that release of DFO can be controlled through the use of an external stimulus [180]. In this case, stimulus consisted of a hyper-thermic pulse of 42° for a period of 1 h. Their experimental model consisted of irradiating hydrogel particles containing 100 µM LTSL-encapsulated DFO with the hyper-thermic pulse on day 2, 6, or 10 [180]. An initial burst release of drug was observed in the first 24 h due to passive DFO diffusion from the LTSLs. A second burst release was observed after application of the hydro-thermic pulse at different time points. With this, they demonstrated that their system is capable of delivering two different doses of drug at different time intervals [180].

3.4.3. Prochelation

Prochelation strategies, in which a chelator is formed in response to a triggering event, are being developed to increase the preferential activity of metal chelators within cancer cells versus healthy cells. One such approach has been undertaken by Tomat et al. using the thiosemicarbazone tridentate scaffold (Figure 9) [186–189]. They have included a redox responsive disulfide bond switch that following reduction in the reducing environment of cells activates the formation of the chelator. This switch is particularly sensitive in cancer cells because of their higher GSH concentrations relative to normal tissue. Tomat et al. have found that the thiosemicarbazone prochelators attenuate the bioavailability of Fe within cancer cells and as a consequence inhibits the activity of the RR enzyme. This finding was demonstrated by the decrease in the g ~ 2 electron paramagnetic resonance (EPR) signal associated with the active enzyme's tyrosyl radical and the growth of a low spin Fe(III) signal in a nearby region (Figure 10). Structural modifications to the prochelators have been made to incorporate functional groups that would allow targeting of specific tumor types [188].

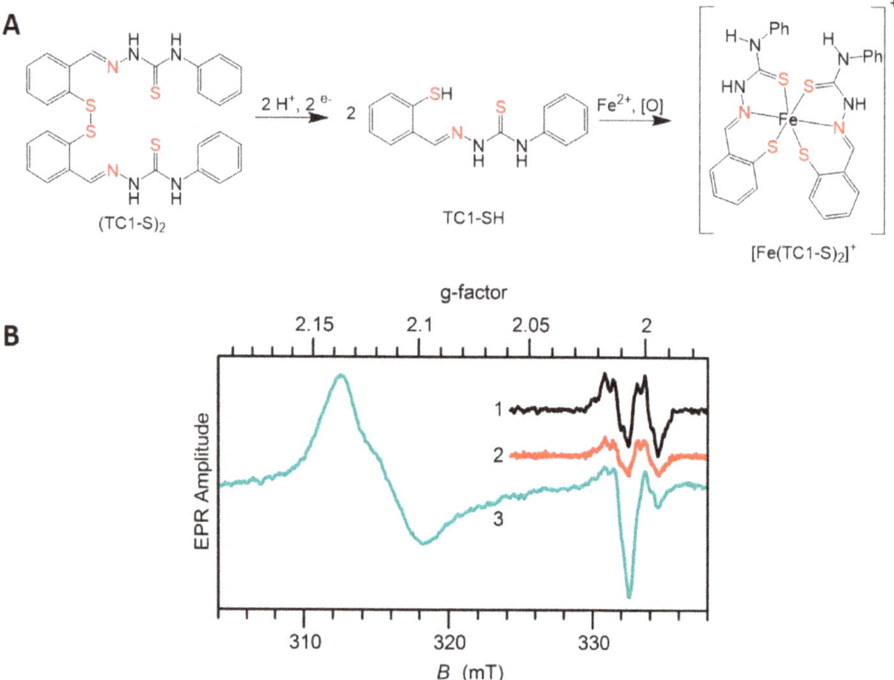

Figure 10. (A) Schematic of the reduction of the prochelator (thiosemicarbazones (TSC)-S)$_2$ for the activation of an antiproliferative thiosemicarbazone chelator (TC1-SH). This chelator would bind Fe from the LIP. (B) Enhanced permeation and retention (EPR) spectra of whole Jurkat cells: (1) untreated cells, (2) after treatment with 50 µM DFO for 3 h, and (3) after treatment with 50 µM (TC1-S)$_2$ for 1 h. Experimental conditions: microwave frequency, 9.338 GHz; microwave power, 2 mW; magnetic field modulation amplitude, 0.5 mT; temperature, 30 K. Reprinted with permission from *Metallomics*, 6, 1905–1912. Copyright 2014 Royal Society of Chemistry [187].

3.4.4. Synergistic Treatment

The clinical use of 3-AP is potentially limited by its low efficacy in some trial studies and possible toxicity related to its dosage [8]. Synergism between compounds has been an area of interest for researchers regarding the anticancer activity of TSC compounds. One way to enhance the ROS formation from TSC compounds is with PDT [190]. This would influence the cellular redox environment of the cells, leading to the induction of oxidative stress. It has been shown that mixtures of PS and TSC increases the production of singlet oxygen, provoking different types of cellular damage, including lipid peroxidation and accumulation of ROS in the mitochondria, due to the labile mitochondrial iron metabolism [191]. Another important characteristic of TSC, is the ability to upregulate the metastasis suppressor NDRG1 [173]. NDRG1 expression is induced by cellular iron depletion, thus effective iron chelators render a promising therapeutic ability through a different pathway. A more potent TSC was generated, di-2-pyridylketone-4-cyclohexyl-4-methyl-3-thiosemicarbazone (DpC), which showed complete inhibition of pancreatic tumor growth [192]. Amongst the targets that DpC affects are: (a) NDRG1; (b) p21CIPI/WAF1 a cyclin-dependent kinase inhibitor; and (c) cyclin D1, which is necessary for the cell cycle progression, by the depletion of iron [193]. Cyclin D1 is known to function as an oncogene in pancreatic cancer, usually being overexpressed in these tumors [194], hence anticancer agents that are able to reduce cyclin D1 levels are beneficial for the treatment of pancreatic cancer. In a study against pancreatic cancer drugs gemcitabine and 5-fluorouracil, DpC's IC_{50} values were at least 4- and 2000-fold lower respectively in four pancreatic cell lines [8]. Dp44mT, a DpC analogue also showed potent anti-cancer activity against various cancer cell lines, with IC_{50} values in the nanomolar range, even against cells lacking the important tumor suppressor protein that prevents tumorigenesis p53 [174]. Moreover, it was found that Dp44mT increased Fe mobilization, resulting in the release of 38% of total cellular Fe, as well as reducing its uptake to less than 10% of the control sample [174].

In summary, there are various drug delivery methods, structural modifications, and synergistic approaches that can mitigate some of the issues with common Fe chelators. Nanoparticle-based therapeutic systems can enhance the bioavailability of the drug, plasma retention time, and produce a stronger anti-tumor effect. A targeting ligand can be attached to carriers in order to increase the selective uptake in tumors and reduce adverse side effects. Responsive systems can be designed for improved drug release in cells and for exclusive activation in cancer cells. Combining chelators with other therapeutic strategies could enhance their cytotoxicity while minimizing their general toxicity.

4. Transmetalation as a New Anticancer Strategy to Target Cu and Fe Chelation

A major limitation in the use of Cu and Fe chelators is that although they have high affinity for these metals and thus are termed selective for them, they still have the capacity to bind other metals and can perturb their homeostasis in the body. The ability of Cu and Fe chelators to bind other metals can be exploited in a therapeutic manner by the judicious choice of transmetalation. In this approach other metals are introduced into the cells in compound form with the Cu or Fe chelators and in the presence of the labile Cu and Fe pool, metal exchange occurs releasing the external metal. The released metal will then synergize with the chelators to enhance the antiproliferative and/or cytotoxic effect. For this approach to be successful, the metal needs to display potent cell killing capacity and also be able to form a stable compound with the chelator that only in the intracellular environment undergoes induced dissociation by the labile Cu and Fe pool at physiologically fast rates [195,196].

In metal-based therapeutics, transmetalation can be a detriment because of the exchange of essential metals in the body for the metals delivered in the drugs [197,198]. For instance, platinum(II) from Pt(II)-based anticancer compounds and gold(I) from Au(I)-based Rheumatoid arthritis compounds have been shown to transmetalate with zinc(II) in undesired parts of the body and decrease the bioavailability of the metal. Pt(II) can displace Zn(II) from human serum albumin [199], one of its main blood transporters, and Au(I) can displace Zn(II) from zinc finger protein sites [200]. With careful consideration of coordination chemistry, these undesirable transmetalation events can be avoided. To this end, Tinoco et al. has sought to manipulate the chemical proximity of titanium(IV) for Fe(III) for a transmetalation anticancer strategy [201,202]. Ti(IV) mimics the coordination chemistry of Fe(III) in being able to bind the same biomolecules though not always in the same coordination modalities [203]. Ti(IV) in chelate form is typically redox inert and this property distinguishes it from Fe because many of its biological functions depend on its redox activity and being able to easy interconvert between Fe(II) and Fe(III). In this capacity, Ti(IV) via transmetalation could work to inhibit molecular mechanisms dependent on Fe and its redox activity [202]. A series of Ti(IV) compounds has been made with a family of Fe(III) chelators termed chemical transferrin mimetics (cTfm). These chelators mimic Fe(III) coordination by the blood transport protein serum transferrin, which is responsible for the delivery of Fe(III) into cells. This protein also serves to transport Ti(IV) and binds it in the same binding site as Fe but with some different ligating atoms [203]. The cTfm chelators form very stable Ti(IV) compounds but have a higher affinity for Fe(III) and this preferential binding facilitates transmetalation. Solution studies that approximate Ti(IV) binding conditions in the blood have demonstrated that these compounds remain intact and that only in the intracellular environment are they dissociated by the labile Fe pool [201,202]. The chelator Def (Figure 9) serves as an excellent cTfm representative. $Ti(Deferasirox)_2{}^{2-}$ exists at pH 7.4 and it is able to decrease the bioavailability of Fe in Jurkat cells and thereby inhibit the activation of RR [202] similar to findings reported by Tomat et al. and their prochelators [187]. In a recent surprise finding, $Ti(Def)_2{}^{2-}$ also transmetalates with Cu(II) possibly leading to the formation of a multinuclear species $[Cu(Def)]_n^{n-}$. A metal competition study between labile citrate sources of Fe(III), Cu(II), and Zn(II) with $Ti(Def)_2^{2-}$ that was monitored by UV-Vis spectroscopy showed that both Cu(II) and Fe(III) underwent transmetalation and Zn(II), which can bind to Def, exhibited no reactivity (Figure 11) [204]. Characteristic ligand to metal charge transfer absorbances were observed for the formation of the $[Cu(Def)]_n^{n-}$ and $Fe(Def)_2^{3-}$ species. $Ti(Def)_2^{2-}$ has been shown to operate by triggering apoptotic cell death [196]. Correlations with possible Cu and Fe chelation and apoptosis are being further studied. This work elucidates a transmetalation strategy to directly impact both Cu and Fe in cancer cells, which could lead to a very promising drug considering the roles Cu and Fe play in cancer progression.

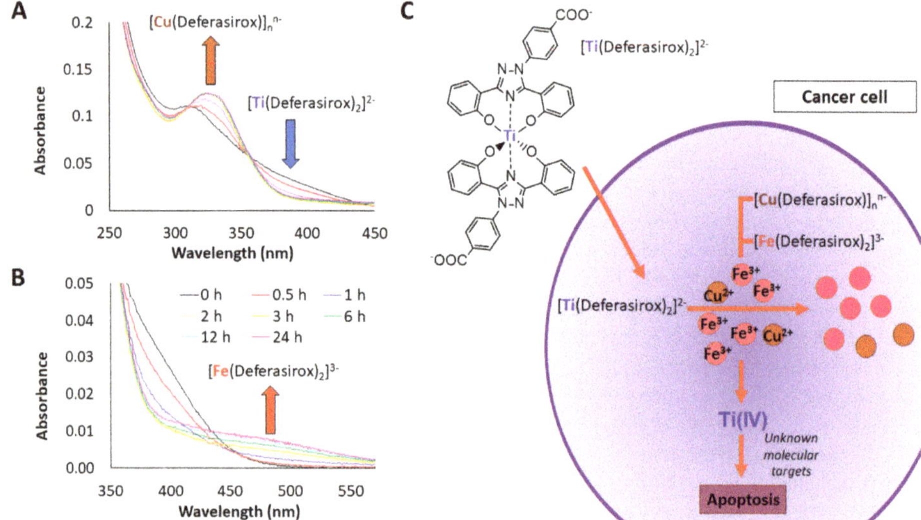

Figure 11. The transmetalation reaction from the simultaneous reaction between $Ti(deferasirox)_2^{2-}$ and labile sources of Fe(III), Cu(II), and Zn(II) monitored by UV-Vis spectroscopy. (**A**) Growth in the ligand to metal charge transfer absorbance for $[Cu(Def)]_n^{n-}$ is observed very rapidly. (**B**) Growth in the ligand to metal charge transfer absorbance for $[Fe(Def)]_2^{3-}$ is observed. (**C**) Proposed model for the intracellular transmetalation of $Ti(deferasirox)_2^{2-}$ by Cu(II) and Fe(III) leading to release of Ti(IV) and subsequently apoptosis.

5. Analytical Tools to Quantify and Track Cu and Fe

The study of trace elements and their functions in living organisms and biological systems requires methods to study metals in cells or tissues [205]. This requirement has spawned the field of metallomics [206] and is providing great insight into the biodistribution of metals in the body to understand biomolecular metal regulation, metal accumulation, and abnormalities that could be indicative of a diseased state [201]. The quantification of metals provides information on the biodegradation of biomolecules containing the metals and biodefense patterns due to changes in the levels of the metals [207]. Studies have shown that some patients infected with a bacterial pathogen experience nutritional immunity wherein their blood Fe levels purposely decrease to deprive the bacteria of their Fe supply [208]. For purposes of understanding the overall impact of applying Cu and Fe chelators for clinical use, the quantification of Cu and Fe in samples, either in vitro, ex vivo (removing tissue from organisms), or in vivo, is quite useful to determine whether the chelators decrease the levels of these metals in the areas of interest and to what extent can the levels be affected in healthy cells, tissue, blood, etc. Given that not all chelators will necessarily work by directly depleting the metal levels, methods have been devised to try to examine tissue localization and biomolecular binding. This information is often lost during metal quantification due to destructive sample preparation and metal extraction steps. This section will briefly discuss tools that are available for quantifying Cu and Fe and for assessing the Cu and Fe metallome in vitro and in the body.

5.1. Techniques to Quantify Cu and Fe Levels

The most common techniques used today for the quantification of Cu and Fe from biological samples in vitro and ex vivo include inductive coupling plasma mass spectrometry (ICP-MS) [209–212], atomic absorption spectrometry (AAS) [212,213], total reflection X-ray fluorescence (TXRF) [214,215], and colorimetric assays by UV-Vis measurements [216,217]. ICP-MS is a highly sensitive tool that allows the precise quantification of metal content down to sub parts per billion (ppb) levels using internal standards [218,219]. These detection limits are possible because highly efficient mass filters and octopolar cells are employed to single out Cu and Fe from possible interferences [220]. The technique is time-consuming, costly, and requires considerable preparation involving sample acid digestion [221]. AAS requires similar sample preparation as ICP-MS but its sensitivity is considerably lower thus requiring higher amounts of samples. In the TXRF technique atoms of the sample (liquids or solids) are excited by a primary X-ray beam and consequently emit low-energy photons that are characteristic of each element [222]. The intensity of each fluorescence radiation is directly proportional to the total amount of the element in the sample after normalization with internal standards [223]. Data collection and processing is fast with this technology and prior sample preparation may not be necessary [224]. The technique is highly sensitive, capable of measuring picograms of Fe [225].

Among the various analytical techniques colorimetry is a simple, reliable, general purpose and cost-effective alternative for routine analysis. While colorimetric assays are easy to perform, sample digestion is often required to release biomolecular bound metal in addition to a specific pH working range. In these assays an absorption spectrum is produced either by a chemical reaction with the metal of interest or from the formation of a colored complex due to ligand binding of the metal. As an example of the latter, the ferrozine-based assay works well to quantify both Fe and Cu [216,217]. Ferrozine (3-(2-pyridyl)-5,6-bis(4-phenylsulfonic acid)-1,2,4-triazine) is an efficient chelator of Fe(II) and Cu(I). To maintain these metals in these oxidations states, reducing agents like hydroxylamine or ascorbic acid are included in the sample preparation [216]. Ferrozine binds to Fe(II) and produces a complex that absorbs strongly at 550 nm [216] whereas the Cu(I) complex absorbs at 470 nm [217]. The ferrozine assay can detect the metals to low ppm values, which is much lower sensitivity compared to ICP-MS (and related techniques), AAS, and TXRF.

5.2. Techniques to Track Cu and Fe in Vitro/ex Vivo and in Vivo

There are several techniques available to help localize metals in living cells and tissue in vitro and ex vivo. X-ray absorption spectroscopy (XAS) approaches such as X-ray absorption near-edge structure (XANES) [226], extended X-ray absorption fine structure (EXAFS) [227], and X-ray fluorescence microscopy [228] are quite useful in determining metal speciation in biological samples. However, these tools can be too specialized for practical purposes of measuring metal level differences. Mass spectrometry techniques are more appropriate for these applications such as secondary ion mass spectrometry (SIMS) and laser ablation inductively coupled plasma mass spectrometry (LA-ICP-MS) [229,230]. Electron spectroscopy imaging (ESI) combined with electron energy loss spectroscopy (EELS) have been very useful in monitoring time dependent changes in metal sub-cellular localization [231–233].

A popular approach for visualizing metals especially at low concentrations is the use of metal binding probes via microscopy [170]. An optical signal can be achieved from several sources including fluorescence, phosphorescence, luminescence, or magnetic resonance imaging (MRI) but fluorescence is the most common one. There are three types of fluorescent probes that differ depending on the effect on the fluorescent property that metal binding induces [170]. Intensiometric sensors are those that alter fluorescence intensity after metal binding either resulting in a "Turn-on" or "Turn-off" probe.

Ratiometric sensors are those that result in a change in the wavelength of excitation or emission and allow for quantification based on the ratio of wavelength intensities between the metal bound and unbound signals. The third type of probe is the class that changes the fluorescence lifetime after metal binding. These probes can come in the small molecule or macromolecule variety and can be fine-tuned for specific organelle or subcellular localization [170]. To identify a subcellular localization, specific dyes are used such as Propidium Iodide (PI), which labels the nucleus and the dye MitoTracker Red, which labels the mitochondria [234]. Due to the predominance of Cu(I) in the reducing environment of cells, most fluorescent probes have been designed to detect Cu(I). Some probes have been designed by conjugating common Cu chelators with highly fluorescent molecules like BODIPY [234]. For studying Fe in cells, fluorescent probes have been created for Fe(II) and Fe(III). Calcein is commonly used to track Fe(II) [169,170] whereas siderophore-containing probes have been synthesized to study Fe(III) in cells [235].

The field is still quite primitive in tracking metals in the body outside of standard blood testing. MRI has been configured to quantify Fe levels in organs [236]. Chang et al. are leading the field in the development of fluorescent and bioluminescent Cu and Fe probes to study these metals in vivo [237–239]. The clinical application of these techniques is pending.

6. Conclusions

Many strategies are being developed to combat the complex set of diseases that constitute cancer. The use of Cu and Fe chelators, while likely not sufficient to combat cancer alone given the status of these metals as essential to the body, holds tremendous promise to be combined with other anticancer approaches. This review has highlighted the numerous ways that Cu and Fe participate in the molecular mechanisms for the onset, proliferation, angiogenesis, and metastasis of cancer. Several Cu and Fe chelators were surveyed to identify important properties that can be exploited in future drug development in addition to structural fine-tuning efforts and analytical tools to improve cellular uptake, cancer cell specificity, and efficacy of the chelators. We anticipate that with the judicial optimization of Cu and Fe chelators and appropriate combination with other drugs that operate in distinct ways, a powerful new treatment will emerge.

Author Contributions: K.G., A.M.V.-S., G.D.-C., I.D.-M., J.A.B.-R., L.F.-V., L.C.S. and A.D.T. participated in the conceptualization, investigation, and writing of the original draft and revised draft, A.C.G., F.P.-D., J.A.M.R., M.V.-C., S.A.L.-R. and X.R.A. participated in the conceptualization, investigation, and writing of the original draft.

Funding: A.M.V-S. was supported by the NIH 5T34GM007821-38, J.A.B.-R. was supported by the NIH 5R25GM061151-16, M.V.-C. was supported by NASA NNX16CA43P, and K.G., L.F-V., S.A.L-R. and A.D.T. were supported by the NIH 5SC1CA190504.

Acknowledgments: We thank all of the funding agencies that made this article possible. L.F.-V. and A.D.T. received additional support from the University of Puerto Rico (UPR) FIPI Grant from the office of the DEGI. We are also thankful to the Departments of Biology and Chemistry at UPR Río Piedras.

Conflicts of Interest: The authors declare no conflict of interest.

References

1. Centers for Disease Control and Prevention: Cancer Prevention and Control. Available online: https://www.cdc.gov/ (accessed on 1 September 2018).
2. Weinberg, R. *The Biology of Cancer*; Garland Science: New York, NY, USA, 2013.
3. Roychowdhury, S.; Chinnaiyan, A.M. Translating cancer genomes and transcriptomes for precision oncology. *CA Cancer J. Clin.* **2016**, *66*, 75–88. [CrossRef] [PubMed]
4. Yu, K.H.; Snyder, M. Omics Profiling in Precision Oncology. *Mol. Cell. Proteom.* **2016**, *15*, 2525–2536. [CrossRef] [PubMed]
5. Mancini, R.J.; Stutts, L.; Ryu, K.A.; Tom, J.K.; Esser-Kahn, A.P. Directing the Immune System with Chemical Compounds. *ACS Chem. Biol.* **2014**, *9*, 1075–1085. [CrossRef] [PubMed]

6. Shao, K.; Singha, S.; Clemente-Casares, X.; Tsai, S.; Yang, Y.; Santamaria, P. Nanoparticle-Based Immunotherapy for Cancer. *ACS Nano.* **2015**, *9*, 16–30. [CrossRef] [PubMed]

7. Shi, T.; Ma, Y.; Yu, L.; Jiang, J.; Shen, S.; Hou, Y.; Wang, T. Cancer Immunotherapy: A Focus on the Regulation of Immune Checkpoints. *Int. J. Mol. Sci.* **2018**, *19*, 1389. [CrossRef] [PubMed]

8. Yu, Y.; Gutierrez, E.; Kovacevic, Z.; Saletta, F.; Obeidy, P.; Suryo Rahmanto, Y.; Richardson, D.R. Iron Chelators for the Treatment of Cancer. *Curr. Med. Chem.* **2012**, *19*, 2689–2702. [CrossRef] [PubMed]

9. Telleria, C.M. Drug Repurposing for Cancer Therapy. *J. Cancer Sci. Ther.* **2012**, *4*, ix–xi. [CrossRef] [PubMed]

10. Mevada, S.T.; AlDhuli, A.S.; Al-Rawas, A.H.; Al-Khabori, M.K.; Nazir, H.; Zachariah, M.; Wali, Y. Liver Enzymes Changes and Safety Profile of Deferasirox Iron Chelator in Omani Children with Thalassemia Major. *Blood* **2014**, *124*, 4903.

11. Grubman, A.; White, A.R. Copper as a key regulator of cell signalling pathways. *Expert Rev. Mol. Med.* **2014**, *16*, E11. [CrossRef] [PubMed]

12. Blockhuys, S.; Wittung-Stafshede, P. Copper chaperone ATOX1 plays role in breast cancer cell migration. *Biochem. Biophys. Res. Commun.* **2017**, *483*, 301–304. [CrossRef] [PubMed]

13. Shen, C.; New, E. What has fluorescent sensing told us about copper and brain malfunction? *Metallomics* **2014**, *7*, 56–65. [CrossRef] [PubMed]

14. Denoyer, D.; Masaldan, S.; La Fontaine, S.; Cater, M.A. Targeting copper in cancer therapy: 'Copper That Cancer'. *Metallomics* **2015**, *7*, 1459–1476. [CrossRef] [PubMed]

15. Blockhuys, S.; Wittung-Stafshede, P. Roles of Copper-Binding Proteins in Breast Cancer. *Int. J. Mol. Sci.* **2017**, *18*, 871. [CrossRef] [PubMed]

16. Li, S.; Zhang, J.; Yang, H.; Wu, C.; Dang, X.; Liu, Y. Copper depletion inhibits $CoCl_2$-induced aggressive phenotype of MCF-7 cells via downregulation of HIF-1 and inhibition of Snail/Twist-mediated epithelial-mesenchymal transition. *Sci. Rep.* **2015**, *5*, 12410. [CrossRef] [PubMed]

17. Shukla, V.; Skuntz, S.; Pant, H.C. Deregulated Cdk5 activity is involved in inducing Alzheimer's disease. *Arch. Med. Res.* **2012**, *43*, 655–662. [CrossRef] [PubMed]

18. Kadowaki, S.; Endoh, D.; Okui, T.; Hayashi, M. Trientine, a copper-chelating agent, induced apoptosis in murine fibrosarcoma cells by activation of the p38 MAPK pathway. *J. Vet. Med. Sci.* **2009**, *71*, 1541–1544. [CrossRef] [PubMed]

19. Wang, J.; Luo, C.; Shan, C.; You, Q.; Lu, J.; Elf, S.; Zhou, Y.; Wen, Y.; Vinkenborg, J.L.; Fan, J. Inhibition of human copper trafficking by a small molecule significantly attenuates cancer cell proliferation. *Nat. Chem.* **2015**, *7*, 968–979. [CrossRef] [PubMed]

20. Hamza, I.; Faisst, A.; Prohaska, J.; Chen, J.; Gruss, P.; Gitlin, J.D. The metallochaperone ATOX1 plays a critical role in perinatal copper homeostasis. *Proc. Natl. Acad. Sci. USA* **2001**, *98*, 6848–6852. [CrossRef] [PubMed]

21. Barresi, V.; Spampinato, G.; Musso, N.; Salinaro, A.T.; Rizzarelli, E.; Condorelli, D.F. ATOX1 gene silencing increases susceptibility to anticancer therapy based on copper ionophores or chelating drugs. *J. Inorg. Biochem.* **2016**, *156*, 145–152. [CrossRef] [PubMed]

22. Ding, P.; Zhang, X.; Jin, S.; Duan, B.; Chu, P.; Zhang, Y.; Chen, Z.-N.; Xia, B.; Song, F. CD147 functions as the signaling receptor for extracellular divalent copper in hepatocellular carcinoma cells. *Oncotarget* **2017**, *8*, 51151. [CrossRef] [PubMed]

23. Ishida, S.; Andreux, P.; Poitry-Yamate, C.; Auwerx, J.; Hanahan, D. Bioavailable copper modulates oxidative phosphorylation and growth of tumors. *Proc. Natl. Acad. Sci. USA* **2013**, *110*, 19507–19512. [CrossRef] [PubMed]

24. Turski, M.L.; Brady, D.C.; Kim, H.J.; Kim, B.-E.; Nose, Y.; Counter, C.M.; Winge, D.R.; Thiele, D.J. A novel role for copper in Ras/mitogen-activated protein kinase signaling. *Mol. Cell. Biol.* **2012**, *32*, 1284–1295. [CrossRef] [PubMed]

25. Brady, D.C.; Crowe, M.S.; Greenberg, D.N.; Counter, C.M. Copper Chelation Inhibits BRAFV600E-driven Melanomagenesis and Counters Resistance to BRAFV600E and MEK1/2 Inhibitors. *Cancer Res.* **2017**, *77*, 6240–6252. [CrossRef] [PubMed]

26. Kohno, T.; Urao, N.; Ashino, T.; Sudhahar, V.; McKinney, R.D.; Hamakubo, T.; Iwanari, H.; Ushio-Fukai, M.; Fukai, T. Novel Role of Copper Transport Protein Antioxidant-1 in Neointimal Formation after Vascular InjurySignificance. *Arterioscler. Thromb. Vasc. Biol.* **2013**, *33*, 805–813. [CrossRef] [PubMed]

27. MacDonald, G.; Nalvarte, I.; Smirnova, T.; Vecchi, M.; Aceto, N.; Doelemeyer, A.; Frei, A.; Lienhard, S.; Wyckoff, J.; Hess, D.; et al. Memo is a copper-dependent redox protein with an essential role in migration and metastasis. *Sci. Signal.* **2014**, *7*, ra56. [CrossRef] [PubMed]

28. Sammons, S.; Brady, D.; Vahdat, L.; Salama, A.K. Copper suppression as cancer therapy: The rationale for copper chelating agents in BRAF V600 mutated melanoma. *Melanoma Manag.* **2016**, *3*, 207–216. [CrossRef] [PubMed]

29. Barresi, V.; Trovato-Salinaro, A.; Spampinato, G.; Musso, N.; Castorina, S.; Rizzarelli, E.; Condorelli, D.F. Transcriptome analysis of copper homeostasis genes reveals coordinated upregulation of SLC31A1, SCO1, and COX11 in colorectal cancer. *FEBS Open Bio* **2016**, *6*, 794–806. [CrossRef] [PubMed]

30. Gupte, A.; Mumper, R.J. Elevated copper and oxidative stress in cancer cells as a target for cancer treatment. *Cancer Treat. Rev.* **2009**, *35*, 32–46. [CrossRef] [PubMed]

31. Ding, W.Q.; Lind, S.E. Metal ionophores—An emerging class of anticancer drugs. *IUBMB Life* **2009**, *61*, 1013–1018. [CrossRef] [PubMed]

32. Denoyer, D.; Pearson, H.B.; Clatworthy, S.A.; Smith, Z.M.; Francis, P.S.; Llanos, R.M.; Volitakis, I.; Phillips, W.A.; Meggyesy, P.M.; Masaldan, S.; et al. Copper as a target for prostate cancer therapeutics: Copper-ionophore pharmacology and altering systemic copper distribution. *Oncotarget.* **2016**, *7*, 37064. [CrossRef] [PubMed]

33. Weekley, C.M.; He, C. Developing drugs targeting transition metal homeostasis. *Curr. Opin. Chem. Biol.* **2017**, *37*, 26–32. [CrossRef] [PubMed]

34. Housecroft, C.E.; Sharpe, A.G. *Inorganic Chemistry*, 4th ed.; Pearson: Cambridge, UK, 2012.

35. Walshe, J. Penicillamine, a new oral therapy for Wilson's disease. *Am. J. Med.* **1956**, *21*, 487–495. [CrossRef]

36. Goldberg, A.; Smith, J.A.; Lochhead, A.C. Treatment of lead-poisoning with oral penicillamine. *Br. Med. J.* **1963**, *1*, 1270. [CrossRef] [PubMed]

37. Walshe, J.M. The story of penicillamine: A difficult birth. *Mov. Disord.* **2003**, *18*, 853–859. [CrossRef] [PubMed]

38. Walshe, J. Disturbances of aminoacid metabolism following liver injury: A study by means of paper chromatography. *Q. J. Med.* **1953**, *22*, 483–505. [PubMed]

39. Weigert, W.M.; Offermanns, H.; Degussa, P.S. D-Penicillamine—Production and Properties. *Angew. Chem. Int. Ed. Engl.* **1975**, *14*, 330–336. [CrossRef] [PubMed]

40. Winterbourn, C.C.; Metodiewa, D. Reactivity of biologically important thiol compounds with superoxide and hydrogen peroxide. *Free Radic. Biol. Med.* **1999**, *27*, 322–328. [CrossRef]

41. Held, K.D.; Biaglow, J.E. Mechanisms for the oxygen radical-mediated toxicity of various thiol-containing compounds in cultured mammalian cells. *Radiat. Res.* **1994**, *139*, 15–23. [CrossRef] [PubMed]

42. Brewer, G.J.; Askari, F.K. Wilson's disease: Clinical management and therapy. *J. Hepatol.* **2005**, *42*, S13–S21. [CrossRef] [PubMed]

43. Khandpur, S.; Jain, N.; Singla, S.; Chatterjee, P.; Behari, M. D-penicillamine induced degenerative dermopathy. *Indian J. Dermatol.* **2015**, *60*, 406. [CrossRef] [PubMed]

44. Levy, R.S.; Fisher, M.; Alter, J.N. Penicillamine: Review and cutaneous manifestations. *J. Am. Acad. Dermatol.* **1983**, *8*, 548–558. [CrossRef]

45. Shiokawa, Y.; Horiuchi, Y.; Honma, M.; Kageyama, T.; Okada, T.; Azuma, T. Clinical evaluation of D-penicillamine by multicentric double-blind comparative study in chronic rheumatoid arthritis. *Arthritis Rheumatol.* **1977**, *20*, 1464–1472. [CrossRef]

46. Brem, S.; Grossman, S.A.; Carson, K.A.; New, P.; Phuphanich, S.; Alavi, J.B.; Mikkelsen, T.; Fisher, J.D. Phase 2 trial of copper depletion and penicillamine as antiangiogenesis therapy of glioblastoma. *Neuro Oncol.* **2005**, *7*, 246–253. [CrossRef] [PubMed]

47. Yoshida, D.; Ikeda, Y.; Nakazawa, S. Suppression of 9L gliosarcoma growth by copper depletion with copper-deficient diet and D-penicillamine. *J. Neurooncol.* **1993**, *17*, 91–97. [CrossRef]

48. Feli, A.; Jazayeri, S.; Bitaraf, M.A.; Dodaran, M.S.; Parastouei, K.; Hosseinzadeh-Attar, M.J. Combination Therapy with Low Copper Diet, Penicillamine and Gamma Knife Radiosurgery Reduces VEGF and IL-8 In Patients with Recurrent Glioblastoma. *Asian Pac. J. Cancer Prev.* **2017**, *18*, 1999. [PubMed]

49. Matsubara, T.; Saura, R.; Hirohata, K.; Ziff, M. Inhibition of human endothelial cell proliferation in vitro and neovascularization in vivo by D-penicillamine. *J. Clin. Investig.* **1989**, *83*, 158–167. [CrossRef] [PubMed]

50. Wadhwa, S.; Mumper, R.J. D-penicillamine and other low molecular weight thiols: Review of anticancer effects and related mechanisms. *Cancer Lett.* **2013**, *337*, 8–21. [CrossRef] [PubMed]

51. Starkebaum, G.; Root, R. D-Penicillamine: Analysis of the mechanism of copper-catalyzed hydrogen peroxide generation. *J. Immunol.* **1985**, *134*, 3371–3378. [PubMed]

52. Gupte, A.; Mumper, R.J. Copper chelation by D-penicillamine generates reactive oxygen species that are cytotoxic to human leukemia and breast cancer cells. *Free Radic. Biol. Med.* **2007**, *43*, 1271–1278. [CrossRef] [PubMed]

53. Gupte, A.; Wadhwa, S.; Mumper, R.J. Enhanced Intracellular Delivery of the Reactive Oxygen Species (ROS)-Generating Copper Chelator D-Penicillamine via a Novel Gelatin—D-Penicillamine Conjugate. *Bioconjug. Chem.* **2008**, *19*, 1382–1388. [CrossRef] [PubMed]

54. Brewer, G.J.; Askari, F.; Lorincz, M.T.; Carlson, M.; Schilsky, M.; Kluin, K.J.; Hedera, P.; Moretti, P.; Fink, J.K.; Tankanow, R.; et al. Treatment of Wilson disease with ammonium tetrathiomolybdate: IV. Comparison of tetrathiomolybdate and trientine in a double-blind study of treatment of the neurologic presentation of Wilson disease. *Arch. Neurol.* **2006**, *63*, 521–527. [CrossRef] [PubMed]

55. Brewer, G.J.; Johnson, V.; Dick, R.D.; Kluin, K.J.; Fink, J.K.; Brunberg, J.A. Treatment of Wilson disease with ammonium tetrathiomolybdate: II. Initial therapy in 33 neurologically affected patients and follow-up with zinc therapy. *Arch. Neurol.* **1996**, *53*, 1017–1025. [CrossRef] [PubMed]

56. Brewer, G.J.; Dick, R.D.; Johnson, V.; Wang, Y.; Yuzbasiyan-Gurkan, V.; Kluin, K.; Fink, J.K.; Aisen, A. Treatment of Wilson's disease with ammonium tetrathiomolybdate: I. Initial therapy in 17 neurologically affected patients. *Arch. Neurol.* **1994**, *51*, 545–554. [CrossRef] [PubMed]

57. Brewer, G.J.; Dick, R.; Ullenbruch, M.R.; Jin, H.; Phan, S.H. Inhibition of key cytokines by tetrathiomolybdate in the bleomycin model of pulmonary fibrosis. *J. Inorg. Biochem.* **2004**, *98*, 2160–2167. [CrossRef] [PubMed]

58. Brewer, G.J. Copper lowering therapy with tetrathiomolybdate as an antiangiogenic strategy in cancer. *Curr. Cancer Drug Targets* **2005**, *5*, 195–202. [CrossRef] [PubMed]

59. Pan, Q.; Kleer, C.G.; Van Golen, K.L.; Irani, J.; Bottema, K.M.; Bias, C.; De Carvalho, M.; Mesri, E.A.; Robins, D.M.; Dick, R.D. Copper deficiency induced by tetrathiomolybdate suppresses tumor growth and angiogenesis. *Cancer Res.* **2002**, *62*, 4854–4859. [PubMed]

60. Brewer, G.J.; Dick, R.D.; Grover, D.K.; LeClaire, V.; Tseng, M.; Wicha, M.; Pienta, K.; Redman, B.G.; Jahan, T.; Sondak, V.K.; et al. Treatment of metastatic cancer with tetrathiomolybdate, an anticopper, antiangiogenic agent: Phase I study. *Clin. Cancer Res.* **2000**, *6*, 1–10. [PubMed]

61. Redman, B.G.; Esper, P.; Pan, Q.; Dunn, R.L.; Hussain, H.K.; Chenevert, T.; Brewer, G.J.; Merajver, S.D. Phase II trial of tetrathiomolybdate in patients with advanced kidney cancer. *Clin. Cancer Res.* **2003**, *9*, 1666–1672. [PubMed]

62. Pass, H.I.; Brewer, G.J.; Dick, R.; Carbone, M.; Merajver, S. A phase II trial of tetrathiomolybdate after surgery for malignant mesothelioma: Final results. *Ann. Thorac. Surg.* **2008**, *86*, 383–390. [CrossRef] [PubMed]

63. Kim, K.K.; Abelman, S.; Yano, N.; Ribeiro, J.R.; Singh, R.K.; Tipping, M.; Moore, R.G. Tetrathiomolybdate inhibits mitochondrial complex IV and mediates degradation of hypoxia-inducible factor-1α in cancer cells. *Sci. Rep.* **2015**, *5*, 14296. [CrossRef] [PubMed]

64. Hayashi, M.; Nishiya, H.; Chiba, T.; Endoh, D.; Kon, Y.; Okui, T. Trientine, a copper-chelating agent, induced apoptosis in murine fibrosarcoma cells in vivo and in vitro. *J. Vet. Med. Sci.* **2007**, *69*, 137–142. [CrossRef] [PubMed]

65. Moriguchi, M.; Nakajima, T.; Kimura, H.; Watanabe, T.; Takashima, H.; Mitsumoto, Y.; Katagishi, T.; Okanoue, T.; Kagawa, K. The copper chelator trientine has an antiangiogenic effect against hepatocellular carcinoma, possibly through inhibition of interleukin-8 production. *Int. J. Cancer* **2002**, *102*, 445–452. [CrossRef] [PubMed]

66. Yoshii, J.; Yoshiji, H.; Kuriyama, S.; Ikenaka, Y.; Noguchi, R.; Okuda, H.; Tsujinoue, H.; Nakatani, T.; Kishida, H.; Nakae, D. The copper-chelating agent, trientine, suppresses tumor development and angiogenesis in the murine hepatocellular carcinoma cells. *Int. J. Cancer* **2001**, *94*, 768–773. [CrossRef] [PubMed]

67. Yin, J.-M.; Sun, L.-B.; Zheng, J.-S.; Wang, X.-X.; Chen, D.-X.; Li, N. Copper chelation by trientine dihydrochloride inhibits liver RFA-induced inflammatory responses in vivo. *Inflamm. Res.* **2016**, *65*, 1009–1020. [CrossRef] [PubMed]

68. European Association for the Study of the Liver. EASL clinical practice guidelines: Wilson's disease. *J. Hepatol.* **2012**, *56*, 671–685. [CrossRef] [PubMed]

69. Gu, J.; Codd, R. Copper (II)-based metal affinity chromatography for the isolation of the anticancer agent bleomycin from *Streptomyces verticillus* culture. *J. Inorg. Biochem.* **2012**, *115*, 198–203. [CrossRef] [PubMed]

70. Ehrenfeld, G.M.; Shipley, J.B.; Heimbrook, D.C.; Sugiyama, H.; Long, E.C.; Van Boom, J.H.; Van der Marel, G.A.; Oppenheimer, N.J.; Hecht, S.M. Copper-dependent cleavage of DNA by bleomycin. *Biochemistry* **1987**, *26*, 931–942. [CrossRef] [PubMed]

71. Solaiman, D.; Rao, E.A.; Antholine, W.; Petering, D.H. Properties of the binding of copper by bleomycin. *J. Inorg. Biochem.* **1980**, *12*, 201–220. [CrossRef]

72. Matsui, H.; Kato, T.; Yamamoto, C.; Takita, T.; Takeuchi, T.; Umezawa, H.; Nagatsu, T. Inhibition of dopamine-β-hydroxylase, a copper enzyme, by bleomycin. *J. Antibiot.* **1980**, *33*, 435–440. [CrossRef] [PubMed]

73. Sugiura, Y.; Ishizu, K.; Miyoshi, K. Studies of metallobleomycins by electronic spectroscopy, electron spin resonance spectroscopy, and potentiometric titration. *J. Antibiot.* **1979**, *32*, 453–461. [CrossRef] [PubMed]

74. Sugiyama, M.; Kumagai, T.; Hayashida, M.; Maruyama, M.; Matoba, Y. The 1.6-Å crystal structure of the copper (II)-bound bleomycin complexed with the bleomycin-binding protein from bleomycin-producing *Streptomyces verticillus*. *J. Biol. Chem.* **2002**, *277*, 2311–2320. [CrossRef] [PubMed]

75. Ahsan, H.; Hadi, S. Strand scission in DNA induced by curcumin in the presence of Cu(II). *Cancer Lett.* **1998**, *124*, 23–30. [CrossRef]

76. Nair, J.; Strand, S.; Frank, N.; Knauft, J.; Wesch, H.; Galle, P.R.; Bartsch, H. Apoptosis and age-dependant induction of nuclear and mitochondrial etheno-DNA adducts in Long-Evans Cinnamon (LEC) rats: Enhanced DNA damage by dietary curcumin upon copper accumulation. *Carcinogenesis* **2005**, *26*, 1307–1315. [CrossRef] [PubMed]

77. Yoshino, M.; Haneda, M.; Naruse, M.; Htay, H.H.; Tsubouchi, R.; Qiao, S.L.; Li, W.H.; Murakami, K.; Yokochi, T. Prooxidant activity of curcumin: Copper-dependent formation of 8-hydroxy-2′-deoxyguanosine in DNA and induction of apoptotic cell death. *Toxicol. In Vitro* **2004**, *18*, 783–789. [CrossRef] [PubMed]

78. Urbina-Cano, P.; Bobadilla-Morales, L.; Ramírez-Herrera, M.A.; Corona-Rivera, J.R.; Mendoza-Magaña, M.L.; Troyo-Sanromán, R.; Corona-Rivera, A. DNA damage in mouse lymphocytes exposed to curcumin and copper. *J. Appl. Genet.* **2006**, *47*, 377–382. [CrossRef] [PubMed]

79. Mishra, S.; Palanivelu, K. The effect of curcumin (turmeric) on Alzheimer's disease: An overview. *Ann. Indian Acad. Neurol.* **2008**, *11*, 13. [CrossRef] [PubMed]

80. Bharti, A.C.; Donato, N.; Singh, S.; Aggarwal, B.B. Curcumin (diferuloylmethane) down-regulates the constitutive activation of nuclear factor-κB and IκBα kinase in human multiple myeloma cells, leading to suppression of proliferation and induction of apoptosis. *Blood* **2003**, *101*, 1053–1062. [CrossRef] [PubMed]

81. Lou, J.R.; Zhang, X.-X.; Zheng, J.; Ding, W.-Q. Transient metals enhance cytotoxicity of curcumin: Potential involvement of the NF-κB and mTOR signaling pathways. *Anticancer Res.* **2010**, *30*, 3249–3255. [PubMed]

82. Zhang, W.; Chen, C.; Shi, H.; Yang, M.; Liu, Y.; Ji, P.; Chen, H.; Tan, R.X.; Li, E. Curcumin is a biologically active copper chelator with antitumor activity. *Phytomedicine* **2016**, *23*, 1–8. [CrossRef] [PubMed]

83. Yu, H.; Zhou, Y.; Lind, S.E.; Ding, W.-Q. Clioquinol targets zinc to lysosomes in human cancer cells. *Biochem. J.* **2009**, *417*, 133–139. [CrossRef] [PubMed]

84. Ding, W.-Q.; Liu, B.; Vaught, J.L.; Yamauchi, H.; Lind, S.E. Anticancer activity of the antibiotic clioquinol. *Cancer Res.* **2005**, *65*, 3389–3395. [CrossRef] [PubMed]

85. Lind, S.E.; Park, J.S.; Drexler, J.W. Pyrithione and 8-hydroxyquinolines transport lead across erythrocyte membranes. *Trans. Res.* **2009**, *154*, 153–159. [CrossRef] [PubMed]

86. Prachayasittikul, V.; Prachayasittikul, S.; Ruchirawat, S.; Prachayasittikul, V. 8-Hydroxyquinolines: A review of their metal chelating properties and medicinal applications. *Drug Des. Dev. Ther.* **2013**, *7*, 1157. [CrossRef] [PubMed]

87. Palanimuthu, D.; Shinde, S.V.; Somasundaram, K.; Samuelson, A.G. In vitro and in vivo anticancer activity of copper bis(thiosemicarbazone) complexes. *J. Med. Chem.* **2013**, *56*, 722–734. [CrossRef] [PubMed]

88. West, D.X.; Liberta, A.E.; Padhye, S.B.; Chikate, R.C.; Sonawane, P.B.; Kumbhar, A.S.; Yerande, R.G. Thiosemicarbazone complexes of copper(II): Structural and biological studies. *Coord. Chem. Rev.* **1993**, *123*, 49–71. [CrossRef]

89. Paterson, B.M.; Donnelly, P.S. Copper complexes of bis(thiosemicarbazones): From chemotherapeutics to diagnostic and therapeutic radiopharmaceuticals. *Chem. Soc. Rev.* **2011**, *40*, 3005–3018. [CrossRef] [PubMed]

90. Cater, M.A.; Pearson, H.B.; Wolyniec, K.; Klaver, P.; Bilandzic, M.; Paterson, B.M.; Bush, A.I.; Humbert, P.O.; La Fontaine, S.; Donnelly, P.S.; et al. Increasing intracellular bioavailable copper selectively targets prostate cancer cells. *ACS Chem. Biol.* **2013**, *8*, 1621–1631. [CrossRef] [PubMed]

91. Park, K.C.; Fouani, L.; Jansson, P.J.; Wooi, D.; Sahni, S.; Lane, D.J.; Palanimuthu, D.; Lok, H.C.; Kovačević, Z.; Huang, M.L.; et al. Copper and conquer: Copper complexes of di-2-pyridylketone thiosemicarbazones as novel anti-cancer therapeutics. *Metallomics* **2016**, *8*, 874–886. [CrossRef] [PubMed]

92. Ruschak, A.M.; Slassi, M.; Kay, L.E.; Schimmer, A.D. Novel proteasome inhibitors to overcome bortezomib resistance. *J. Natl. Cancer Inst.* **2011**, *103*, 1007–1017. [CrossRef] [PubMed]

93. Wang, F.; Zhai, S.; Liu, X.; Li, L.; Wu, S.; Dou, Q.P.; Yan, B. A novel dithiocarbamate analogue with potentially decreased ALDH inhibition has copper-dependent proteasome-inhibitory and apoptosis-inducing activity in human breast cancer cells. *Cancer Lett.* **2011**, *300*, 87–95. [CrossRef] [PubMed]

94. Helsel, M.E.; Franz, K.J. Pharmacological activity of metal binding agents that alter copper bioavailability. *Dalton Trans.* **2015**, *44*, 8760–8770. [CrossRef] [PubMed]

95. Caragounis, A.; Du, T.; Filiz, G.; Laughton, K.M.; Volitakis, I.; Sharples, R.A.; Cherny, R.A.; Masters, C.L.; Drew, S.C.; Hill, A.F.; et al. Differential modulation of Alzheimer's disease amyloid β-peptide accumulation by diverse classes of metal ligands. *Biochem. J.* **2007**, *407*, 435–450. [CrossRef] [PubMed]

96. White, A.R.; Du, T.; Laughton, K.M.; Volitakis, I.; Sharples, R.A.; Xilinas, M.E.; Hoke, D.E.; Holsinger, R.D.; Evin, G.; Cherny, R.A.; et al. Degradation of the Alzheimer disease amyloid β-peptide by metal-dependent up-regulation of metalloprotease activity. *J. Biol. Chem.* **2006**, *281*, 17670–17680. [CrossRef] [PubMed]

97. Allensworth, J.L.; Evans, M.K.; Bertucci, F.; Aldrich, A.J.; Festa, R.A.; Finetti, P.; Ueno, N.T.; Safi, R.; McDonnell, D.P.; Thiele, D.J. Disulfiram (DSF) acts as a copper ionophore to induce copper-dependent oxidative stress and mediate anti-tumor efficacy in inflammatory breast cancer. *Mol. Oncol.* **2015**, *9*, 1155–1168. [CrossRef] [PubMed]

98. Safi, R.; Nelson, E.R.; Chitneni, S.K.; Franz, K.J.; George, D.J.; Zalutsky, M.R.; McDonnell, D.P. Copper signaling axis as a target for prostate cancer therapeutics. *Cancer Res.* **2014**, *74*, 5819–5831. [CrossRef] [PubMed]

99. Ranganathan, R.; Madanmohan, S.; Kesavan, A.; Baskar, G.; Krishnamoorthy, Y.R.; Santosham, R.; Ponraju, D.; Rayala, S.K.; Venkatraman, G. Nanomedicine: Towards development of patient-friendly drug-delivery systems for oncological applications. *Int. J. Nanomed.* **2012**, *7*, 1043–1060.

100. Dabrowiak, J.C. *Metals in Medicine*, 2nd ed.; Wiley: Hoboken, NJ, USA, 2017.

101. Wadhwa, S.; Mumper, R.J. Intracellular delivery of the reactive oxygen species generating agent D-penicillamine upon conjugation to poly-L-glutamic acid. *Mol. Pharm.* **2010**, *7*, 854–862. [CrossRef] [PubMed]

102. Norum, O.-J.; Fremstedal, A.S.V.; Weyergang, A.; Golab, J.; Berg, K. Photochemical delivery of bleomycin induces T-cell activation of importance for curative effect and systemic anti-tumor immunity. *J. Control. Release* **2017**, *268*, 120–127. [CrossRef] [PubMed]

103. Luo, C.-Q.; Xing, L.; Cui, P.-F.; Qiao, J.-B.; He, Y.-J.; Chen, B.-A.; Jin, L.; Jiang, H.-L. Curcumin-coordinated nanoparticles with improved stability for reactive oxygen species-responsive drug delivery in lung cancer therapy. *Int. J. Nanomed.* **2017**, *12*, 855. [CrossRef] [PubMed]

104. Liu, H.; Zhang, Y.; Zheng, S.; Weng, Z.; Ma, J.; Li, Y.; Xie, X.; Zheng, W. Detention of copper by sulfur nanoparticles inhibits the proliferation of A375 malignant melanoma and MCF-7 breast cancer cells. *Biochem. Biophys. Res. Commun.* **2016**, *477*, 1031–1037. [CrossRef] [PubMed]

105. Ishida, S.; McCormick, F.; Smith-McCune, K.; Hanahan, D. Enhancing tumor-specific uptake of the anticancer drug cisplatin with a copper chelator. *Cancer Cell* **2010**, *17*, 574–583. [CrossRef] [PubMed]

106. Lambert, I.; Sørensen, B. Facilitating the Cellular Accumulation of Pt-Based Chemotherapeutic Drugs. *Int. J. Mol. Sci.* **2018**, *19*, 2249. [CrossRef] [PubMed]

107. Voss, F.K.; Ullrich, F.; Münch, J.; Lazarow, K.; Lutter, D.; Mah, N.; Andrade-Navarro, M.A.; von Kries, J.P.; Stauber, T.; Jentsch, T.J. Identification of LRRC8 heteromers as an essential component of the volume-regulated anion channel VRAC. *Science* **2014**, *344*, 634–638. [CrossRef] [PubMed]

108. Gradogna, A.; Gavazzo, P.; Boccaccio, A.; Pusch, M. Subunit-dependent oxidative stress sensitivity of LRRC8 volume-regulated anion channels. *J. Physiol.* **2017**, *595*, 6719–6733. [CrossRef] [PubMed]

109. Jentsch, T.J.; Lutter, D.; Planells-Cases, R.; Ullrich, F.; Voss, F.K. VRAC: Molecular identification as LRRC8 heteromers with differential functions. *Pflug. Arch. Eur. J. Phys.* **2016**, *468*, 385–393. [CrossRef] [PubMed]

110. Cairo, G.; Bernuzzi, F.; Recalcati, S. A precious metal: Iron, an essential nutrient for all cells. *Genes Nutr.* **2006**, *1*, 25–39. [CrossRef] [PubMed]

111. Lane, D.J.R.; Merlot, A.M.; Huang, M.L.H.; Bae, D.H.; Jansson, P.J.; Sahni, S.; Kalinowski, D.S.; Richardson, D.R. Cellular iron uptake, trafficking and metabolism: Key molecules and mechanisms and their roles in disease. *Biochim. Biophys. Acta Mol. Cell Res.* **2015**, *1853*, 1130–1144. [CrossRef] [PubMed]

112. Bertini, I.; Gray, H.B.; Stiefel, E.I.; Valentine, J.S. *Biological Inorganic Chemistry: Structure and Reactivity*; University Science Books: Mill Valley, CA, USA, 2007.

113. Wang, J.; Pantopoulos, K. Regulation of cellular iron metabolism. *Biochem. J.* **2011**, *434*, 365–381. [CrossRef] [PubMed]

114. Torti, S.V.; Torti, F.M. Iron and cancer: More ore to be mined. *Nat. Rev. Cancer* **2013**, *13*, 342. [CrossRef] [PubMed]

115. Huang, X. Iron overload and its association with cancer risk in humans: Evidence for iron as a carcinogenic metal. *Mutat. Res. Fund. Mol. Mech. Mutagen.* **2003**, *533*, 153–171. [CrossRef]

116. Vashchenko, G.; MacGillivray, R. Multi-copper oxidases and human iron metabolism. *Nutrients* **2013**, *5*, 2289–2313. [CrossRef] [PubMed]

117. Askwith, C.; Kaplan, J. Iron and copper transport in yeast and its relevance to human disease. *Trends Biochem. Sci.* **1998**, *23*, 135–138. [CrossRef]

118. Brissot, P.; Ropert, M.; Le Lan, C.; Loréal, O. Non-transferrin bound iron: A key role in iron overload and iron toxicity. *Biochim. Biophys. Acta Gen. Subj.* **2012**, *1820*, 403–410. [CrossRef] [PubMed]

119. Batey, R.; Fong, P.L.C.; Shamir, S.; Sherlock, S. A non-transferrin-bound serum iron in idiopathic hemochromatosis. *Dig. Dis. Sci.* **1980**, *25*, 340–346. [CrossRef] [PubMed]

120. Porter, J.; Abeysinghe, R.; Marshall, L.; Hider, R.; Singh, S. Kinetics of removal and reappearance of non-transferrin-bound plasma iron with deferoxamine therapy. *Blood* **1996**, *88*, 705–713. [PubMed]

121. Luria-Pérez, R.; Helguera, G.; Rodríguez, J.A. Antibody-mediated targeting of the transferrin receptor in cancer cells. *Bol. Med. Hosp. Infant. Mex.* **2016**, *73*, 372–379. [PubMed]

122. Eckenroth, B.E.; Steere, A.N.; Chasteen, N.D.; Everse, S.J.; Mason, A.B. How the binding of human transferrin primes the transferrin receptor potentiating iron release at endosomal pH. *Proc. Natl. Acad. Sci. USA* **2011**, *108*, 13089–13094. [CrossRef] [PubMed]

123. Steere, A.N.; Byrne, S.L.; Chasteen, N.D.; Mason, A.B. Kinetics of iron release from transferrin bound to the transferrin receptor at endosomal pH. *Biochim. Biophys. Acta Gen. Subj.* **2012**, *1820*, 326–333. [CrossRef] [PubMed]

124. Kraiter, D.C.; Zak, O.; Aisen, P.; Crumbliss, A.L. A determination of the reduction potentials for diferric and C- and N-lobe monoferric transferrins at endosomal pH (5.8). *Inorg. Chem.* **1998**, *37*, 964–968. [CrossRef]

125. Dhungana, S.; Taboy, C.H.; Zak, O.; Larvie, M.; Crumbliss, A.L.; Aisen, P. Redox properties of human transferrin bound to its receptor. *Biochemistry* **2004**, *43*, 205–209. [CrossRef] [PubMed]

126. Bou-Abdallah, F. Does iron release from transferrin involve a reductive process? *Bioenergetics* **2012**, *1*, e1111.

127. Clardy, S.L.; Connor, J.R.; Beard, J. *Restless Legs Syndrome*; Elsevier Inc.: Amsterdam, The Netherlands, 2009; pp. 50–60.

128. Wolff, N.A.; Garrick, M.D.; Zhao, L.; Garrick, L.M.; Ghio, A.J.; Thévenod, F. A role for divalent metal transporter (DMT1) in mitochondrial uptake of iron and manganese. *Sci. Rep.* **2018**, *8*, 211. [CrossRef] [PubMed]

129. Wiseman, H.; Halliwell, B. Damage to DNA by reactive oxygen and nitrogen species: Role in inflammatory disease and progression to cancer. *Biochem. J.* **1996**, *313*, 17–29. [CrossRef] [PubMed]

130. Hider, R.C.; Kong, X.L. Iron speciation in the cytosol: An overview. *Dalton Trans.* **2013**, *42*, 3220–3229. [CrossRef] [PubMed]

131. Philpott, C.C.; Ryu, M.-S. Special delivery: Distributing iron in the cytosol of mammalian cells. *Front. Pharmacol.* **2014**, *5*. [CrossRef] [PubMed]

132. Fenton, H.J.H. LXXIII.-Oxidation of tartaric acid in presence of iron. *J. Chem. Soc. Trans.* **1894**, *65*, 899–910. [CrossRef]

133. Jang, S.; Imlay, J.A. Micromolar intracellular hydrogen peroxide disrupts metabolism by damaging iron-sulfur enzymes. *J. Biol. Chem.* **2007**, *282*, 929–937. [CrossRef] [PubMed]

134. Keyer, K.; Imlay, J.A. Superoxide accelerates DNA damage by elevating free-iron levels. *Proc. Natl. Acad. Sci. USA* **1996**, *93*, 13635–13640. [CrossRef] [PubMed]

135. Richardson, D.R.; Kalinowski, D.S.; Lau, S.; Jansson, P.J.; Lovejoy, D.B. Cancer cell iron metabolism and the development of potent iron chelators as anti-tumour agents. *Biochim. Biophys. Acta Gen. Subj.* **2009**, *1790*, 702–717. [CrossRef] [PubMed]

136. Min Pang, B.S.; Connor, J.R. Role of Ferritin in Cancer Biology. *J. Cancer Sci. Ther.* **2015**, *7*, 155–160.

137. Kabat, G.C.; Miller, A.B.; Jain, M.; Rohan, T.E. Dietary iron and haem iron intake and risk of endometrial cancer: A prospective cohort study. *Br. J. Cancer.* **2007**, *98*, 194. [CrossRef] [PubMed]

138. Toyokuni, S. Iron-induced carcinogenesis: The role of redox regulation. *Free Radic. Biol. Med.* **1996**, *20*, 553–566. [CrossRef]

139. Richmond, H.G. Induction of Sarcoma in the Rat by Iron—Dextran Complex. *Br. Med. J.* **1959**, *1*, 947. [CrossRef] [PubMed]

140. Hanahan, D.; Weinberg, R.A. Hallmarks of Cancer: The Next Generation. *Cell* **2011**, *144*, 646–674. [CrossRef] [PubMed]

141. Corcé, V.; Gouin, S.G.; Renaud, S.; Gaboriau, F.; Deniaud, D. Recent advances in cancer treatment by iron chelators. *Bioorg. Med. Chem. Lett.* **2016**, *26*, 251–256. [CrossRef] [PubMed]

142. Prutki, M.; Poljak-Blazi, M.; Jakopovic, M.; Tomas, D.; Stipancic, I.; Zarkovic, N. Altered iron metabolism, transferrin receptor 1 and ferritin in patients with colon cancer. *Cancer Lett.* **2006**, *238*, 188–196. [CrossRef] [PubMed]

143. Shinohara, H.; Fan, D.; Ozawa, S.; Yano, S.; Van Arsdell, M.; Viner, J.L.; Beers, R.; Pastan, I.; Fidler, I.J. Site-specific expression of transferrin receptor by human colon cancer cells directly correlates with eradication by antitransferrin recombinant immunotoxin. *Int. J. Oncol.* **2000**, *17*, 643–694. [CrossRef] [PubMed]

144. Shindelman, J.E.; Ortmeyer, A.E.; Sussman, H.H. Demonstration of the transferrin receptor in human breast cancer tissue. Potential marker for identifying dividing cells. *Int. J. Cancer* **1981**, *27*, 329–334. [CrossRef] [PubMed]

145. Daniels, T.R.; Delgado, T.; Rodriguez, J.A.; Helguera, G.; Penichet, M.L. The transferrin receptor part I: Biology and targeting with cytotoxic antibodies for the treatment of cancer. *Clin. Immunol.* **2006**, *121*, 144–158. [CrossRef] [PubMed]

146. Sutherland, R.; Delia, D.; Schneider, C.; Newman, R.; Kemshead, J.; Greaves, M. Ubiquitous cell-surface glycoprotein on tumor cells is proliferation-associated receptor for transferrin. *Proc. Natl. Acad. Sci. USA* **1981**, *78*, 4515–4519. [CrossRef] [PubMed]

147. Pinnix, Z.K.; Miller, L.D.; Wang, W.; D'Agostino, R.; Kute, T.; Willingham, M.C.; Hatcher, H.; Tesfay, L.; Sui, G.; Di, X.; et al. Ferroportin and Iron Regulation in Breast Cancer Progression and Prognosis. *Sci. Transl. Med.* **2010**, *2*, 43ra56. [CrossRef] [PubMed]

148. Liotta, L.A.; Kohn, E.C. The microenvironment of the tumour–host interface. *Nature* **2001**, *411*, 375. [CrossRef] [PubMed]

149. Agarwal, D.; Goodison, S.; Nicholson, B.; Tarin, D.; Urquidi, V. Expression of matrix metalloproteinase 8 (MMP-8) and tyrosinase-related protein-1 (TYRP-1) correlates with the absence of metastasis in an isogenic human breast cancer model. *Differentiation* **2003**, *71*, 114–125. [CrossRef] [PubMed]

150. Liou, G.-Y.; Storz, P. Reactive oxygen species in cancer. *Free Radic. Res.* **2010**, *44*. [CrossRef] [PubMed]

151. Richardson, D.R. Iron chelators as therapeutic agents for the treatment of cancer. *Crit. Rev. Oncol. Hematol.* **2002**, *42*, 267–281. [CrossRef]

152. Buss, J.L.; Torti, F.M.; Torti, S.V. The role of iron chelation in cancer therapy. *Curr. Med. Chem.* **2003**, *10*, 1021–1034. [CrossRef] [PubMed]

153. Kalinowski, D.S.; Richardson, D.R. The Evolution of Iron Chelators for the Treatment of Iron Overload Disease and Cancer. *Pharmacol. Rev.* **2005**, *57*, 547–583. [CrossRef] [PubMed]

154. Ninomiya, T.; Ohara, T.; Noma, K.; Katsura, Y.; Katsube, R.; Kashima, H.; Kato, T.; Tomono, Y.; Tazawa, H.; Kagawa, S.; et al. Iron depletion is a novel therapeutic strategy to target cancer stem cells. *Oncotarget* **2017**, *8*, 98405–98416. [CrossRef] [PubMed]

155. Nurchi, V.M.; Crisponi, G.; Lachowicz, J.I.; Medici, S.; Peana, M.; Zoroddu, M.A. Chemical features of in use and in progress chelators for iron overload. *J. Trace Elem. Med. Biol.* **2016**, *38*, 10–18. [CrossRef] [PubMed]

156. Mobarra, N.; Shanaki, M.; Ehteram, H.; Nasiri, H.; Sahmani, M.; Saeidi, M.; Goudarzi, M.; Pourkarim, H.; Azad, M. A Review on Iron Chelators in Treatment of Iron Overload Syndromes. *Int. J. Hematol. Oncol. Stem Cell Res.* **2016**, *10*, 239–247. [PubMed]

157. Neilands, J.B. Siderophores-Structure and function of microbial iron transport compounds. *J. Biol. Chem.* **1995**, *270*, 26723–26726. [CrossRef] [PubMed]

158. Kontoghiorghes, G.J. New chelation therapies and emerging chelating drugs for the treatment of iron overload. *Expert Opin. Emerg. Drugs* **2006**, *11*, 1–5. [CrossRef] [PubMed]

159. Salis, O.; Bedir, A.; Kilinc, V.; Alacam, H.; Gulten, S.; Okuyucu, A. The anticancer effects of desferrioxamine on human breast adenocarcinoma and hepatocellular carcinoma cells. *Cancer Biomark.* **2014**, *14*, 419–426. [CrossRef] [PubMed]

160. Kicic, A.; Chua, A.C.; Baker, E. Desferrithiocin is a more potent antineoplastic agent than desferrioxamine. *Br. J. Pharmacol.* **2002**, *135*, 1393–1402. [CrossRef] [PubMed]

161. Zhou, T.; Ma, Y.; Kong, X.; Hider, R.C. Design of iron chelators with therapeutic application. *Dalton Trans.* **2012**, *41*, 6371–6389. [CrossRef] [PubMed]

162. Yan, X.; Yu, Y.; Ji, P.; He, H.; Qiao, C. Antitumor activity of endoperoxide-iron chelator conjugates—Design, synthesis and biological evaluation. *Eur. J. Med. Chem.* **2015**, *102*, 180–187. [CrossRef] [PubMed]

163. Steinhauser, S.; Heinz, U.; Bartholomä, M.; Weyhermüller, T.; Nick, H.; Hegetschweiler, K. Complex formation of ICL670 and related ligands with Fe(III) and Fe(II). *Eur. J. Inorg. Chem.* **2004**, *2004*, 4177–4192. [CrossRef]

164. Ohyashiki, J.H.; Kobayashi, C.; Hamamura, R.; Okabe, S.; Tauchi, T.; Ohyashiki, K. The oral iron chelator deferasirox represses signaling through the mTOR in myeloid leukemia cells by enhancing expression of REDD1. *Cancer Sci.* **2009**, *100*, 970–977. [CrossRef] [PubMed]

165. Chantrel-Groussard, K.; Gaboriau, F.; Pasdeloup, N.; Havouis, R.; Nick, H.; Pierre, J.-L.; Brissot, P.; Lescoat, G. The new orally active iron chelator ICL670A exhibits a higher antiproliferative effect in human hepatocyte cultures than O-trensox. *Eur. J. Pharmacol.* **2006**, *541*, 129–137. [CrossRef] [PubMed]

166. Jeon, S.-R.; Lee, J.-W.; Jang, P.-S.; Chung, N.-G.; Cho, B.; Jeong, D.-C. Anti-leukemic properties of deferasirox via apoptosis in murine leukemia cell lines. *Blood Res.* **2015**, *50*, 33–39. [CrossRef] [PubMed]

167. Theerasilp, M.; Chalermpanapun, P.; Ponlamuangdee, K.; Sukvanitvichai, D.; Nasongkla, N. Imidazole-modified deferasirox encapsulated polymeric micelles as pH-responsive iron-chelating nanocarrier for cancer chemotherapy. *RSC Adv.* **2017**, *7*, 11158–11169. [CrossRef]

168. Hrušková, K.; Potůčková, E.; Hergeselová, T.; Liptáková, L.; Hašková, P.; Mingas, P.; Kovaříková, P.; Šimůnek, T.; Vávrová, K. Aroylhydrazone iron chelators: Tuning antioxidant and antiproliferative properties by hydrazide modifications. *Eur. J. Med. Chem.* **2016**, *120*, 97–110. [CrossRef] [PubMed]

169. Ma, Y.; Abbate, V.; Hider, R.C. Iron-sensitive fluorescent probes: Monitoring intracellular iron pools. *Metallomics* **2015**, *7*, 212–222. [CrossRef] [PubMed]

170. Dean, K.M.; Qin, Y.; Palmer, A.E. Visualizing metal ions in cells: An overview of analytical techniques, approaches, and probes. *Biochim. Biophys. Acta Mol. Cell Res.* **2012**, *1823*, 1406–1415. [CrossRef] [PubMed]

171. Sun, D.; Melman, G.; LeTourneau, N.J.; Hays, A.M.; Melman, A. Synthesis and antiproliferating activity of iron chelators of hydroxyamino-1,3,5-triazine family. *Bioorg. Med. Chem. Lett.* **2010**, *20*, 458–460. [CrossRef] [PubMed]

172. Yu, Y.; Kalinowski, D.S.; Kovacevic, Z.; Siafakas, A.R.; Jansson, P.J.; Stefani, C.; Lovejoy, D.B.; Sharpe, P.C.; Bernhardt, P.V.; Richardson, D.R. Thiosemicarbazones from the old to new: Iron chelators that are more than just ribonucleotide reductase inhibitors. *J. Med. Chem.* **2009**, *52*, 5271–5294. [CrossRef] [PubMed]

173. Mrozek-Wilczkiewicz, A.; Serda, M.; Musiol, R.; Malecki, G.; Szurko, A.; Muchowicz, A.; Golab, J.; Ratuszna, A.; Polanski, J. Iron chelators in photodynamic therapy revisited: Synergistic effect by novel highly active thiosemicarbazones. *ACS Med. Chem. Lett.* **2014**, *5*, 336–339. [CrossRef] [PubMed]

174. Serda, M.; Kalinowski, D.S.; Mrozek-Wilczkiewicz, A.; Musiol, R.; Szurko, A.; Ratuszna, A.; Pantarat, N.; Kovacevic, Z.; Merlot, A.M.; Richardson, D.R.; et al. Synthesis and characterization of quinoline-based thiosemicarbazones and correlation of cellular iron-binding efficacy to anti-tumor efficacy. *Bioorg. Med. Chem. Lett.* **2012**, *22*, 5527–5531. [CrossRef] [PubMed]

175. Finch, R.A.; Liu, M.-C.; Cory, A.H.; Cory, J.G.; Sartorelli, A.C. Triapine (3-aminopyridine-2-carboxaldehyde thiosemicarbazone; 3-AP): An inhibitor of ribonucleotide reductase with antineoplastic activity. *Adv. Enzyme Regul.* **1999**, *39*, 3–12. [CrossRef]

176. Finch, R.A.; Liu, M.-C.; Grill, S.P.; Rose, W.C.; Loomis, R.; Vasquez, K.M.; Cheng, Y.-C.; Sartorelli, A.C. Triapine (3-aminopyridine-2-carboxaldehyde-thiosemicarbazone): A potent inhibitor of ribonucleotide reductase activity with broad spectrum antitumor activity. *Biochem. Pharmacol.* **2000**, *59*, 983–991. [CrossRef]

177. Koppenol, W.; Hider, R. Iron and redox cycling. Do's and don'ts. *Free Radic. Biol. Med.* **2018**, in press. [CrossRef] [PubMed]

178. Yuan, J.; Lovejoy, D.B.; Richardson, D.R. Novel di-2-pyridyl-derived iron chelators with marked and selective antitumor activity: In vitro and in vivo assessment. *Blood.* **2004**, *104*, 1450–1458. [CrossRef] [PubMed]

179. Kang, Y.J.; Kuo, C.F.; Majd, S. Nanoparticle-based delivery of an anti-proliferative metal chelator to tumor cells. In Proceedings of the 2017 39th Annual International Conference of the IEEE Engineering in Medicine and Biology Society (EMBC), Seogwipo, Korea, 11–15 July 2017; pp. 309–312.

180. O'Neill, H.S.; Herron, C.C.; Hastings, C.L.; Deckers, R.; Lopez Noriega, A.; Kelly, H.M.; Hennink, W.E.; McDonnell, C.O.; O'Brien, F.J.; Ruiz-Hernández, E.; et al. A stimuli responsive liposome loaded hydrogel provides flexible on-demand release of therapeutic agents. *Acta Biomater.* **2017**, *48*, 110–119. [CrossRef] [PubMed]

181. Abayaweera, G.S.; Wang, H.; Shrestha, T.B.; Yu, J.; Angle, K.; Thapa, P.; Malalasekera, A.P.; Maurmann, L.; Troyer, D.L.; Bossmann, S.H. Synergy of Iron Chelators and Therapeutic Peptide Sequences Delivered via a Magnetic Nanocarrier. *J. Funct. Biomater.* **2017**, *8*, 23. [CrossRef] [PubMed]

182. Agemy, L.; Friedmann-Morvinski, D.; Kotamraju, V.R.; Roth, L.; Sugahara, K.N.; Girard, O.M.; Mattrey, R.F.; Verma, I.M.; Ruoslahti, E. Targeted nanoparticle enhanced proapoptotic peptide as potential therapy for glioblastoma. *Proc. Natl. Acad. Sci. USA* **2011**, *108*, 17450–17455. [CrossRef] [PubMed]

183. Croy, S.R.; Kwon, G.S. Polymeric Micelles for Drug Delivery. *Curr. Pharm. Des.* **2006**, *12*, 4669–4684. [CrossRef] [PubMed]

184. Hume, D.A. The mononuclear phagocyte system. *Curr. Opin. Immunol.* **2006**, *18*, 49–53. [CrossRef] [PubMed]

185. Samad, A.; Sultana, Y.; Aqil, M. Liposomal drug delivery systems: An update review. *Curr. Drug Deliv.* **2007**, *4*, 297–305. [CrossRef] [PubMed]

186. Chang, T.M.; Tomat, E. Disulfide/thiol switches in thiosemicarbazone ligands for redox-directed iron chelation. *Dalton Trans.* **2013**, *42*, 7846–7849. [CrossRef] [PubMed]

187. Akam, E.A.; Chang, T.M.; Astashkin, A.V.; Tomat, E. Intracellular reduction/activation of a disulfide switch in thiosemicarbazone iron chelators. *Metallomics* **2014**, *6*, 1905–1912. [CrossRef] [PubMed]

188. Akam, E.A.; Tomat, E. Targeting Iron in Colon Cancer via Glycoconjugation of Thiosemicarbazone Prochelators. *Bioconjug. Chem.* **2016**, *27*, 1807–1812. [CrossRef] [PubMed]

189. Akam, E.A.; Utterback, R.D.; Marcero, J.R.; Dailey, H.A.; Tomat, E. Disulfide-masked iron prochelators: Effects on cell death, proliferation, and LTA hemoglobin production. *J. Inorg. Biochem.* **2018**, *180*, 186–193. [CrossRef] [PubMed]

190. MacCormack, M.A. Photodynamic therapy. *Adv. Dermatol.* **2006**, *22*, 219–258. [CrossRef] [PubMed]

191. Mrozek-Wilczkiewicz, A.; Malarz, K.; Rams-Baron, M.; Serda, M.; Bauer, D.; Montforts, F.-P.; Ratuszna, A.; Burley, T.; Polanski, J.; Musiol, R. Iron chelators and exogenic photosensitizers. Synergy through oxidative stress gene expression. *J. Cancer.* **2017**, *8*, 1979. [CrossRef] [PubMed]

192. Kovacevic, Z.; Chikhani, S.; Lovejoy, D.B.; Richardson, D.R. Novel thiosemicarbazone iron chelators induce up-regulation and phosphorylation of the metastasis suppressor, NDRG1: A new strategy for the treatment of pancreatic cancer. *Mol. Pharmacol.* **2011**, mol-111. [CrossRef]

193. Nurtjahja-Tjendraputra, E.; Fu, D.; Phang, J.M.; Richardson, D.R. Iron chelation regulates cyclin D1 expression via the proteasome: A link to iron deficiency–mediated growth suppression. *Blood* **2007**, *109*, 4045–4054. [CrossRef] [PubMed]

194. Kornmann, M.; Ishiwata, T.; Itakura, J.; Tangvoranuntakul, P.; Beger, H.G.; Korc, M. Increased cyclin D1 in human pancreatic cancer is associated with decreased postoperative survival. *Oncology* **1998**, *55*, 363–369. [CrossRef] [PubMed]

195. Parks, T.B.; Cruz, Y.M.; Tinoco, A.D. Applying the Fe(III) Binding Property of a Chemical Transferrin Mimetic to Ti(IV) Anticancer Drug Design. *Inorg. Chem.* **2014**, *53*, 1743–1749. [CrossRef] [PubMed]

196. Loza-Rosas, S.A.; Vázquez-Salgado, A.M.; Rivero, K.I.; Negrón, L.J.; Delgado, Y.; Benjamín-Rivera, J.A.; Vázquez-Maldonado, A.L.; Parks, T.B.; Munet-Colón, C.; Tinoco, A.D. Expanding the therapeutic potential of the iron chelator deferasirox in the development of aqueous stable Ti(IV) anticancer complexes. *Inorg. Chem.* **2017**, *56*, 7788–7802. [CrossRef] [PubMed]

197. Diaz-Garcia, J.D.; Gallegos-Villalobos, A.; Gonzalez-Espinoza, L.; Sanchez-Nino, M.D.; Villarrubia, J.; Ortiz, A. Deferasirox nephrotoxicity—The knowns and unknowns. *Nat. Rev. Nephrol.* **2014**, *10*, 574–586. [CrossRef] [PubMed]

198. Crisponi, G.; Nurchi, V.M.; Crespo-Alonso, M.; Sanna, G.; Zoroddu, M.A.; Alberti, G.; Biesuz, R. A Speciation Study on the Perturbing Effects of Iron Chelators on the Homeostasis of Essential Metal Ions. *PLoS ONE* **2015**, *10*, e0133050. [CrossRef] [PubMed]

199. Hu, W.; Luo, Q.; Wu, K.; Li, X.; Wang, F.; Chen, Y.; Ma, X.; Wang, J.; Liu, J.; Xiong, S.; et al. The anticancer drug cisplatin can cross-link the interdomain zinc site on human albumin. *Chem. Commun.* **2011**, *47*, 6006–6008. [CrossRef] [PubMed]

200. Larabee, J.L.; Hocker, J.R.; Hanas, J.S. Mechanisms of Aurothiomalate–Cys2His2 Zinc Finger Interactions. *Chem. Res. Toxicol.* **2005**, *18*, 1943–1954. [CrossRef] [PubMed]

201. Loza-Rosas, S.A.; Saxena, M.; Delgado, Y.; Gaur, K.; Pandrala, M.; Tinoco, A.D. A ubiquitous metal, difficult to track: Towards an understanding of the regulation of titanium(IV) in humans. *Metallomics* **2017**, *9*, 346–356. [CrossRef] [PubMed]

202. Saxena, M.; Loza Rosas, S.; Gaur, K.; Sharma, S.; Perez Otero, S.C.; Tinoco, A.D. Exploring titanium(IV) chemical proximity to iron(III) to elucidate a function for Ti(IV) in the human body. *Coord. Chem. Rev.* **2018**, *363*, 109–125. [CrossRef] [PubMed]

203. Tinoco, A.D.; Saxena, M.; Sharma, S.; Noinaj, N.; Delgado, Y.; Gonzalez, E.P.Q.; Conklin, S.E.; Zambrana, N.; Loza-Rosas, S.A.; Parks, T.B. Unusual synergism of transferrin and citrate in the regulation of Ti(IV) speciation, transport, and toxicity. *J. Am. Chem. Soc.* **2016**, *138*, 5659–5665. [CrossRef] [PubMed]

204. Loza Rosas, S.A.; Zayas-Ortiz, A.; Vazquez-Salgado, A.M.; Benjamin-Rivera, J.A.; Gaur, K.; Perez Otero, S.C.; Mendez-Fernandez, A.P.; Vazquez-Maldonado, A.L.; Alicea, N.; Tinoco, A.D. Transmetalation with a titanium(IV) compound transforms biofunctional copper into a cytotoxic agent. 2018; in preparation.

205. Deda, D.K.; Cardoso, R.M.; Uchiyama, M.K.; Pavani, C.; Toma, S.H.; Baptista, M.S.; Araki, K. A reliable protocol for colorimetric determination of iron oxide nanoparticle uptake by cells. *Anal. Bioanal. Chem.* **2017**, *409*, 6663–6675. [CrossRef] [PubMed]

206. Mounicou, S.; Szpunar, J.; Lobinski, R. Metallomics: The concept and methodology. *Chem. Soc. Rev.* **2009**, *38*, 1119–1138. [CrossRef] [PubMed]

207. Patil, U.; Adireddy, S.; Jaiswal, A.; Mandava, S.; Lee, B.; Chrisey, D. In Vitro/In Vivo Toxicity Evaluation and Quantification of Iron Oxide Nanoparticles. *Int. J. Mol. Sci.* **2015**, *16*, 24417–24450. [CrossRef] [PubMed]

208. Weiss, G.; Carver, P.L. Role of divalent metals in infectious disease susceptibility and outcome. *Clin. Microbiol. Infect.* **2018**, *24*, 16–23. [CrossRef] [PubMed]

209. Nunes, J.A.; Batista, B.L.; Rodrigues, J.L.; Caldas, N.M.; Neto, J.A.G.; Barbosa, F. A Simple Method Based on ICP-MS for Estimation of Background Levels of Arsenic, Cadmium, Copper, Manganese, Nickel, Lead, and Selenium in Blood of the Brazilian Population. *J. Toxicol. Environ. Health A* **2010**, *73*, 878–887. [CrossRef] [PubMed]

210. Meng, H.; Chen, Z.; Xing, G.; Yuan, H.; Chen, C.; Zhao, F.; Zhang, C.; Zhao, Y. Ultrahigh reactivity provokes nanotoxicity: Explanation of oral toxicity of nano-copper particles. *Toxicol. Lett.* **2007**, *175*, 102–110. [CrossRef] [PubMed]

211. Sanz-Medel, A.; Montes-Bayón, M.; Luisa Fernández Sánchez, M. Trace element speciation by ICP-MS in large biomolecules and its potential for proteomics. *Anal. Bioanal. Chem.* **2003**, *377*, 236–247. [CrossRef] [PubMed]

212. Herrera, C.; Pettiglio, M.A.; Bartnikas, T.B. Investigating the role of transferrin in the distribution of iron, manganese, copper, and zinc. *J. Biol. Inorg. Chem.* **2014**, *19*, 869–877. [CrossRef] [PubMed]

213. Singh, B.P.; Dwivedi, S.; Dhakad, U.; Murthy, R.C.; Choubey, V.K.; Goel, A.; Sankhwar, S.N. Status and Interrelationship of Zinc, Copper, Iron, Calcium and Selenium in Prostate Cancer. *Indian J. Clin. Biochem.* **2016**, *31*, 50–56. [CrossRef] [PubMed]

214. Matusiak, K.; Skoczen, A.; Setkowicz, Z.; Kubala-Kukus, A.; Stabrawa, I.; Ciarach, M.; Janeczko, K.; Jung, A.; Chwiej, J. The elemental changes occurring in the rat liver after exposure to PEG-coated iron oxide nanoparticles: Total reflection X-ray fluorescence (TXRF) spectroscopy study. *Nanotoxicology* **2017**, *11*, 1225–1236. [CrossRef] [PubMed]

215. Kubala-Kukuś, A.; Banaś, D.; Braziewicz, J.; Majewska, U.; Pajek, M.; Wudarczyk-Moćko, J.; Antczak, G.; Borkowska, B.; Góźdź, S.; Smok-Kalwat, J. Analysis of Copper Concentration in Human Serum by Application of Total Reflection X-ray Fluorescence Method. *Biol. Trace Elem. Res.* **2014**, *158*, 22–28. [CrossRef] [PubMed]

216. Riemer, J.; Hoepken, H.H.; Czerwinska, H.; Robinson, S.R.; Dringen, R. Colorimetric ferrozine-based assay for the quantitation of iron in cultured cells. *Anal. Biochem.* **2004**, *331*, 370–375. [CrossRef] [PubMed]

217. Kundra, S.K.; Katyal, M.; Singh, R.P. Spectrophotometric determination of copper(I) and cobalt(II) with ferrozine. *Anal. Chem.* **1974**, *46*, 1605–1606. [CrossRef]

218. Shimberg, G.D.; Pritts, J.D.; Michel, S.L.J. *Methods Enzymology*; David, S.S., Ed.; Academic Press: Cambridge, MA, USA, 2018; Volume 599, pp. 101–137.

219. Abou-Shakra, F.R. *Handbook of Analytical Separations*; Wilson, I.D., Ed.; Elsevier Science B.V.: Amstedam, The Netherlands, 2003; Volume 4, pp. 351–371.

220. Ammerman, J.; Huang, C.; Sailstad, J.; Wieling, J.; Whitmire, M.L.; Wright, D.; de Lisio, P.; Keenan, F.; McCurdy, E.; Woods, B.; et al. Technical aspects of inductively coupled plasma bioanalysis techniques. *Bioanalysis* **2013**, *5*, 1831–1841. [CrossRef] [PubMed]

221. Godfrey, L.V.; Glass, J.B. Methods Enzymology. Klotz, M.G., Ed.; Academic Press: Cambridge, MA, USA, 2011; Volume 486, pp. 483–506.

222. Zhang, R.; Li, L.; Sultanbawa, Y.; Xu, Z.P. X-ray fluorescence imaging of metals and metalloids in biological systems. *Am. J. Nucl. Med. Mol. Imaging* **2018**, *8*, 169–188. [PubMed]

223. Antosz, F.J.; Xiang, Y.; Diaz, A.R.; Jensen, A.J. The use of total reflectance X-ray fluorescence (TXRF) for the determination of metals in the pharmaceutical industry. *J. Pharm. Biomed. Anal.* **2012**, *62*, 17–22. [CrossRef] [PubMed]

224. Marcó, L.M.; Greaves, E.D.; Alvarado, J. Analysis of human blood serum and human brain samples by total reflection X-ray fluorescence spectrometry applying Compton peak standardization. *Spectrochim. Acta Part B At. Spectrosc.* **1999**, *54*, 1469–1480. [CrossRef]

225. Wobrauschek, P. Total reflection X-ray fluorescence analysis—A review. *X-Ray Spectrom.* **2007**, *36*, 289–300. [CrossRef]

226. Terzano, R.; Al Chami, Z.; Vekemans, B.; Janssens, K.; Miano, T.; Ruggiero, P. Zinc distribution and speciation within rocket plants (*Eruca vesicaria* L. Cavalieri) grown on a polluted soil amended with compost as determined by XRF microtomography and Micro-XANES. *J. Agric. Food Chem.* **2008**, *56*, 3222–3231. [CrossRef] [PubMed]

227. Montarges-Pelletier, E.; Chardot, V.; Echevarria, G.; Michot, L.J.; Bauer, A.; Morel, J.L. Identification of nickel chelators in three hyperaccumulating plants: An X-ray spectroscopic study. *Phytochemistry* **2008**, *69*, 1695–1709. [CrossRef] [PubMed]

228. Fahrni, C.J. Biological applications of X-ray fluorescence microscopy: Exploring the subcellular topography and speciation of transition metals. *Curr. Opin. Chem. Biol.* **2007**, *11*, 121–127. [CrossRef] [PubMed]

229. Qin, Z.Y.; Caruso, J.A.; Lai, B.; Matusch, A.; Becker, J.S. Trace metal imaging with high spatial resolution: Applications in biomedicine. *Metallomics* **2011**, *3*, 28–37. [CrossRef] [PubMed]

230. Becker, J.S.; Kumtabtim, U.; Wu, B.; Steinacker, P.; Otto, M.; Matusch, A. Mass spectrometry imaging (MSI) of metals in mouse spinal cord by laser ablation ICP-MS. *Metallomics* **2012**, *4*, 284–288. [CrossRef] [PubMed]

231. Köpf-Maier, P. Electron-spectroscopic imaging—A method for analysing the distribution of light elements in mammalian cells and tissues. *Acta Histochem.* **1991**, *91*, 25–37. [CrossRef]

232. Kapp, N.; Studer, D.; Gehr, P.; Geiser, M. *Electron Microscopy: Methods and Protocols*; Kuo, J., Ed.; Humana Press: Totowa, NJ, USA, 2007; pp. 431–447.

233. Morello, M.; Canini, A.; Mattioli, P.; Sorge, R.P.; Alimonti, A.; Bocca, B.; Forte, G.; Martorana, A.; Bemardi, G.; Sancesario, G. Sub-cellular localization of manganese in the basal ganglia of normal and manganese-treated rats—An electron spectroscopy imaging and electron energy-loss spectroscopy study. *Neurotoxicology* **2008**, *29*, 60–72. [CrossRef] [PubMed]

234. Mukherjee, N.; Podder, S.; Mitra, K.; Majumdar, S.; Nandi, D.; Chakravarty, A.R. Targeted photodynamic therapy in visible light using BODIPY-appended copper(II) complexes of a vitamin B6 Schiff base. *Dalton Trans.* **2018**, *47*, 823–835. [CrossRef] [PubMed]

235. Que, E.L.; Domaille, D.W.; Chang, C.J. Metals in neurobiology: Probing their chemistry and biology with molecular imaging. *Chem. Rev.* **2008**, *108*, 1517–1549. [CrossRef] [PubMed]

236. Hernando, D.; Levin, Y.S.; Sirlin, C.B.; Reeder, S.B. Quantification of Liver Iron With MRI: State of the Art and Remaining Challenges. *J. Magn. Reson. Imaging* **2014**, *40*, 1003–1021. [CrossRef] [PubMed]
237. Heffern, M.C.; Park, H.M.; Au-Yeung, H.Y.; Van de Bittner, G.C.; Ackerman, C.M.; Stahl, A.; Changa, C.J. In vivo bioluminescence imaging reveals copper deficiency in a murine model of nonalcoholic fatty liver disease. *Proc. Natl. Acad. Sci. USA* **2016**, *113*, 14219–14224. [CrossRef] [PubMed]
238. Hirayama, T.; Van de Bittner, G.C.; Gray, L.W.; Lutsenko, S.; Chang, C.J. Near-infrared fluorescent sensor for in vivo copper imaging in a murine Wilson disease model. *Proc. Natl. Acad. Sci. USA* **2012**, *109*, 2228–2233. [CrossRef] [PubMed]
239. Aron, A.T.; Heffern, M.C.; Lonergan, Z.R.; Vander Wal, M.N.; Blank, B.R.; Spangler, B.; Zhang, Y.; Park, H.M.; Stahl, A.; Renslo, A.R.; et al. In vivo bioluminescence imaging of labile iron accumulation in a murine model of *Acinetobacter baumannii* infection. *Proc. Natl. Acad. Sci. USA* **2017**, *114*, 12669–12674. [CrossRef] [PubMed]

 *inorganics* 

Article

Synthesis, Reactivity Studies, and Cytotoxicity of Two *trans*-Iodidoplatinum(II) Complexes. Does Photoactivation Work?

Leticia Cubo [1], Thalia Parro [1], Amancio Carnero [2], Luca Salassa [3], Ana I. Matesanz [1] and Adoracion G. Quiroga [1,*]

[1]   Departamento de Química Inorgánica, Universidad Autónoma de Madrid, 28049-Madrid, Spain;
      leticia.cubo@uam.es (L.C.); thalia.pd20@gmail.com (T.P.); ana.matesanz@uam.es (A.I.M.)
[2]   IBIS, Campus Hospital Universitario Virgen del Rocío, 41013-Sevilla, Spain; acarnero@us.es
[3]   Donostia International Physics Center, 20018-Donostia, Spain; lsalassa@dipc.org
*   Correspondence: adoracion.gomez@uam.es; Tel.: +34-914974050

Received: 19 October 2018; Accepted: 23 November 2018; Published: 3 December 2018

Abstract: *trans*-Platinum complexes have been the landmark in unconventional drugs prompting the development of innovative structures that might exhibit chemical and biological profiles different to cisplatin. Iodido complexes signaled a new turning point in the platinum drug design field when their cytotoxicity was reevaluated and reported. In this new study, we have synthesized and evaluated diiodidoplatinum complexes *trans*-[PtI$_2$(amine)(pyridine)] bearing aliphatic amines (isopropylamine and methylamine) and pyridines in *trans* configuration. X-ray diffraction data support the structural characterization. Their cytotoxicity has been evaluated in tumor cell lines such as SAOS-2, A375, T-47D, and HCT116. Moreover, we report their solution behavior and reactivity with biological models. Ultraviolet-a (UVA) irradiation induces an increase in their reactivity towards model nucleobase 5′-GMP in early stages, and promotes the release of the pyridine ligand (spectator ligand) at longer reaction times. Density Functional calculations have been performed and the results are compared with our previous studies with other iodido derivatives.

Keywords: platinum iodido complexes; cytotoxicity; photoactivation

---

1. Introduction

Antitumor platinum drug design has been and still is a major project in metallodrug research. The latest reviews and the large number of contributions therein identify new complex designs to face the known side effect problems of these antitumor drugs [1]. One of the latest contributions to this field has been the results that we published when studying the impact of the leaving group in the reactivity and cytotoxicity of the iodido complexes [2,3]. Looking for a new design, our group of researchers reevaluated *cis* platinum iodido complexes; their reactivity turned out to be quite unexpected versus sulfur donor biomolecules [4]. Our studies proved that the iodido groups stayed in the adducts formed in the reaction with the protein cytochrome c or lysozime while the aliphatic amines acted as leaving groups [5]. This peculiar reactivity was not detected versus DNA, with which they showed classical reactivity cisplatin like.

Following these results, we extended our studies to include the reactivity of *trans* diiodido diamine platinum(II) with different aliphatic amines versus some selected models of biomolecules [6]. These reactivity studies revealed a very similar profile for the *cis* and *trans* complexes upon binding to model nucleobases (DNA). The adduct formation occurs with retention of the amine spectator ligands. Additionally, *trans*-type complexes manifested a lower propensity to form adducts with peptide and a more classical reactivity, the iodido ligands release upon protein binding. Morevoer, these last *trans* series seemed to be affected by the size of the amine ligands showing differences in their reactivity versus S-donor models and in their cytotoxicity [6]. The reconsideration of these iodido derivatives have shown a great impact in the field, and the trend and behavior of these complexes is being reevaluated [7].

Within the design of novel metallodrugs, the use of activation therapies such as the use of UVA light offers a wide fan of possibilities for metallic complexes such as activation of reactivity [8], change in conformation [9], heterobimetallic complexes with PDT ligands [10] and singlet oxygen producing complexes [11]. Photochemical studies on a selected number of *trans* iodido complexes initially showed that irradiation induces a faster reaction with CT DNA and a higher amount of Pt bound to DNA. These complexes also reacted faster with 5′-GMP under irradiation and even showed noticeable improvements in their cytotoxicity when treatment was combined with UVA light [12]. These results support the idea that UVA light could be used to increase the activity of the diiodido platinum complexes, and might even make it more selective, in a similar way than those observed for chlorido derivatives [8].

Within this frame, we have looked at the role of the spectator ligand within the iodido complex structures and replaced one of the aliphatic amines with an aromatic planar amine, widely used in the *trans* type of complexes with highly satisfying results [13]. In particular, we prepared two *trans* configured iodido-platinum complexes with a pyridine and isopropylamine/methylamine ligands. X-ray structure studies of both complexes complete the structural characterization. The cytotoxicity was evaluated and the possible mechanism investigated, looking for a possible photochemical activation. The interaction studies of these new complexes with a representative model biomolecules such as: model nucleobase 5′-GMP and *N*-Methylimidazol (MeIm) have been performed and photochemical studies on such interaction analyzed and discussed together with the DFT calculations.

2. Results

2.1. Synthesis and Characterization

The *trans*-[PtI$_2$(amine)(pyridine)] complexes were prepared according to published procedures with slight variations, using the *cis*-[PtI$_2$(amine)$_2$] complexes with the corresponding aliphatic amine: isopropylamine or methylamine as the starting material [6]. The reaction was carried out in water with an excess of pyridine and without isolating the tetraamine species, giving the desired *trans* complexes by slow evaporation of the solvent at high temperature.

The structure and numbering of the complexes are depicted in Figure 1. Their characterization performed by usual techniques is nicely in agreement with the proposed structure and detailed data have been collected in the experimental section.

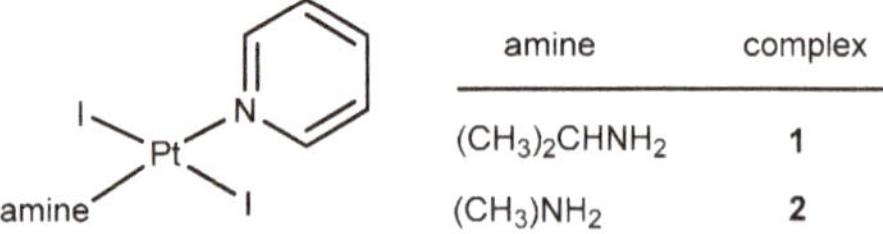

Figure 1. Structure and numbering of the complexes studied in this manuscript.

The molecular structure of complexes **1** and **2** are shown in Figure 2. Crystal data are listed in Table S1. Selected bond lengths and angles are shown in Table 1. The platinum(II) atom has a square planar coordination geometry and is coordinated to two nitrogen atoms of the methylamine or isopropylamine and pyridine ligands and two iodido ligands in *trans*-arrangement.

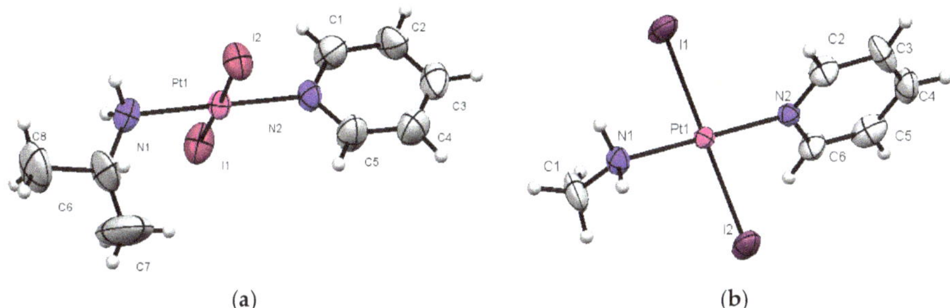

Figure 2. Molecular views: (**a**) complex **1** (**b**) complex **2**.

Table 1. Selected distances (Å) and angles (°) in complexes **1** and **2**.

Distances	Complex 1	Complex 2
Pt–I1	2.5962(9)	2.5906(11)
Pt–I2	2.5920(8)	2.59861(19)
Pt–N1	2.050(9)	2.077(13)
Pt–N2	2.014(8)	2.025(13)
Angles	**Complex 1**	**Complex 2**
N2–Pt–N1	179.24(4)	179.3(5)
N1–Pt–I1	90.1(3)	90.2(4)
N1–Pt–I2	90.0(3)	89.1(4)
N2–Pt–I1	89.7(3)	90.1(4)
N2–Pt–I2	90.02(3)	90.6(4)
I1–Pt–I2	179.05(3)	177.24(4)

2.2. Cytotoxicity of trans-[PtI$_2$(amine)(py)] Complexes

The cytotoxicity of complexes **1** and **2** was determined in comparison to cisplatin. The cancer cell lines selected for this study are SAOS-2 (human osteosarcoma), A375 (melanoma), T-47D (breast carcinoma), HCT116++ (human colon carcinoma with the presence of p53) and HCT116– (human colon carcinoma in the absence of p53). The IC$_{50}$ and standard deviation values for complexes **1** and **2** are shown in Table 2.

Table 2. IC$_{50}$ values for complexes **1** and **2** and cisplatin in five cancer cell lines. Data were collected after 96 h of exposure to the drugs. Standard deviation is shown in brackets.

Cell line	Complex 1	Complex 2	Cisplatin
SAOS-2	32.9 (22.6) *	53.7(15.8)	5.9(1.5)
A375	18.9(3.4)	27.9(5.1)	9.2(1.7)
T-47D	30.9(5.6)	45.0(2.5)	10.2(3.4)
HCT116++	18.1(3.6)	29.8(4.4)	8.3(2.7)
HCT116− −	56.4(10.3)	63.4(17.3)	62.2(14.3)

* This value could be an overestimated number, because of the high standard deviation.

2.3. Reactivity of trans-[PtI$_2$(amine)(py)] Complexes with 5′-GMP and MeIm

One of the main cellular targets of metallodrugs is DNA [14,15]. There are many techniques to study the interaction with DNA, since our objective in this section is to compare the reactivity of these compounds with the data available in the references, we have used a small model of DNA, such as 5′-GMP to carry out the experiments. This model has proved to be an excellent approach in many systems by monitoring the changes at its H8 signal by ^{1}H NMR [16,17].

Thus, the study on the reactivity of complexes **1** and **2** with the model nucleobase 5′-GMP was carried out. The complexes under investigation were incubated at a molar ratio of complex: 5′-GMP of 1:2 as described in the experimental section and the reactivity toward this nucleobase was monitored by NMR spectroscopy. The ^{1}H NMR spectra of the interaction of complex **1** with 5′-GMP monitored in a mixture of acetone-d_6 and D_2O (ratio 2:1) from 1 h to 7 h are shown in Figure 3. When following the changes at the H8 peak signal of 5′-GMP, a new signal arises at 9.1 ppm that is assignable to a H8 nucleotide adduct and the intensity increases during the first 3 h of reaction while the H8-signal of the free nucleotide decreases. At longer reaction times, the first monoadduct intensity does not increase and simultaneously, a new adduct is detected at 8.2 ppm. The monitoring of the reaction provides evidence for the presence of 5′-GMP adducts in the sample (label S at Figure 3 and Figure S1). The reaction of the *trans* complex **2** with 5′-GMP was also studied and the results are similar to the reactivity described for complex **1**. However, the reactivity of compound **2** seems to be slower since the signals assigned to the adduct species are weaker in complex **2** than for complex **1** after 3 h of reaction (Figure S2). Furthermore, a smaller amount of the model nucleotide adducts are detected for complex **2** over longer periods of time.

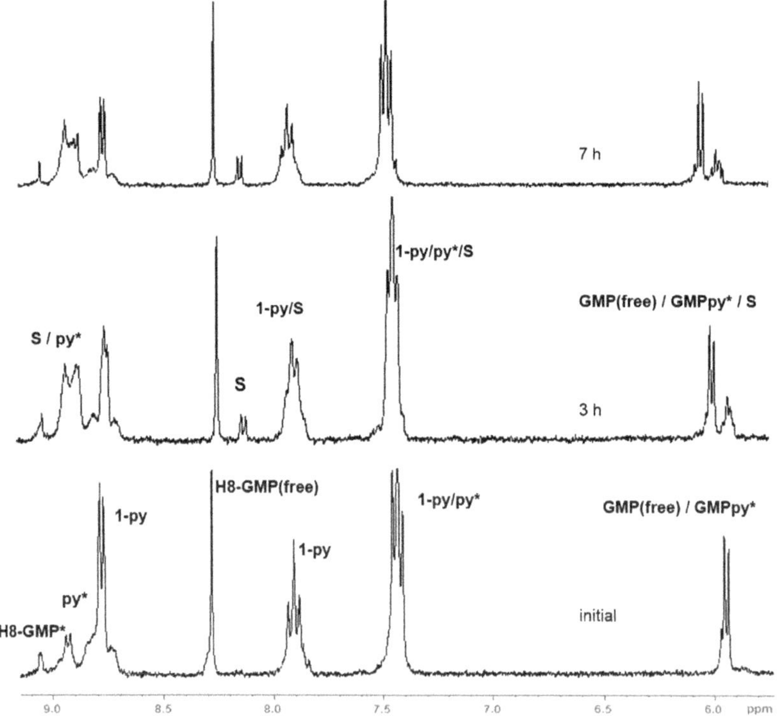

Figure 3. Progress of the reaction between complex **1** and 5′-GMP at 37 °C monitored by ^{1}H NMR (acetone-d_6 and D_2O in a ratio 2:1) showing changes in the aromatic area. Labels: GMP(free), free 5′-GMP; H8-GMP*: H8-5′-GMP coordinated adduct signal; 1-py, pyridine ligand signals of complex **1**; py*, adduct pyridine signal, and S: speciation.

When similar experiments were repeated under irradiation, the reactivity of complex **1** with 5′-GMP showed some important differences compared to the non-irradiated sample. In the first hour of irradiation, the reported monoadduct is formed in a higher amount (integrals shows a higher formation), but after 1 h the spectra shows a new signal at 8.6 ppm (Figure 4) corresponding to free pyridine and its intensity increases along the time. Moreover, we can clearly observe new species in the high-field region of the spectra and there are also signals corresponding to the isopropylamine group that appear to split into several sets.

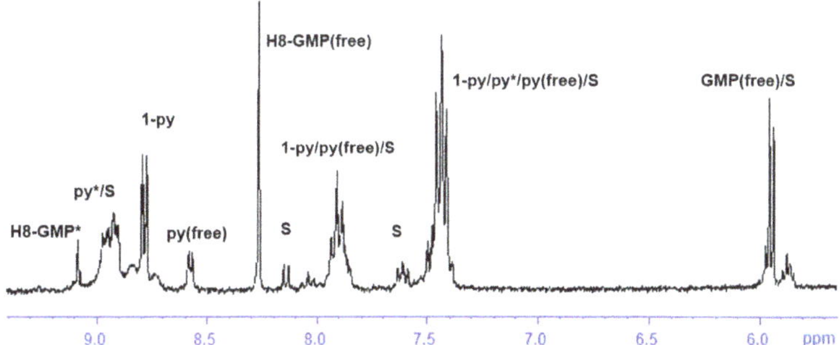

Figure 4. ^{1}H NMR spectrum of the reaction between complex **1** and 5′-GMP (acetone-d_6 and D$_2$O in a ratio 2:1) showing changes in the aromatic and aliphatic area after 1 h under irradiation at 37 °C. Labels: GMP(free), free 5′-GMP; H8-GMP*: H8-5′-GMP coordinated adduct signal; 2-py, pyridine ligand signals of complex **2**; py*, adduct pyridine signal, and **S**: speciation.

In addition, we extended our NMR study of the complexes' coordination to the heterocyclic ligand *N*-methylImidazole, MeIm. It has been reported that this model can provide information about the possibilities of a complex to bind into a lopsided configuration like those presented by biological molecules such as proteins or DNA [18]. Moreover, its structure is different to 5′-GMP (already used) but similar to those presented by proteins at, for example, histidine's sites. We monitored the reactivity with this model at 37 °C (Figure 5a) and detected monoadduct species formation within the first hour, becoming almost a major product only after 3 h. The speciation is clearly formed quickly and the starting material has fully reacted after 7 h. The photoactivation of the sample (Figure 5b) enhances the reactivity but more importantly, produces the release of the pyridine ligand like in its reaction with 5′-GMP.

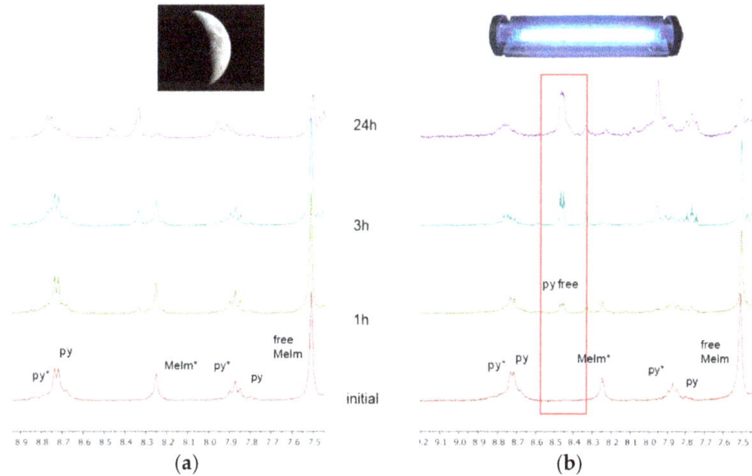

Figure 5. Progress of the reaction between complex **1** and MeIm monitored by ^{1}H NMR (acetone-d_6 and D$_2$O in a ratio 2:1) showing changes in the aromatic area. Labels: py, pyridine ligand complex **1**; py*, pyridine signal adducts, MeIm*, *N*-Methylimidazol signal adducts, free MeIm, free *N*-Methylimidazol. (**a**) reaction at 37 °C and (**b**) reaction at 37 °C and irradiating.

2.4. DFT Calculations

For a better understanding of the photochemistry of these *trans*-iodido platinum(II) complexes, we performed a set of Density Functional Theory (DFT) and time-dependent DFT (TD-DFT) calculations on complexes **1** and **2** and their 9-EtG adducts. The DNA basis mimic, 9-EtG, was used instead of 5′-GMP to ease the computational calculations. After geometry optimization (Table S2a–d), we analyzed the single-singlet transitions on the complexes and gauged their excited-state chemistry (Tables S3 and S4). Consistently with experimental results, the parent complexes **1** and **2** display electronic transition that are mostly dissociative towards the iodido ligands (Table S3). The calculated absorption spectrum of the mono 9-EtG adducts of the two complexes displays a band at ca. 350 nm, which can be excited under UVA excitation (Table S5). Both mono and bis 9-EtG adducts present a dissociative transition towards the four ligands, as shown in the electron difference density maps (EDDMs, Table S6). As an example, Figure 6 (and Table S7) reports two EDDMS for complexes **1** and **2** corresponding to the calculated lowest-energy electronic transition in which the antibonding orbitals LUMO+4 and 5 have significant contributions. According to this scenario, the 9-EtG adducts are more likely to prompt the release of a pyridine under light irradiation compared to the parent 1 and 2. A summary of the most relevant bond distances for **1** and **2** and their 9Et-G derivatives are reported in Table S8.

Figure 6. Selected electron difference density maps (EDDMs, transition 1, Table S7) and corresponding LUMOs for the bis 9-EtG adducts of complexes **1** (**a**) and **2** (**b**) in water at the CAM-B3LYP/LANL08/6-31G** level. In the EDDMs (top), gray indicates a decrease in electron density, while blue–gray indicates an increase.

3. Discussion

The complexes were designed based on the positive results achieved with previous complexes of nonconventional structure and using iodidos, aliphatic amines, and pyridine as ligands. The synthesis was performed following our published procedure, although the final yields obtained for complexes **1** and **2** are lower than usual. Many attempts for varying concentrations, temperature or longer reaction time did not improve the results. The characterization by usual techniques was in accordance with the X-Ray study and allowed for a detailed knowledge of the structural features. The mentioned structures are compared with those of published Pt complexes with aliphatic amines, iodide, and pyridine ligands. The Pt–N (aliphatic amine) [19,20], Pt–I [21,22] and Pt–N (pyridine) [23,24] distances fall in the typical ranges. The interactions between the methyl groups and the iodido ligands result in significant strain in the complexes. This is reflected in the deviation of the Pt–N–C angles (116°) from the ideal tetrahedral values (109°). Cytotoxicity was evaluated versus SAOS-2 human osteosarcoma, A375 melanoma, T-47D breast cancer, SF-268 glioblastoma, NCI-H460 lung cancer and colorectal carcinoma HCT116, and matched p53-deficient HCT116 (-/-) cell lines. The antiproliferative effects are found to be moderate, with IC_{50} values generally being in the micromolar range and the differences between the cytotoxic potencies of complexes **1** and **2** are comparatively small. Substitution of one aliphatic amine by pyridine seems to decrease cytotoxic potency; the steric demand could be responsible for these poor specific antitumor effects. We believe that irradiation of the complexes will produce the enhancement of the antitumoral activity, as discussed in the following results.

- The interaction of these complexes with 5′-GMP shows the formation of the expected DNA adduct after hydrolysis of the iodido complexes as reported for similar compounds with no release of the spectator ligands.
- However, once we irradiate the samples, we can only observe a promotion of the reactivity in the first hour, and then the release of the pyridine ligand becomes clear at longer reaction times under irradiation.
- This release of the spectator ligand, enhanced by irradiation, has not been observed studying *trans* isomers with iodido ligands versus DNA [6], but only with *cis* isomers and versus proteins [5]. In addition, when the results of the interaction of compounds **1** and **2** with MeIm are evaluated, they are very similar to the results obtained with DNA.
- Our interpretation of the data is that after irradiation, the reactivity of complexes **1** and **2** is enhanced, forming active species. However, at longer reaction times, the compounds lose the pyridine, affording new adducts and species that are more similar to those reported from the cytotoxic compounds than their original aqua species/adducts.
- DFT calculations justify the quicker formation of the bis/monoadduct species (Figure 6) along the first hour, but once the irradiation applies for longer periods of time, the pyridine is released and the compound is no longer the structure proposed but similar to those bearing only aliphatic amines [12].

The potential photoactivation of these poor pharmacological complexes is apparently not a simple mechanism. In the short term the irradiation will afford the DNA adducts but the most effective photoactivation will take place in the long term, as the species will be the same as the most active iodido series reported previously (Figure 7).

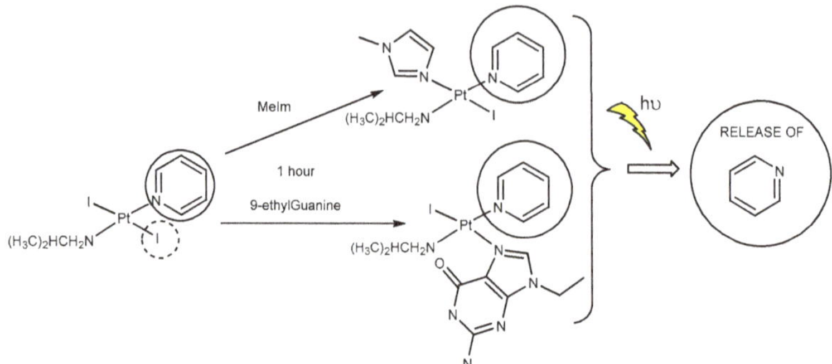

Figure 7. Reactivity versus 9-EtG and MeIm proposed in the discussion.

4. Materials and Methods

Complexes *cis*-PtI$_2$(amine)$_2$ (where amine = isopropylamine or methylamine) were synthesized as described in the previous communications based on reported procedures [12]. Characterization was in agreement with our previous data. Chemical starting materials; K$_2$PtCl$_4$, amines, pyridine and biological molecules—were purchased from VWR (Madrid, Spain).

4.1. Method for the Synthesis of trans-[PtI$_2$(amine)(pyridine)]

We followed published methods with slight variations, briefly: 500 mg of *cis*-PtI$_2$(amine)$_2$ (where amine is ipa for **1** and ma for **2**) and 20 equivalents of pyridine were mixed together in 30 mL of water and heated at reflux temperature for 3–6 h. The suspension turned to an almost clear solution that was filtered over celite. The resulting solution was concentrated very slowly at high temperature (100 °C) until the detection of an orange solid, which was allowed to stand overnight at 4 °C until complete precipitation. Then, the solid was filtered off, washed with warm water, and vacuum dried overnight at 60 °C in a drying oven. Afterwards, recrystallization in chloroform/ether was required. Single crystals were achieved by slow evaporation of a chloroform solution.

trans-[PtI$_2$(ipa)(py)] **1** (orange solid). Yield: 18%. Elemental analysis found, C, 16.23; H, 2.44; N, 4.63; C$_6$H$_{10}$I$_2$N$_2$Pt requires C, 16.37; H, 2.40; N, 4.77. NMR (acetone-d_6): δ (1H): 1.33 (d, *J* = 6.46 Hz, 6H, 2CH$_3$-ipa), 3.50 (sept, *J* = 6.7 Hz, 1H, CH-ipa), 4.16 (b.s., 2H, NH$_2$), 7.36 (dd, *J* = 6.6, 1.5 Hz, 2H, CH-py), 7.83 (td, *J* = 7.7, 1.7 Hz, 1H, CH-py), 8.81(dd, *J* = 6.6, 1.8 Hz, 2H, CH-py).

trans-[PtI$_2$(ma)(py)] **2** (orange solid). Yield: 15%. Elemental analysis found, C, 13.20; H, 1.94; N, 5.03; C$_8$H$_{14}$I$_2$N$_2$Pt requires C, 12.89; H, 1.80; N, 5.01. NMR (acetone-d_6): δ (1H): 2.57 (t, *J* = 6.5 Hz, 3H, CH$_3$-ma), 4.20 (b.s., 2H, NH$_2$), 7.40 (dd, *J* = 7.1, 1.51 Hz, 2H, CH-py), 7.88 (td, *J* = 7.1, 1.9 Hz, 1H, CH-py), 8.84 (dd, *J* = 6.6, 1.84 Hz, 2H, CH-py).

4.2. Nuclear Magnetic Resonance Spectroscopy

1D ^{1}H NMR spectra were recorded on a 300 MHz Bruker Advance III HD (Bruker, Rivas-Vaciamadrid, Madrid, Spain) and 500 MHz DRX spectrometers (Bruker, Rivas-Vaciamadrid, Madrid, Spain). Solutions were prepared in acetone-d_6 for the characterization spectra and in a mixture of acetone-d_6 and D$_2$O in a ratio 2:1 for the interaction studies with the 5′-GMP and MeIm with a final concentration of [Pt] = 6.5 mM. ^{1}H chemical shifts were internally referenced to sodium 3-(trimethylsilyl)propionate (TSP).

4.3. Irradiation

The light source used in the photoactivation experiments was a Photoreactor LZC-ICH2 from Luzchem (Ottawa, ON, Canada) fitted with UVA lamps (2 mW/cm^2, λ = 365 nm). The temperature in the light chamber during irradiation was kept at 37 °C and the sample preparation was performed as describe in the Section 4.2.

4.4. X-ray Diffraction

The structural features of the new complexes **1** and **2** were unambiguously proven by X-ray diffraction. Data collection was performed on a Bruker Kappa Apex II (X8 APEX, Bruker, Rivas-Vaciamadrid, Madrid, Spain) area detector X-ray diffractometer using a graphite-monochromated Mo radiation. Crystal data and the structure refinement parameters are listed in Table S1. CCDC 913250 (complex **1**) and 913249 (complex **2**) contains the supplementary crystallographic data for this paper. These data can be obtained free of charge via http://www.ccdc.cam.ac.uk/conts/retrieving.html (or from the CCDC, 12 Union Road, Cambridge CB2 1EZ, UK; Fax: +44 1223 336033; E-mail: deposit@ccdc.cam.ac.uk).

4.5. Cytotoxicity

SAOS-2 human osteosarcoma, A375 melanoma, T-47D breast cancer, SF-268 glioblastoma, NCI-H460 lung cancer cell lines were purchased from ATCC (Barcelona. Spain). Colorectal carcinoma cell lines HCT116 and matched p53-deficient HCT116 ($-/-$) were a kind gift of Bert Vogelstein (John-Hopkins University, Baltimore, MD, USA). These cell lines were cultured with DMEN (A375 and NCI-H460) or RPMI (HCT116) (Sigma, Madrid, Spain) with 10% fetal bovine serum, 2 mM L-glutamine, 100 U/mL penicillin and 100 µg/mL streptomycin in humidified air with 5% CO_2 at 37 °C.

The cytotoxicity was performed in Dr. Amancio Carnero's laboratory, following our published methodology, which is fully described in reference [6]. The compounds were dissolved in DMSO, diluted with PBS phosphate buffered saline to reach 10 mM solutions.

4.6. DFT Calculations

All calculations on complexes **1** and **2** and their 9-EtG adducts were performed with the Gaussian 09 (G09) program [25] employing the DFT and TD-DFT [26,27] methods. Basis sets, ECPs for Pt and I and functionals were benchmarked (not shown) and the best combination in terms of performance and computational demand was the PBE1PBE:LANL08/6-31G** [28] for geometry optimization and CAM-B3LYP/LANL08/6-31G** [29] for electronic transition calculations (see below). The PCM solvent model [30] was adopted in all DFT and TD-DFT calculations with water as solvent. The nature of all stationary points was confirmed by normal mode analysis.

Thirty-two singlet excited states with the corresponding oscillator strengths were determined for the complexes at the ground-state geometry by TD-DFT. Theoretical UV-Vis spectra were obtained using GAUSSSUM 2.2 [31].

Molecular graphics images were produced using the UCSF Chimera package from the Resource for Biocomputing, Visualization, and Informatics at the University of California, San Francisco (supported by NIH P41 RR001081) [32].

5. Conclusions

We have presented two new diiodido complexes with aliphatic amines (ipa, **1** and ma, **2**) in *trans* to a pyridine ligand. Although the cytotoxicity of the complexes was not as encouraging as cisplatin and other similar compounds, their reactivity in the presence of UVA light and upon binding to model biological molecules revealed a new and interesting profile. We observed that irradiation at 365 nm enhances the DNA adduct formation at early stages (up to 1 h). Our photoactivation experiments suggest that longer irradiation times produce the release of the pyridine ligand and the subsequence generation of species that have previously proved to be active and interact with biomolecular targets.

Since this is the first time that we have observed the release of the spectator ligand in *trans*-type complexes, these results suggest that the aromatic nature of the amine ligand could play an important role in the photochemistry of *trans* diiodido platinum complexes. We have identified two potential photoactivable compounds that showed two different photoactivation pathways.

Supplementary Materials: The following are available online at http://www.mdpi.com/2304-6740/6/4/127/s1, Cif and cifchecked files. Table S1. Crystal Data for complex **1** and **2**; Figure S1. Progress of the reaction between complex **1** and 5′-GMP at 37 °C monitored by ^{1}H NMR (acetone-d_6 and D_2O in a ratio 2:1) showing changes in the aromatic area and showing above the ^{1}H NMR (acetone-d_6 and D_2O in a ratio 2:1) of non coordinated 5′-GMP and pyridine. Figure S2. Progress of the reaction between complex **2** and 5′-GMP at 37 °C monitored by ^{1}H NMR showing changes in the aromatic area. Table S2a. Selected bond distances (Å) for complexes **1** and **2** optimized with the DFT method at the PBE1PBE:LANL08/6-31G** level using the PCM solvent (water) model. Table S2b. Selected bond distances (Å) for complex **2** optimized with the DFT method at the PBE1PBE:ECP/6-31G** level using different ECPs for the Pt atom and the PCM solvent (water) model. Table S2c. Selected bond distances (Å) for complex **2** optimized with the DFT method at the PBE1PBE:LANL08/BS level using different basis sets (BS) for the non-Pt atoms and the PCM solvent (water) model. Table S2d. Selected bond distances (Å) for complex **2** optimized with the DFT method using different functionals, the LANL08/6-31G** ECP/basis set and the PCM solvent (water) model. Table S3. Experimental and theoretical absorption spectra for complexes **1** and **2** in water at the CAM-B3LYP/LANL08/6-31G** level. Table S4. Selected TD-DFT singlet-singlet transitions and corresponding electron difference density maps (EDDMs) for complexes **1** and **2** in water at the CAM-B3LYP/LANL08/6-31G** level. Table S5. Experimental and theoretical absorption spectra for mono and bis 9-EtG adducts (complexes **3** to **6**) of complexes **1** and **2** in water at the CAM-B3LYP/LANL08/6-31G** level. Table S6. Selected TDDFT singlet-singlet transitions and corresponding electron difference density maps (EDDMs) for mono and bis 9-EtG adducts of complexes **1** and **2** in water at the CAM-B3LYP/LANL08/6-31G** level. In the EDDMs gray indicates a decrease in electron density, while blue-gray indicates an increase. Table S7. Selected frontier molecular orbitals for complexes **3** to **6**. Table S8. Selected bond distances (Å) for the mono and bis 9-EtG adducts of complexes **1** and **2** optimized with the DFT method at the PBE1PBE:LANL08/6-31G** level using the PCM solvent (water) model.

Author Contributions: Conceptualization: L.C. and A.G.Q.; Software, L.S.; Validation: L.C.; A.C., L.S., A.G.Q and A.I.M.; Formal Analysis and Data Curation: A.G.Q., A.I.M., L.S., A.C.; Investigation: L.C., T.P, A.G.Q., A.C., L.S., and A.I.M.; Resources: A.G.Q, A.C., L.S.; Writing-Original Draft Preparation: A.G.Q.; Writing—Review & Editing, A.G.Q., A.I.M., L.S., L.C.; Visualization: A.G.Q. and L.C.; Supervision, A.G.Q. and A.I.M.; Project Administration: A.G.Q.; Funding Acquisition: A.G.Q.

Funding: This research was funded by MINECO grant number CTQ-2015-68779R.

Acknowledgments: The MetDrugs network is acknowledged for providing opportunities for discussion.

Conflicts of Interest: The authors declare no conflict of interest.

References

1.   Johnstone, T.C.; Suntharalingam, K.; Lippard, S.J. The Next Generation of Platinum Drugs: Targeted Pt(II) Agents, Nanoparticle Delivery, and Pt(IV) Prodrugs. *Chem. Rev.* **2016**, *116*, 3436–3486. [CrossRef] [PubMed]

2.   Medrano, A.; Dennis, S.M.; Alvarez-Valdes, A.; Perles, J.; McGregor Mason, T.; Quiroga, A.G. Synthesis, cytotoxicity, DNA interaction and cell cycle studies of *trans*-diiodophosphine Pt(II) complexes. *Dalton Trans.* **2015**, *44*, 3557–3562. [CrossRef] [PubMed]

3.   Quiroga, A.G. Understanding *trans* platinum complexes as potential antitumor drugs beyond targeting DNA. *J. Inorg. Biochem.* **2012**, *114*, 106–112. [CrossRef] [PubMed]

4.   Messori, L.; Casini, A.; Gabbiani, C.; Michelucci, E.; Cubo, L.; Rios-Luci, C.; Padron, J.M.; Navarro-Ranninger, C.; Quiroga, A.G. Cytotoxic Profile and Peculiar Reactivity with Biomolecules of a Novel "Rule-Breaker" Iodidoplatinum(II) Complex. *ACS Med. Chem. Lett.* **2010**, *1*, 381–385. [CrossRef] [PubMed]

5.   Messori, L.; Cubo, L.; Gabbiani, C.; Alvarez-Valdes, A.; Michelucci, E.; Pieraccini, G.; Rios-Luci, C.; Leon, L.G.; Padron, J.M.; Navarro-Ranninger, C.; et al. Reactivity and Biological Properties of a Series of Cytotoxic $Pt_{I2}(amine)_2$ Complexes, Either *cis* or *trans* Configured. *Inorg. Chem.* **2012**, *51*, 1717–1726. [CrossRef] [PubMed]

6.   Parro, T.; Medrano, M.A.; Cubo, L.; Munoz-Galvan, S.; Carnero, A.; Navarro-Ranninger, C.; Quiroga, A.G. The second generation of iodido complexes: *trans*-[PtI_2(amine)(amine′)] bearing different aliphatic amines. *J. Inorg. Biochem.* **2013**, *127*, 182–187. [CrossRef] [PubMed]

7.   Starha, P.; Vanco, J.; Travnicek, Z.; Hosek, J.; Klusakova, J.; Dvorak, Z. Platinum(II) Iodido Complexes of 7-Azaindoles with Significant Antiproliferative Effects: An Old Story Revisited with Unexpected Outcomes. *PLoS ONE* **2016**, *11*, e0165062. [CrossRef]

8. Cubo, L.; Pizarro, A.M.; Quiroga, A.G.; Salassa, L.; Navarro-Ranninger, C.; Sadler, P.J. Photoactivation of *trans* diamine platinum complexes in aqueous solution and effect on reactivity towards nucleotides. *J. Inorg. Biochem.* **2010**, *104*, 909–918. [CrossRef]

9. Presa, A.; Brissos, R.F.; Caballero, A.B.; Borilovic, I.; Korrodi-Gregório, L.; Pérez-Tomás, R.; Roubeau, O.; Gamez, P. Photoswitching the Cytotoxic Properties of Platinum(II) Compounds. *Angew. Chem. Int. Ed.* **2014**, *54*, 4561–4565. [CrossRef]

10. Quental, L.; Raposinho, P.; Mendes, F.; Santos, I.; Navarro-Ranninger, C.; Alvarez-Valdes, A.; Huang, H.; Chao, H.; Rubbiani, R.; Gasser, G.; et al. Combining imaging and anticancer properties with new heterobimetallic Pt(II)/M(I) (M = Re, 99 mTc) complexes. *Dalton Trans.* **2017**, *46*, 14523–14536. [CrossRef]

11. Frei, A.; Rubbiani, R.; Tubafard, S.; Blacque, O.; Anstaett, P.; FelgentrÃger, A.; Maisch, T.; Spiccia, L.; Gasser, G. Synthesis, Characterization, and Biological Evaluation of New Ru(II) Polypyridyl Photosensitizers for Photodynamic Therapy. *J. Med. Chem.* **2014**, *57*, 7280–7292. [CrossRef] [PubMed]

12. Navas, F.; Perfahl, S.; Garino, C.; Salassa, L.; Novakova, O.; Navarro-Ranninger, C.; Bednarski, P.J.; Malina, J.; Quiroga, A.N.G. Increasing DNA reactivity and in vitro antitumor activity of *trans* diiodido Pt(II) complexes with UVA light. *J. Inorg. Biochem.* **2015**, *153*, 211–218. [CrossRef] [PubMed]

13. Aris, S.M.; Farrell, N.P. Towards Antitumor Active *trans*-Platinum Compounds13. *Eur. J. Inorg. Chem.* **2009**, *2009*, 1293–1302. [CrossRef]

14. Farrell, N.P. Multi-platinum anti-cancer agents. Substitution-inert compounds for tumor selectivity and new targets. *Chem. Soc. Rev.* **2015**, *44*, 8773–8785. [CrossRef] [PubMed]

15. Brabec, V. DNA Modifications by antitumor platinum and ruthenium compounds: Their recognition and repair. In *Progress in Nucleic Acid Research and Molecular Biology*; Academic Press: New York, NY, USA, 2002; Volume 71, pp. 1–68.

16. Mellish, K.J.; Qu, Y.; Scarsdale, N.; Farrell, N. Effect of Geometric Isomerism in Dinuclear Platinum Antitumour Complexes on the Rate of Formation and Structure of Intrastrand Adducts with Oligonucleotides. *Nucleic Acids Res.* **1997**, *25*, 1265–1271. [CrossRef] [PubMed]

17. Zhao, Y.; Woods, J.A.; Farrer, N.J.; Robinson, K.S.; Pracharova, J.; Kasparkova, J.; Novakova, O.; Li, H.; Salassa, L.; Pizarro, A.M.; et al. Diazido Mixed-Amine Platinum(IV) Anticancer Complexes Activatable by Visible-Light Form Novel DNA Adducts. *Chem. A Eur. J.* **2013**, *19*, 9578–9591. [CrossRef] [PubMed]

18. Velders, A.H.; Quiroga, A.G.; Haasnoot, J.G.; Reedijk, J. Orientation- and Temperature-Dependent Rotational Behavior of Imidazole Ligands (L) in β-[Ru(azpy)$_2$(L)$_2$](PF$_6$)$_2$ Complexes. *Eur. J. Inorg. Chem.* **2003**, *2003*, 713–719. [CrossRef]

19. Perez, J.M.; Montero, E.I.; Gonzalez, A.M.; Solans, X.; Font-Bardia, M.; Fuertes, M.A.; Alonso, C.; Navarro-Ranninger, C. X-ray Structure of Cytotoxic *trans*-[PtCl$_2$(dimethylamine)(isopropylamine)]: Interstrand Cross-Link Efficiency, DNA Sequence Specificity, and Inhibition of the B–Z Transition. *J. Med. Chem.* **2000**, *43*, 2411–2418. [CrossRef]

20. Rochon, F.D.; Buculei, V. Multinuclear NMR study and crystal structures of complexes of the types cis- and *trans*-Pt(amine)$_2$I$_2$. *Inorg. Chim. Acta* **2004**, *357*, 2218–2230. [CrossRef]

21. Thiele, G.; Wagner, D. Über die Reaktion von Platiniodiden mit Pyridin und über die Molekül- und Kristallstruktur von *trans*-Diiodobis(pyridin)platin(II). *Chem. Ber.* **1978**, *111*, 3162–3170. [CrossRef]

22. Berger, I.; Nazarov, A.A.; Hartinger, C.G.; Groessl, M.; Valiahdi, S.-M.; Jakupec, M.A.; Keppler, B.K. A glucose derivative as natural alternative to the cyclohexane-1,2-diamine ligand in the anticancer drug oxaliplatin? *ChemMedChem* **2007**, *2*, 505–514. [CrossRef] [PubMed]

23. Tessier, C.; Rochon, F.D. Multinuclear NMR study and crystal structures of complexes of the types *cis*- and *trans*-Pt(Ypy)$_2$X$_2$, where Ypy=pyridine derivative and X=Cl and I. *Inorg. Chim. Acta* **1999**, *295*, 25–38. [CrossRef]

24. Rochon, F.D.; Tessier, C. Pt(II) compounds with sulfoxide ligands and crystal structures of complexes of the types I(R$_2$SO)Pt($Î\frac{1}{4}$-I)$_2$Pt(R$_2$SO)I and *trans*-Pt(R$_2$SO)(L)X$_2$ (L = amine, pyridine and pyrimidine). *Inorg. Chim. Acta* **2008**, *361*, 2591–2600. [CrossRef]

25. Frisch, M.J.; Schlegel, G.E.; Scuseria, M.A.; Robb, J.R.; Cheeseman, G.; Scalmani, V.; Barone, B.; Mennucci, G.A.; Petersson, H.; Nakatsuji, M.; et al. *Gaussian 09, Revision B.01*; Gaussian, Inc.: England, UK, 2009.

26. Stratmann, R.E.; Gustavo, E.S.; Michael, J.F. An efficient implementation of time-dependent density-functional theory for the calculation of excitation energies of large molecules. *J. Chem. Phys.* **1998**, *109*, 8218–8224. [CrossRef]

27. Mark, E.C.; Christine, J.; Kim, C.C.; Dennis, R.S. Molecular excitation energies to high-lying bound states from time-dependent density-functional response theory: Characterization and correction of the time-dependent local density approximation ionization threshold. *J. Chem. Phys.* **1998**, *108*, 4439–4449. [CrossRef]
28. Perdew, J.P.; Burke, K.; Ernzerhof, M. Generalized Gradient Approximation Made Simple. *Phys. Rev. Lett.* **1996**, *77*, 3865–3868. [CrossRef] [PubMed]
29. Yanai, T.; Tew, D.P.; Handy, N.C. A new hybrid exchange-correlation functional using the Coulomb-attenuating method (CAM-B3LYP). *Chem. Phys. Lett.* **2004**, *393*, 51–57. [CrossRef]
30. Miertu, S.; Scrocco, E.; Tomasi, J. Electrostatic interaction of a solute with a continuum. A direct utilizaion of AB initio molecular potentials for the prevision of solvent effects. *Chem. Phys.* **1981**, *55*, 117–129. [CrossRef]
31. O'Boyle, N.M.; Tenderholt, A.L.; Langner, K.M. cclib: A library for package-independent computational chemistry algorithms. *J. Comput. Chem.* **2008**, *29*, 839–845. [CrossRef] [PubMed]
32. Pettersen, E.F.; Goddard, T.D.; Huang, C.C.; Couch, G.S.; Greenblatt, D.M.; Meng, E.C.; Ferrin, T.E. UCSF ChimeraA visualization system for exploratory research and analysis. *J. Comput. Chem.* **2004**, *25*, 1605–1612. [CrossRef]

Article

Development and Validation of Liquid Chromatography-Based Methods to Assess the Lipophilicity of Cytotoxic Platinum(IV) Complexes

Matthias H. M. Klose [1,2], Sarah Theiner [3], Hristo P. Varbanov [1,4], Doris Hoefer [1], Verena Pichler [1,5], Markus Galanski [1], Samuel M. Meier-Menches [2,3,*] and Bernhard K. Keppler [1,2,*]

[1]   Institute of Inorganic Chemistry, University of Vienna, 1090 Vienna, Austria;
      matthias.klose@univie.ac.at (M.H.M.K.); hristo.varbanov@uni-graz.at (H.P.V.);
      doris.hoefer@univie.ac.at (D.H.); verena.pichler@meduniwien.ac.at (V.P.);
      markus.galanski@univie.ac.at (M.G.)
[2]   Research Cluster 'Translational Cancer Therapy Research', University of Vienna, 1090 Vienna, Austria
[3]   Department of Analytical Chemistry, University of Vienna, 1090 Vienna, Austria; sarah.theiner@univie.ac.at
[4]   Institute of Chemistry—Inorganic Chemistry, University of Graz, 8010 Graz, Austria
[5]   Division of Nuclear Medicine, Department of Biomedical Imaging and Image-Guided Therapy,
      Medical University of Vienna, 1090 Vienna, Austria
*   Correspondence: samuel.meier@univie.ac.at (S.M.M.-M.); bernhard.keppler@univie.ac.at (B.K.K.);
     Tel.: +43-1-4277-52373 (S.M.M.-M.); +43-1-4277-52602 (B.K.K.)

Received: 2 October 2018; Accepted: 29 November 2018; Published: 4 December 2018

Abstract: Lipophilicity is a crucial parameter for drug discovery, usually determined by the logarithmic partition coefficient (Log P) between octanol and water. However, the available detection methods have restricted the widespread use of the partition coefficient in inorganic medicinal chemistry, and recent investigations have shifted towards chromatographic lipophilicity parameters, frequently without a conversion to derive Log P. As high-performance liquid chromatography (HPLC) instruments are readily available to research groups, a HPLC-based method is presented and validated to derive the partition coefficient of a set of 19 structurally diverse and cytotoxic platinum(IV) complexes exhibiting a dynamic range of at least four orders of magnitude. The chromatographic lipophilicity parameters ϕ_0 and Log k_w were experimentally determined for the same set of compounds, and a correlation was obtained that allows interconversion between the two lipophilicity scales, which was applied to an additional set of 34 platinum(IV) drug candidates. Thereby, a $\phi_0 = 58$ corresponds to Log P = 0. The same approaches were successfully evaluated to determine the distribution coefficient (Log D) of five ionisable platinum(IV) compounds to sample pH-dependent effects on the lipophilicity. This study provides straight-forward HPLC-based methods to determine the lipophilicity of cytotoxic platinum(IV) complexes in the form of Log P and ϕ_0 that can be interconverted and easily expanded to other metal-based compound classes.

Keywords: ϕ_0; anticancer agents; chromatographic lipophilicity parameter; distribution coefficient; HPLC; lipophilicity; Log k_w; Log P; partition coefficient; platinum(IV)

---

1. Introduction

Optimising lipophilicity is a crucial process in drug discovery [1,2], because it significantly affects the diffusion of drugs through cell membranes and thus, plays a role in pharmacokinetic processes, including absorption, distribution, metabolism, and excretion [2,3]. Some authors have also suggested that an optimal lipophilicity might increase the chances of success during drug development [1].

Lipophilicity is typically determined by means of a compound's partition coefficient between octanol and water on a logarithmic scale, commonly abbreviated as Log $P_{o/w}$, henceforth called

Log P [4–6]. The Organisation for Economic Cooperation and Development (OECD) has outlined the standard method for experimentally obtaining partition coefficients of investigational compounds by the shake flask method [7], which is followed by an appropriate detection technique such as photometry, gas chromatography, or high-performance liquid chromatography (HPLC) [7–9]. The partition coefficient obtained by the shake flask method is typically found between $-2 <$ Log P < 4 [7]. Hydrophilic compounds are characterised by negative values and lipophilic compounds by positive values.

Purely chromatographic alternatives to the partition coefficient between octanol and water have emerged as well because of the potential for automation, higher throughput, and minimising sample preparation efforts [10,11]. In particular, reversed phase-HPLC (RP-HPLC) has been suggested to provide a suitable means to directly assess the lipophilic property of an investigational compound. As these methods do not involve the shake flask procedure, they have the additional advantage of being independent of the concentration effects. In this setup, RP-HPLC is performed by using a C8- or C18-bonded stationary phase and a polar mobile phase, the latter being a mixture of water and methanol or water and acetonitrile. Chromatographic retention results from the partition of analytes between the two phases and can thus directly relate to the lipophilicity of an analyte [10]. For example, high capacity factors are indicative of a strong interaction with the lipophilic stationary phase and, thus, the strong lipophilic character of an analyte.

The corresponding chromatographic parameters Log k_w and ϕ_0 are independent of the flow rate and column length and can indeed provide relevant information about the lipophilic property of an analyte featuring its own scale even without conversion into Log P values [11]. On the one hand, Log k_w is defined as the logarithmic capacity factor of a compound in a mobile phase containing pure water. It is usually obtained by extrapolation using the linear Soczewinski–Snyder relationship [12,13], defined as Log k = Log $k_w - S\phi$, where k is the capacity factor of a compound in a specific mobile phase composition, ϕ is the volume percentage of the organic modifier in the eluent, and S is a constant for a given analyte and HPLC system. On the other hand, ϕ_0 corresponds to the volume percentage of the organic modifier in the mobile phase at which the analyte is equally distributed in the mobile and the stationary phase [14].

The estimation of lipophilicity has been mostly implemented in the discovery process of organic drugs. However, lipophilicity determinations have also become of growing importance for the optimisation of metal-based anticancer drugs, among which is the class of platinum compounds [15–21], as they must accumulate efficiently in cells to execute their DNA-binding properties [22]. The lipophilicity of platinum anticancer agents was most commonly assessed by the shake flask method, followed by the detection of a platinum isotope by atomic absorption spectroscopy or inductively coupled plasma mass spectrometry (ICP-MS) [16–18,21]. Hyphenation of microemulsion electrokinetic chromatography to ICP-MS is a promising alternative but is experimentally demanding [23]. Due to restrictions in accessing such instruments, the chromatographically determined lipophilicity parameter Log k_w has recently become more popular [24–26]. In addition, quantitative structure property relationships have been developed for modelling the lipophilicity of platinum complexes, and a free access database for the prediction of their Log P values is available [20,25,27–30]. However, an in depth analysis of representatives of the platinum(IV) class has not been reported to date. Also, a direct conversion of the chromatographic lipophilicity parameter ϕ_0 into partition coefficients has not been established, which would allow the interconversion between the two lipophilicity scales of ϕ_0 and Log P.

In this study, a set of 19 structurally diverse cytotoxic platinum(IV) complexes was used for evaluating the shake flask method by chromatographic detection and deriving lipophilicity parameters directly by chromatographic approaches (Table 1). The findings were validated by ICP-MS detection and the approach was extended to the distribution coefficient that accounts for ionisable compounds (e.g., containing carboxylic acids). Finally, we provide for the first time an empirical equation to convert the chromatography-based lipophilicity parameter ϕ_0 (and Log k_w) of platinum(IV) complexes into

Log P. This equation was then used to calculate Log P values for an additional set of 34 cytotoxic platinum(IV) complexes from experimentally determined ϕ_0.

Table 1. List of 19 platinum(IV) complexes included in the standard set. The lipophilicity of each compound was assessed by the shake flask method using chromatographic (the logarithmic partition coefficient (Log P, HPLC) and element specific (Log P, inductively coupled plasma mass spectrometry, ICP-MS) detection. The chromatographic lipophilicity parameters Log k_w and ϕ_0 were determined using potassium iodide (KI) as a dead time marker. The compounds were sorted according to increasing Log P (HPLC).

Nr.	Structure	Log k_w (KI)	ϕ_0 (KI)	Log P (HPLC)	Log P (ICP-MS)
1	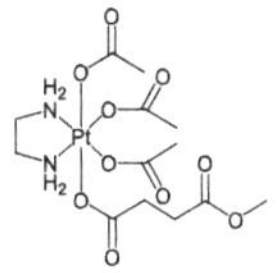	1.14	25.6	−2.07	−1.70
2	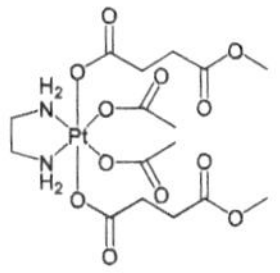	1.60	36.1	−1.93	−2.23
3	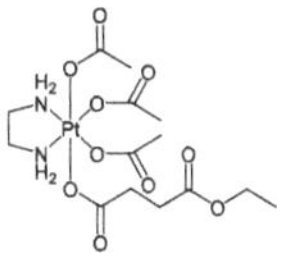	1.47	36.6	−1.88	−1.41
4	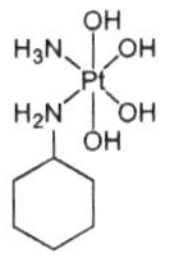	0.76	22.5	−1.87	−1.70
5	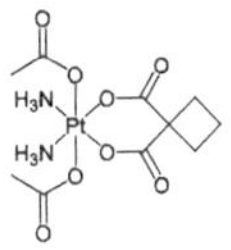	1.07	27.1	−1.23	−1.42
6	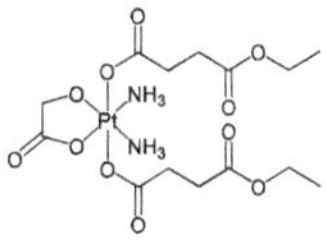	1.39	41.0	−1.17	−1.24
7	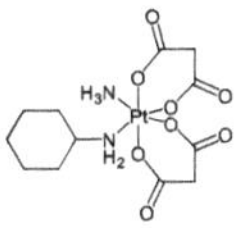	1.18	32.7	−0.94	−1.00
8	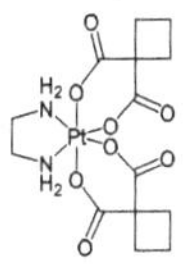	1.77	44.7	−0.57	−0.66

Table 1. *Cont.*

Nr.	Structure	Log k_w (KI)	ϕ_0 (KI)	Log P (HPLC)	Log P (ICP-MS)
9	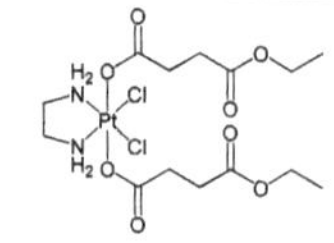	1.86	50.5	−0.34	−0.35
10	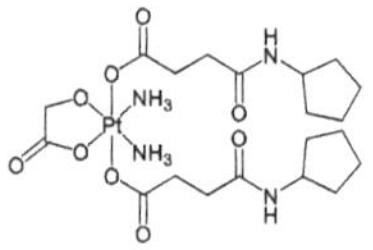	2.28	56.7	−0.30	−0.39
11	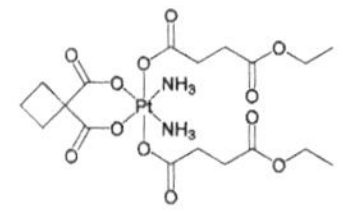	2.67	57.5	0.09	0.06
12	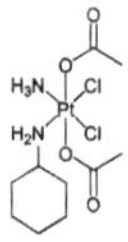	2.20	55.4	0.26	0.23
13	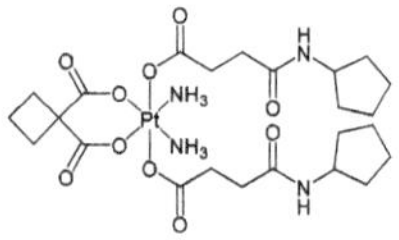	3.14	66.3	0.89	0.77
14	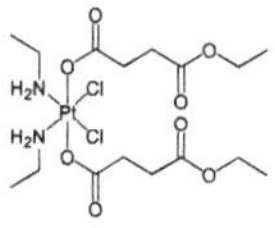	3.03	63.2	0.94	0.71
15	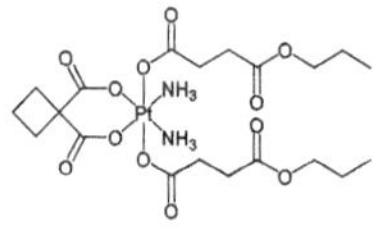	3.26	69.0	1.19	0.95
16	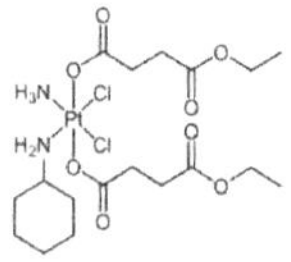	4.04	72.8	1.66	1.20
17	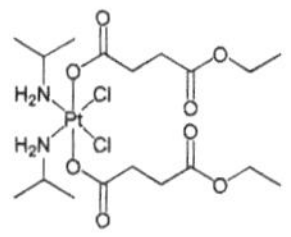	3.48	70.3	1.68	1.30
18	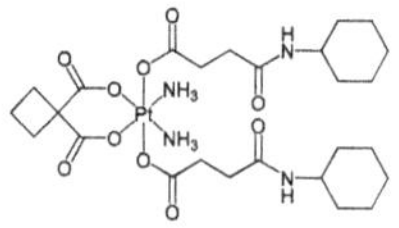	3.86	74.2	1.83	1.33
19	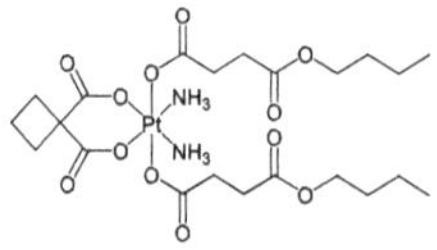	4.18	76.5	2.37	1.06

2. Results and Discussion

Typically, lipophilicity is determined by the shake flask method as the logarithmic partition coefficient (Log P) of the concentration of an analyte between 1-octanol and water [7]. The concentration of each phase is derived by spectroscopic or spectrometric detection techniques [7–9]. Alternatively, lipophilicity can be directly obtained from chromatographic experiments, such as RP-HPLC, where the retention time of the analyte depends on the partition between the lipophilic stationary and the hydrophilic mobile phase. Here, the analyte is detected online by either spectroscopic or mass spectrometric techniques. The chromatographic lipophilicity parameters are calculated from capacity factors that are independent of the void volume, flow rate, and column length.

Several studies have found correlations between the lipophilicity of metal-based anticancer agents, usually determined by the shake flask method with element-specific detection, and their cytotoxic activity [21,30–35]. However, a restricted access to such instrumentation may have favoured the recent emergence of chromatographic lipophilicity parameters for metallodrugs [24,26,31,32]. So far, correlations between Log P and chromatographic parameters have been derived only for relatively homogenous series [20,25,35]. Thus, such approaches are presented and validated here to directly obtain Log P from shake flask experiments and indirectly by converting the chromatographic lipophilicity parameter ϕ_0 into Log P (Scheme 1). The current investigation was performed with a set of 19 structurally diverse cytotoxic platinum(IV) complexes (Table 1). An additional set of 34 platinum(IV) complexes was used to test the conversion of the chromatographic lipophilicity parameter ϕ_0 into Log P. A detailed list of the additionally employed compounds can be found in Tables S1 and S2.

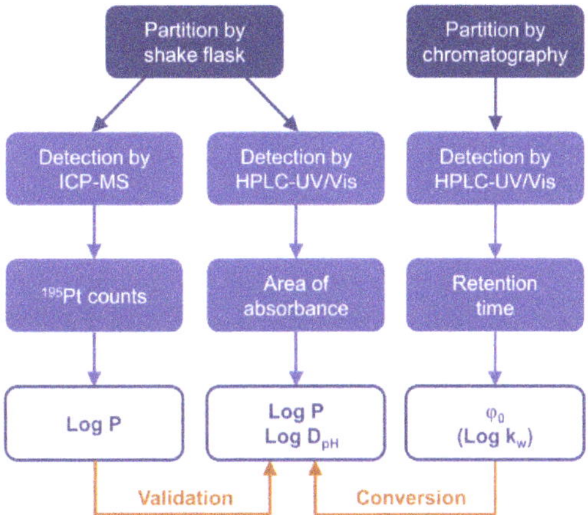

Scheme 1. The lipophilicity of platinum(IV) complexes was evaluated by chromatographic workflows and validated by ICP-MS detection.

2.1. Determining Partition Coefficients by the Shake Flask Method

The platinum complexes were individually dissolved in water, saturated with 1-octanol, and the classical shake flask method was performed according to the OECD guidelines (Scheme 1) [7]. A stock solution of 0.5 mmol·L^{-1} was prepared of each compound. Then, equal volumes of the stock solution and 1-octanol pre-saturated with water were mixed for 60 min, and after phase separation, the individual phases were analysed. It must be noted that the metal-based anticancer agents must be inert with respect to ligand exchange reactions during this time period in order to be amenable to the shake flask procedure.

Of each sample, the stock solution and the resulting aqueous and 1-octanol phases were analysed by RP-HPLC using isocratic methods. The Log P values of the platinum compounds were calculated based on the areas of absorbance determined in the aqueous and 1-octanol phases that are directly proportional to the concentration according to the Lambert–Beer law. The sum of the area values of the aqueous and 1-octanol phase was compared with the stock solution and considered valid if the two values matched. After each run of a 1-octanol sample, the column was washed with 95% methanol before equilibrating to the original mobile phase composition. This was necessary to remove any remaining 1-octanol bound to the stationary phase of the column, which would increase the back pressure of the system, alter the properties of the stationary phase, and thus, influence the partition of the analytes between the mobile and the stationary phase.

Under these conditions, the lipophilicity of the representative platinum(IV) complexes was found between $-2.0 <$ Log P < 2.4, spanning at least four orders of magnitude. In general, the complexes featuring equatorial acetates **1–3** showed the lowest lipophilicity, while the examples containing carboplatin- (**11**, **13**, **15**, **18**, and **19**) or cisplatin/satraplatin-derived cores (**12**, **14**, **16**, and **17**) were the most tuneable and lipophilic compounds in our series (Table 1).

Aliquots of the stock solution and the aqueous phase of the same samples were serially diluted for ICP-MS analysis by acquiring the ^{195}Pt isotope signal. Because the injection of the organic phase containing 1-octanol into the ICP-MS is problematic, the Log P was calculated by assuming $c_{octanol} = c_{total} - c_{water}$. The lipophilicity of the same compounds was then found between $-2.2 <$ Log P < 1.3, spanning roughly three orders of magnitude.

The Log P values obtained by the two methods correlated well with a regression coefficient of $R^2 = 0.957$ (Figure 1). Importantly, the slope of the fitting line was close to one and the intercept close to zero, thus validating the chromatographic approach to determine Log P of the cytotoxic platinum(IV) complexes. Some variation was observed at the hydrophilic and hydrophobic ends, which probably stems from the uncertainty of the assumption of $c_{octanol} = c_{total} - c_{water}$ to calculate Log P by ICP-MS. A linear relationship seemed to exist over three orders of magnitudes between $-1.5 <$ Log P < 1.8 ($R^2 = 0.991$) between the two approaches.

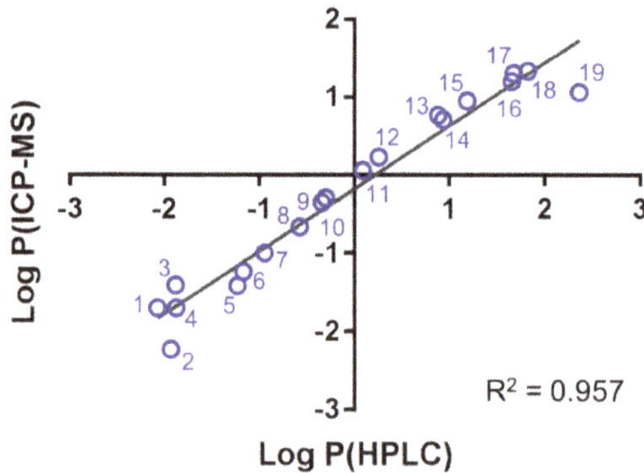

Figure 1. The correlation of the Log P values obtained by the shake flask method using HPLC–UV/Vis and ICP-MS detection was linear according to Log P (ICP-MS) = 0.805·Log P (HPLC) − 0.177.

2.2. Extending the Approach to Determine Distribution Coefficients

For ionisable compounds, the partition between an aqueous and 1-octanol phase is a function of the pH value and is described as the logarithmic distribution coefficient (Log D_{pH}, abbreviated as Log D) [36]. The distribution coefficient includes all species of an analyte in each of the two phases (e.g., both the ionised and neutral species) and also the potential hydrolysis products. Lipophilic properties at different pH values are essential for orally administered platinum(IV) anticancer agents that must additionally permeate several cellular membranes [37].

In order to extend the previous experiments to Log D values, five acidic platinum(IV) compounds (**20–24**, Table S1) featuring free carboxylic groups were analysed at different pH values (i.e., pH = 1.7, 2.5, 3.7, 4.2, 4.7, 5.6, 6.2, and 7.4 (Table 2)). Again, the shake flask method was used, followed by RP-HPLC–UV/Vis detection. Compound **10** was used as a non-ionisable control, as its lipophilicity should be independent of the pH value of the solution. Indeed, it showed an average Log P = -0.27 ± 0.02 over the entire tested pH range. Compound **20** did not yield a clear UV/Vis signal during Log D determination and, thus, was not analysed. In general, the acidic platinum(IV) representatives showed a clear dependence of the Log D on the pH value that is indicative of their titration curves. Increasing the pH above the pK_a of the acidic groups significantly increased the hydrophilicity by deprotonation. Thus, the distribution coefficient can also be obtained by the shake flask method using chromatography-based methods with UV/Vis detection and can provide important insight into the lipophilicity of different species of an analyte in solution depending on the pH.

Table 2. The distribution coefficient (Log D) values of the platinum(IV) compounds at different pH values, determined by the shake flask method using RP-HPLC–UV/Vis detection. The colour code is determined by blue = +0.5, white = 0, red = -3. n.d. = not detected.

Complex	pH							
	1.7	2.5	3.7	4.2	4.7	5.6	6.2	7.4
20	n.d.	n.d.	n.d.	n.d.	n.d.	n.d.	n.d.	n.d.
21	−1.56	−1.54	−1.82	−2.16	−2.91	n.d.	n.d.	n.d.
22	−1.19	−1.26	−1.42	−1.69	−2.03	n.d.	n.d.	n.d.
23	−0.35	−0.39	−0.56	−0.91	−1.36	−2.73	n.d.	n.d.
24	0.46	0.45	0.37	0.15	−0.23	−1.34	−2.17	n.d.
10	−0.25	−0.28	−0.24	−0.27	−0.27	−0.27	−0.29	−0.27

2.3. Determination of the Chromatographic Lipophilicity Parameter ϕ_0

The chromatographic lipophilicity parameter ϕ_0 is intuitive, as it represents the percentage of the organic phase in the eluent at which the partition of an analyte between the mobile and stationary phase is equal. Higher values indicate a stronger lipophilic character of the analyte, and ϕ_0 ranges between 0–100. The parameter is obtained by determining the capacity factors of an analyte with at least three different mobile phase compositions and then solving the linear Soczewinski–Snyder relationship for Log k = 0 (see Section 4).

In a first step, uracil and potassium iodide (KI) were investigated as system dead time (t_0) markers, and their retention times were determined in the range of 5–90% organic modifier (Figure S1). These were used to calculate ϕ_0 values for the same set of 19 platinum(IV) complexes. The ϕ_0 values derived from the two different dead time markers correlated linearly with R^2 = 0.999 and covered the range of ϕ = 20–80. The retention time of uracil increased with increasing aqueous phase from 70–95% (Figure S1) and consequently, KI may be more appropriate for hydrophilic compounds, as its retention time remained relatively constant over the entire range of mobile phase compositions. As metal-based anticancer agents may undergo ligand exchange reactions, it is advised to use potassium iodide as an external marker if necessary. Of the platinum(IV) compounds, **4** displayed the lowest ϕ_0 = 22.5, being the most hydrophilic, and **19** was the most lipophilic, with the highest ϕ_0 = 76.5.

The correlation of ϕ_0 and Log k_w for the same set of compounds appeared to be a quadratic polynomial, with an acceptable fit of R^2 = 0.982 for potassium iodide as the dead time marker (Figure S2). The quadratic dependence of these parameters was already observed by Schoenmakers [38]. In a more recent publication, it was noted that the dependence of ϕ_0 and Log k_w was linear for narrow ranges of ϕ_0 and generally quadratic when considering wide ranges of ϕ_0. Also, the choice of organic modifier has a considerable impact on the Log k_w value of purely organic molecules with respect to the regression model used, and methanol was suggested as the most suitable [39].

2.4. Converting Chromatographic Lipophilicity to Shake Flask Lipophilicity

According to the OECD guidelines, the Log k_w values of compounds with known Log P can be used to create a calibration curve to evaluate partition coefficients from chromatographic data [7].

In this study, the ϕ_0 values were plotted against the chromatographically detected Log P values of the 19 platinum(IV) complexes. The resulting correlation curve was a quadratic polynomial with an $R^2 = 0.951$ (Figure 2). Thus, the calculated Log P values (cLog P) of platinum(IV) complexes can be obtained from ϕ_0-values according to Equation (1):

$$\text{cLog P} = 0.001\phi_0^2 \text{ (KI)} - 0.027\phi_0 \text{ (KI)} - 1.758. \tag{1}$$

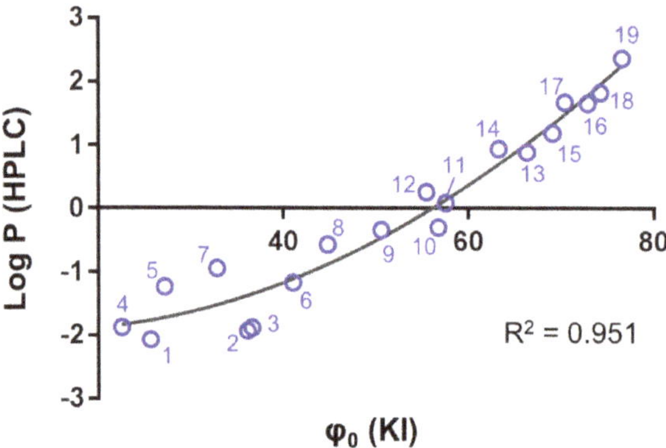

Figure 2. The overall correlation of the chromatographically determined ϕ_0 (using potassium iodide as a dead time marker) and Log P by the shake flask method (using HPLC–UV/Vis) is shown.

The corresponding linear correlation between Log k_w and Log P featured $R^2 = 0.918$ (Figure S2). However, the hydrophilic region showed increased variance with both chromatographic approaches reflecting the uncertainty of the experimental determination of Log P < -1 directly by HPLC. As a rule of thumb, a $\phi_0 = 58$ yields a Log P $= 0$, and consequently, lipophilic platinum(IV) compounds are characterised by $\phi_0 > 58$, while hydrophilic platinum(IV) compounds display $\phi_0 < 58$. The calibration curve was applied to the set of 34 platinum(IV) complexes, and their cLog P values were calculated based on their experimentally determined ϕ_0 (Table S2). The conversion worked well for the lipophilic representatives. For example, **54** featured a cLog P $= 0.6$, which was very similar to the reported Log P $= 0.7$ [21]. The quadratic approximation implied that ϕ_0-values < 25 would yield constant cLog P values of -1.9. In fact, the uncertainty in the hydrophilic region is also reflected by the OECD guidelines that restrain the range of experimental Log P values determined by the shake flask method to a lower end of -2 [7]. The increased uncertainty of the correlation in the hydrophilic region is further exemplified by compound **34**, which displayed a cLog P $= -1.5$, while its reported value was Log P $= -0.8$ [21].

As the antiproliferative activity is known for most of the complexes reported herein, we tried to directly correlate the concentration to inhibit 50% of cell proliferation (IC$_{50}$) of the set of 19 platinum(IV) complexes (except **4–6**) in a particular cell line with their lipophilicity parameters (Figure S4). As expected, the two parameters did not directly correlate, and the regression coefficients were uniformly below 0.1. This was indicative that active cellular accumulation and stability towards reduction may have strongly contributed to the cytotoxic property of these platinum(IV) complexes in addition to the lipophilicity [33,40,41].

3. Materials and Methods

Ultrapure water (18.2 MΩ cm, Milli-Q Advantage, Darmstadt, Germany) was used for the RP-HPLC experiments, as well as for all the dilutions in the ICP-MS measurements. Nitric acid (TraceSELECT®, ≥65%, p.a., Fluka, Buchs, Switzerland) was used without further purification. The platinum and rhenium standards for the ICP-MS measurements were obtained from CPI

International (Amsterdam, The Netherlands). All the other reagents and solvents were obtained from standard commercial sources and were used without further purification.

NMR spectra were recorded with a Bruker Avance III 500 MHz NMR spectrometer (Bruker Daltonics, Bremen, Germany) at 500.32 MHz (^{1}H). Mass spectra were measured with a Bruker maXis electrospray-ionisation-quadrupole-time-of-flight mass spectrometer (Bruker Daltonics, Bremen, Germany) using ACN/MeOH + 1% H_2O as the solvent or a Bruker Amazon SL ion trap using water/MeOH. The elemental analysis was performed at the Microanalytical Laboratory of the University of Vienna with a Eurovector EA3000 elemental analyser (Eurovector Srl, Pavia, Italy) for CHNS analysis.

3.1. Platinum(IV) Complexes

The platinum(IV) compounds used in this investigation were prepared at the Institute of Inorganic Chemistry, University of Vienna. A complete list of the complexes, together with their structures and determined lipophilicity parameters can be found in Tables 1, S1 and S2. The complexes **1–15**, **17–19** [24,26,30,33,35,40–42], the ionisable complexes **20–24** [24,43], and the extended set of platinum(IV) complexes **25–26**, **28**, **30**, and **32–58** [24,26,32,33,35,40–42,44–47] were synthesised according to procedures in the literature.

3.2. (OC-6-43)-Amminedichlorido(cyclohexaneamine)bis[(4-ethoxy)-4-oxobutanoato]platinum(IV) (16)

Complex **16** was obtained upon the reaction of (OC-6-43)-amminedichlorido(cyclohexaneamine) dihydroxidoplatinum(IV) with 3-(ethoxycarbonyl)propanoic anhydride (3 eq) in *N,N*-dimethylformamide (DMF) under anhydrous conditions and light protection, as described for analogous compounds in reference [41]. Yield = 45% (55 mg). Elemental analysis: $C_{18}H_{34}Cl_2N_2O_8Pt$ (672.46); calcd. C 32.15, H 5.10, N 4.17; found C 32.17, H 5.12, N 3.97. ^{1}H-NMR (DMF-d_7): δ= 7.49 (bs, 2H, NH$_2$), 6.96 (m, 3H, NH$_3$), 4.10 (m, 4H, H9), 3.01 (bs, 1H, H1), 2.58 (m, 4H, H6), 2.50 (m, 4H, H7), 2.16 (m, 2H, H2), 1.72 (m, 2H, H3), 1.60 (m, 1H, H4), 1.36 (m, 4H, H2 and H3), 1.22 (t, $^3J_{H,H}$ = 7.1 Hz, 6H, H10), 1.17 (m, 1H, H4) ppm. ESI-MS (positive): m/z 673.1397 (m$_{theor}$ 673.1411) [M + H]$^+$, 695.1215 (m$_{theor}$ 695.1230) [M + Na]$^+$. ESI-MS (negative): m/z 671.1245 (671.1265) [M − H]$^-$.

3.3. (OC-6-54)-Dichlorido(N,N-dimethyl-ethane-1,2-diamine)hydroxido[4-(2-propyn-1-ylamino)-4-oxo-butanoato]platinum(IV) (27)

Complex **27** was synthesised by dissolving the precursor (0.20 g, published in [46]), 0.41 mmol) in 4 mL of DMF under an argon atmosphere and heating it to 50 °C. Then, 1,1-carbonyldiimidazol (0.07 g, 0.45 mmol) dissolved in 2 mL of DMF was added dropwise. After stirring for 10 min, the reaction mixture was cooled to room temperature, and the remaining CO_2 was removed by a flush of argon stream through the solution. Propargylamine (29 µL) was added, and the reaction was stirred overnight. DMF was removed under reduced pressure, and the crude mixture was purified by column chromatography (mobile phase EtOAc/MeOH/triethylamine = 3:1:0.1) to yield a white powder. Yield = 33% (70 mg). Elemental analysis: $C_{11}H_{21}Cl_2N_3O_4Pt$ (525.29); calcd. C 25.12, H 4.03, N 8.00; found C 25.06, H 4.26, N 7.65. ^{1}H-NMR (d_6-DMSO): δ = 9.45 (bs, 1H, NH5a); 8.25 (t, 1H, NH10, *J* = 5.3 Hz); 7.10 (bs, 1H, NH5b); 3.81 (dd, 2H, H11, *J* = 2.5, 5.4 Hz); 3.07 (t, 1H, H13, *J* = 2.5 Hz); 2.86–2.75 (bm, 4H, H3/H4); 2.68 (bs (s+d), 3H, H2a or H2b); 2.59 (bs (s+d), 3H, H2a or H2b); 2.50–2.25 (bm, 4H, H7/H8); 1.48 (bs, 1H, OH1) ppm. ESI-MS (positive): m/z 547.6 [M + Na$^+$]$^+$, 563.6 [M + K$^+$]$^+$. ESI-MS (negative): m/z 523.5 [M − H$^+$]$^+$.

3.4. (OC-6-33)-Dichlorido(ethane-1,2-diamine)bis[(4-(2-propyn-1-ylamino)-4-oxobutanoato]platinum(IV) (29)

Complex **29** was synthesised by dissolving **20** (0.40 g, 1.11 mmol) in 8 mL DMF under an argon atmosphere and heating it to 60 °C. Then, 1,1-carbonyldiimidazol (0.24 g, 1.48 mmol) dissolved in 16 mL of DMF was added dropwise. After stirring for 10 min, the reaction mixture was cooled to room temperature and the remaining CO_2 was removed by a flush of argon stream through the solution. Propargylamine (110 µL) dissolved in 24 mL of DMF was added, and the reaction was stirred overnight. DMF was removed under reduced pressure, and the crude mixture was purified by column chromatography (mobile phase EtOAc/MeOH = 7:1) to yield a white powder. Yield = 23% (110 mg). Elemental analysis: $C_{16}H_{24}Cl_2N_4O_6Pt$ (634.40); calcd. $C_{16}H_{24}Cl_2N_4O_6Pt \cdot 0.5H_2O$ (644.39) C 29.87, H 3.92, N 8.71; found C 29.56, H 3.58, N 8.61. ^{1}H-NMR (d_6-DMSO): δ = 8.42 (bs, 4H, $NH_2$2), 8.27 (t, 2H, NH7, J = 5.4 Hz), 3.84 (dd, 4H, H8, J = 2.5, 5.4 Hz), 3.09 (t, 2H, H10, J = 2.5 Hz), 2.67 (bs, 4H, H1), 2.48 (t, 4H, H4 or H5, J = 7.1 Hz), 2.31 (t, 4H, H4 or H5, J = 7.1 Hz) ppm. ESI-MS (positive): m/z 657.1 [M + Na$^+$]$^+$.

3.5. (OC-6-33)-Dichlorido(ethane-1,2-diamine)bis[(4-dimethylamino)-4-oxobutanoato]platinum(IV) (31)

Complex **31** was synthesised by dissolving 1-(dimethylcarbamoyl)-3-methyl-1H-imidazol-3-ium iodide (0.41 g, 2.95 mmol) and **20** (0.40 g, 1.11 mmol) in 8 mL of DMF under an argon atmosphere. Trimethylamine (0.15 g) was added, and the clear brown reaction mixture was stirred at room temperature for 24 h. DMF was removed under reduced pressure, and the crude mixture was purified by column chromatography (mobile phase EtOAc/MeOH = 1:1). A white powder was obtained after crystallisation in methanol. Yield = 35% (150 mg). Elemental analysis: $C_{14}H_{28}Cl_2N_4O_6Pt$ (614.4); calcd. $C_{14}H_{28}Cl_2N_4O_6Pt \cdot H_2O$ (632.4) C 26.59, H 4.78, N 8.86; found C 26.47, H 4.64, N 8.61. ^{1}H-NMR (d_6-DMSO): δ = 8.44 (bs, 4H, 2NH_2), 2.95 (s, 6H, H7 or H7′), 2.79 (s, 6H, H7 or H7′), 2.68 (bs, 4H, H1), 2.47 (bs, 8H, H4/H5) ppm. ESI-MS (positive): m/z 637.1 [M + Na$^+$]$^+$. ESI-MS (negative): m/z 613.0 [M − H$^+$]$^-$.

3.6. Instrumentation

HPLC. The RP-HPLC experiments were carried out on a Dionex Ultimate 3000 RS system (Dionex, Germering, Germany), controlled by the Dionex Chromeleon 6.80 software. The following chromatographic conditions were used: Agilent Poroshell 120 SB-C18 column (2.1 × 150 mm, 2.7 µm); injection volume: 1 µL; flow rate: 0.2 mL·min^{-1}; collection rate: 5–100 Hz; temperature of the column and autosampler: 25°C; UV/Vis detection set up at 210–256 nm. Mobile phases consisted of (A) water with 0.1% FA and (B) methanol with 0.1% FA.

ICP-MS. The ICP-MS measurements were carried out on an ICP-quadrupole MS instrument Agilent 7500ce (Agilent Technologies, Waldbronn, Germany) equipped with a CETAC ASX-520 autosampler (Omaha, NE, USA) and a MicroMist nebuliser at a sample uptake rate of approx. 0.25 mL min^{-1}. The instrument was tuned on a daily basis in order to achieve the maximum sensitivity. Rhenium (^{185}Re) served as internal standard for platinum to account for instrumental fluctuations, and the platinum isotopes ^{194}Pt and ^{195}Pt were acquired. The ICP-MS was equipped with nickel cones and operated at an RF power of 1500 W. Argon was used as the plasma gas (15 L·min^{-1}), carrier gas (~1 L·min^{-1}), and make up gas (~0.2 L·min^{-1}). The dwell time was set to 0.1 s, and the measurement was performed in 10 replicates. The Agilent MassHunter software package (Workstation Software, Version B.01.01, 2012, Agilent, Santa Clara, CA, USA) was used for data processing.

3.7. Shake Flask Procedure

The logarithmic partition coefficient Log P between 1-octanol and water phases was determined for platinum complexes using the shake flask method according to the OECD guidelines with slight modifications [7]. The platinum compounds were dissolved in water (Milli-Q, pre-saturated with 1-octanol) to yield stock solutions of 0.5 mmol·L^{-1}, which were filtered through a 0.45-µm filter (Minisart RC 25, Sartorius AG, Göttingen, Germany). An equal volume of 1-octanol (pre-saturated with ultrapure water) was added to the platinum-containing aqueous solutions, and the mixtures were shaken mechanically for 60 min at room temperature. The samples were centrifuged to assist with bilayer formation. The experiments were performed at least in duplicates.

3.8. Analysis of Samples Obtained from the Shake Flask Procedure

RP-HPLC. The two phases obtained by the shake flask method were analysed by RP-HPLC for each complex, using an isocratic mode in a suitable water/methanol ratio ranging from 95:5 to 10:90. However, after elution of each sample containing 1-octanol, the methanol percentage was increased to 95% for 20 min before returning to the initial isocratic conditions in order to remove any remaining 1-octanol from the column. Additionally, the total platinum concentration was determined by analysing an aliquot of each stock solution. The concentration of the platinum compounds was detected by UV/Vis spectroscopy. According to the Lambert–Beer law, the area of absorbance (A) of an eluting analyte is directly proportional to its concentration (c). The Log P was calculated according to Equation (2):

$$\text{Log P(HPLC)} = \text{Log} \left(c_{octanol} / c_{water} \right) = \text{Log} \left(A_{octanol} / A_{water} \right). \tag{2}$$

ICP-MS. The total platinum content in the initial stock solution and in the final aqueous phase obtained by the shake flask method was determined by ICP-MS. Each aliquot was diluted gravimetrically with 1% HNO_3 to match the linear range of the ICP-MS system. By assuming $c_{octanol} = c_{total} - c_{water}$, the Log P was calculated according to Equation (3):

$$\text{Log P(ICP-MS)} = \text{Log} \left[\left(c_{total} - c_{water} \right) / c_{water} \right] \tag{3}$$

3.9. Determining the Distribution Coefficient by the Shake Flask Method with HPLC–UV/Vis Detection

Compared with the partition coefficient (Log P), the distribution coefficient (Log D_{pH}, henceforth called Log D) takes into account all the species of an analyte at a given pH and is suitable for ionisable compounds. Thus, the Log D of 5 platinum(IV) compounds (**20–24**) featuring an acidic moiety was determined at different pH values. The platinum(IV) complex **10**, bearing non-ionisable functional groups, was evaluated as a control.

In this case, the aqueous phase consisted of phosphate or acetic acid buffer solutions at 10 mmol·L^{-1}, saturated with octanol. The phosphate buffers were prepared at pH = 1.7, 2.5, 6.2, and 7.4, and the acetic acid buffers were prepared at pH = 3.7, 4.2, 4.7, and 5.6. Thus, the platinum complexes were dissolved in those buffer solutions, and the shake flask method was performed as described above using HPLC–UV/Vis detection.

3.10. Determining the Lipophilicity from Chromatographic Parameters

The platinum complexes (0.5 mmol·L^{-1}) were dissolved in 1 mL of water/methanol (1:1) and the solutions were filtered through a 0.45-μm filter (Minisart RC 25, Sartorius AG, Göttingen, Germany). Uracil was employed as an internal standard (1 μmol·L^{-1}) and potassium iodide (KI) as an external standard for determining the system dead time (t_0). The obtained samples were directly analysed by RP-HPLC. The RP-HPLC experiments were performed in isocratic modes at least in duplicates, employing at least 3 different water/methanol compositions per complex. The capacity factors $k = (t_r - t_0)/t_0$, where t_r is the retention time of the analyte, were determined for all the investigated eluent compositions. From these, the capacity factor at 100% aqueous mobile phase (Log k_w) was extrapolated using the linear Soczewinski–Snyder relationship [12,13], Log k = Log k_w − Sϕ, where Log k is the capacity factor in a specific mobile phase composition, ϕ is the volume fraction of the methanol in the eluent, and S is a constant for a given analyte and HPLC system (slope). Then, the chromatographic hydrophobicity parameter $ϕ_0$ was derived, defined as the volume percentage of the organic modifier required to achieve an equal distribution of a compound between the mobile and stationary phase, corresponding to Log k = 0 [10,14].

4. Conclusions

Lipophilicity is a crucial parameter in drug discovery and also for metal-based drug candidates. However, experimentally determining lipophilicity by the shake flask method may have been somewhat impeded by restricted access to suitable detection methods besides element-specific spectroscopy or spectrometry. Because high-performance liquid chromatography (HPLC) instruments are more widely available to research groups, chromatographic lipophilicity parameters have recently become more popular but are in different scales compared with the classically obtained partition coefficient. Here, a straight-forward HPLC-based method was validated to derive the partition coefficient (Log P) of cytotoxic platinum(IV) complexes from shake flask experiments, which exhibited a dynamic range of at least four orders of magnitude from $-2 <$ Log P < 2.4. In addition, chromatographic partition by means of the parameter ϕ_0 was assessed and calibrated against the corresponding Log P values so that the chromatographic lipophilicity of cytotoxic platinum(IV) complexes may be directly converted into Log P, which is valid for $\phi_0 > 25$. The same methods were extended to determine the distribution coefficient (Log D) of ionisable platinum(IV) representatives, which showed pH-dependent lipophilicity changes. These approaches can be applied to other classes of metal complexes that may ultimately enable the estimation of lipophilicity for metal-based anticancer agents in general.

Supplementary Materials: The following are available online at http://www.mdpi.com/2304-6740/6/4/130/s1: Table S1: List of five platinum(IV) complexes bearing ionisable groups for determining distribution coefficients; Table S2: List of 35 platinum(IV) complexes for converting ϕ_0 into calculated Log P (cLog P) based on the calibration curve described in the main text; Figure S1: Comparison of the retention times of uracil (detection at 256 nm) and potassium iodide (KI, detection at 230 nm) at different percentages of methanol (ϕ) in the eluent; Figure S2: The correlation between the chromatographic lipophilicity parameters ϕ_0 and Log k_w is a quadratic polynomial (using potassium iodide as a dead time marker) with $R^2 = 0.982$ for the standard set of 19 platinum compounds; Figure S3: The correlation between Log P by the shake flask method and the chromatographic Log k_w is linear (using potassium iodide as a dead time marker) with $R^2 = 0.918$ for the standard set of 19 platinum compounds. Log P (HPLC) = $1.250 \cdot$ Log $k_w - 3.000$.

Author Contributions: Conceptualisation, M.H.M.K., S.T., S.M.M.-M., and B.K.K.; data curation, M.H.M.K. and S.T.; formal analysis, S.M.M.-M.; investigation, M.H.M.K. and S.T.; resources, H.P.V., D.H., V.P., M.G., and B.K.K.; validation, M.H.M.K. and S.T.; visualisation, S.M.M.-M.; original draft preparation, M.H.M.K. and S.M.M.-M.; and editing of manuscript, S.T., H.P.V., D.H., V.P., M.G., S.M.M.-M., and B.K.K.

Funding: M.H.M.K. is grateful to the *"Stipendium der Monatshefte für Chemie"* for financial support.

Conflicts of Interest: The authors declare no conflicts of interest.

References

1. Waring, M.J. Lipophilicity in Drug Discovery. *Expert Opin. Drug Discov.* **2010**, *5*, 235–248. [CrossRef] [PubMed]

2. Varma, M.V.S.; Radi, Z.A.; Rotter, C.J.; Litchfield, J.; El-Kattan, A.F.; Lyubimov, A.V. Pharmacokinetics and Toxicokinetics in Drug Discovery and Development. In *Encyclopedia of Drug Metabolism and Interactions*; John Wiley & Sons, Inc.: Hoboken, NJ, USA, 2011.

3. Mälkiä, A.; Murtomäki, L.; Urtti, A.; Kontturi, K. Drug Permeation in Biomembranes: In Vitro and In Silico Prediction and Influence of Physicochemical Properties. *Eur. J. Pharm. Sci.* **2004**, *23*, 13–47. [CrossRef] [PubMed]

4. Hansch, C.; Maloney, P.P.; Fujita, T. Correlation of Biological Activity of Phenoxyacetic Acids with Hammett Substituent Constants and Partition Coefficients. *Nature* **1962**, *194*, 178–180. [CrossRef]

5. Hansch, C.; Fujita, T. p-s-p Analysis. A Method for the Correlation of Biological Activity and Chemical Structure. *J. Am. Chem. Soc.* **1964**, *86*, 1616–1626. [CrossRef]

6. Leo, A.; Hansch, C.; Elkins, D. Partition Coefficients and Their Uses. *Chem. Rev.* **1971**, *71*, 525–616. [CrossRef]

7. OECD. *Test No. 107: Partition Coefficient (n-Octanol/Water): Shake Flask Method*; OECD Publishing: Paris, France, 1995.

8. Albert, A. *Selective Toxicity: The Physicochemical Basis of Therapy*; Chapman and Hall: London, UK, 1979.

9. Kerns, E.H. High Throughput Physicochemical Profiling for Drug Discovery. *J. Pharm. Sci.* **2001**, *90*, 1838–1858. [CrossRef] [PubMed]

10. Nasal, A.; Siluk, D.; Kaliszan, R. Chromatographic Retention Parameters in Medicinal Chemistry and Molecular Pharmacology. *Curr. Med. Chem.* **2003**, *10*, 381–426. [CrossRef] [PubMed]

11. Valkó, K. Application of High-Performance Liquid Chromatography Based Measurements of Lipophilicity to Model Biological Distribution. *J. Chromatogr. A* **2004**, *1037*, 299–310. [CrossRef] [PubMed]

12. Soczewinski, E.; Wachtmeister, C.A. The Relation between the Composition of Certain Ternary Two-Phase Solvent Systems and RM Values. *J. Chromatogr. A* **1962**, *7*, 311–320. [CrossRef]

13. Snyder, L.R.; Dolan, J.W.; Gant, J.R. Gradient Elution in High-Performance Liquid Chromatography: I. Theoretical Basis for Reversed-Phase Systems. *J. Chromatogr. A* **1979**, *165*, 3–30. [CrossRef]

14. Valkó, K.; Slégel, P. New Chromatographic Hydrophobicity Index (ϕ_0) Based on the Slope and the Intercept of the log k versus Organic Phase Concentration Plot. *J. Chromatogr. A* **1993**, *631*, 49–61. [CrossRef]

15. Souchard, J.P.; Ha, T.T.; Cros, S.; Johnson, N.P. Hydrophobicity Parameters for Platinum Complexes. *J. Med. Chem.* **1991**, *34*, 863–864. [CrossRef] [PubMed]

16. Screnci, D.; McKeage, M.J.; Galettis, P.; Hambley, T.W.; Palmer, B.D.; Baguley, B.C. Relationships Between Hydrophobicity, Reactivity, Accumulation and Peripheral Nerve Toxicity of a Series of Platinum Drugs. *Br. J. Cancer* **2000**, *82*, 966–972. [CrossRef] [PubMed]

17. Robillard, M.S.; Galanski, M.; Zimmermann, W.; Keppler, B.K.; Reedijk, J. (Aminoethanol)dichloroplatinum(II) Complexes: Influence of the Hydroxyethyl Moiety on 5′-GMP and DNA Binding, Intramolecular Stability, the Partition Coefficient and Anticancer Activity. *J. Inorg. Biochem.* **2002**, *88*, 254–259. [CrossRef]

18. Hall, M.D.; Amjadi, S.; Zhang, M.; Beale, P.J.; Hambley, T.W. The Mechanism of Action of Platinum(IV) Complexes in Ovarian Cancer Cell Lines. *J. Inorg. Biochem.* **2004**, *98*, 1614–1624. [CrossRef] [PubMed]

19. Ghezzi, A.; Aceto, M.; Cassino, C.; Gabano, E.; Osella, D. Uptake of Antitumor Platinum(II)-Complexes by Cancer Cells, Assayed by Inductively Coupled Plasma Mass Spectrometry (ICP-MS). *J. Inorg. Biochem.* **2004**, *98*, 73–78. [CrossRef] [PubMed]

20. Platts, J.A.; Oldfield, S.P.; Reif, M.M.; Palmucci, A.; Gabano, E.; Osella, D. The RP-HPLC Measurement and QSPR Analysis of Log Po/w Values of Several Pt(II) Complexes. *J. Inorg. Biochem.* **2006**, *100*, 1199–1207. [CrossRef] [PubMed]

21. Reithofer, M.R.; Bytzek, A.K.; Valiahdi, S.M.; Kowol, C.R.; Groessl, M.; Hartinger, C.G.; Jakupec, M.A.; Galanski, M.; Keppler, B.K. Tuning of Lipophilicity and Cytotoxic Potency by Structural Variation of Anticancer Platinum(IV) Complexes. *J. Inorg. Biochem.* **2011**, *105*, 46–51. [CrossRef] [PubMed]

22. Gibson, D. The Mechanism of Action of Platinum Anticancer Agents—What Do We Really Know About it? *Dalton Trans.* **2009**, 10681–10689. [CrossRef]

23. Bytzek, A.K.; Reithofer, M.R.; Galanski, M.; Groessl, M.; Keppler, B.K.; Hartinger, C.G. The First Example of MEEKC-ICP-MS Coupling and its Application for the Analysis of Anticancer Platinum Complexes. *Electrophoresis* **2010**, *31*, 1144–1150. [CrossRef]

24. Varbanov, H.P.; Valiahdi, S.M.; Kowol, C.R.; Jakupec, M.A.; Galanski, M.; Keppler, B.K. Novel Tetracarboxylato-platinum(IV) Complexes as Carboplatin Prodrugs. *Dalton Trans.* **2012**, *41*, 14404–14415. [CrossRef] [PubMed]

25. Ermondi, G.; Caron, G.; Ravera, M.; Gabano, E.; Bianco, S.; Platts, J.A.; Osella, D. Molecular interaction fields vs. quantum-mechanical-based descriptors in the modelling of lipophilicity of platinum(IV) complexes. *Dalton Trans.* **2013**, *42*, 3482–3489. [CrossRef] [PubMed]

26. Varbanov, H.P.; Göschl, S.; Heffeter, P.; Theiner, S.; Roller, A.; Jensen, F.; Jakupec, M.A.; Berger, W.; Galanski, M.; Keppler, B.K. A Novel Class of Bis- and Tris-Chelate Diam(m)inebis(dicarboxylato)platinum(IV) Complexes as Potential Anticancer Prodrugs. *J. Med. Chem.* **2014**, *57*, 6751–6764. [CrossRef] [PubMed]

27. Oldfield, S.P.; Hall, M.D.; Platts, J.A. Calculation of Lipophilicity of a Large, Diverse Dataset of Anticancer Platinum Complexes and the Relation to Cellular Uptake. *J. Med. Chem.* **2007**, *50*, 5227–5237. [CrossRef] [PubMed]

28. Tetko, I.V.; Jaroszewicz, I.; Platts, J.A.; Kuduk-Jaworska, J. Calculation of Lipophilicity for Pt(II) Complexes: Experimental Comparison of Several Methods. *J. Inorg. Biochem.* **2008**, *102*, 1424–1437. [CrossRef] [PubMed]

29. Platts, J.A.; Ermondi, G.; Caron, G.; Ravera, M.; Gabano, E.; Gaviglio, L.; Pelosi, G.; Osella, D. Molecular and Statistical Modeling of Reduction Peak Potential and Lipophilicity of Platinum(IV) Complexes. *J. Biol. Inorg. Chem.* **2011**, *16*, 361–372. [CrossRef] [PubMed]

30. Tetko, I.V.; Varbanov, H.P.; Galanski, M.; Talmaciu, M.; Platts, J.A.; Ravera, M.; Gabano, E. Prediction of LogP for Pt(II) and Pt(IV) Complexes: Comparison of Statistical and Quantum-Chemistry Based Approaches. *J. Inorg. Biochem.* **2016**, *156*, 1–13. [CrossRef] [PubMed]

31. Meier, S.M.; Hanif, M.; Adhireksan, Z.; Pichler, V.; Novak, M.; Jirkovsky, E.; Jakupec, M.A.; Arion, V.B.; Davey, C.A.; Keppler, B.K.; et al. Novel Metal(II) Arene 2-Pyridinecarbothioamides: A Rationale to Orally Active Organometallic Anticancer Agents. *Chem. Sci.* **2013**, *4*, 1837–1846. [CrossRef]

32. Pichler, V.; Goschl, S.; Meier, S.M.; Roller, A.; Jakupec, M.A.; Galanski, M.; Keppler, B.K. Bulky (*N*,*N*)-(Di) alkylethane-1,2-diamineplatinum(II) Compounds as Precursors for Generating Unsymmetrically Substituted Platinum(IV) Complexes. *Inorg. Chem.* **2013**, *52*, 8151–8162. [CrossRef]

33. Varbanov, H.P.; Jakupec, M.A.; Roller, A.; Jensen, F.; Galanski, M.; Keppler, B.K. Theoretical Investigations and Density Functional Theory Based Quantitative Structure-Activity Relationships Model for Novel Cytotoxic Platinum(IV) Complexes. *J. Med. Chem.* **2013**, *56*, 330–344. [CrossRef]

34. Meier-Menches, S.M.; Gerner, C.; Berger, W.; Hartinger, C.G.; Keppler, B.K. Structure-Activity Relationships for Ruthenium and Osmium Anticancer Agents—Towards Clinical Development. *Chem. Soc. Rev.* **2018**, *47*, 909–928. [CrossRef] [PubMed]

35. Varbanov, H.; Valiahdi, S.M.; Legin, A.A.; Jakupec, M.A.; Roller, A.; Galanski, M.; Keppler, B.K. Synthesis and Characterization of Novel Bis(carboxylato)dichloridobis(ethylamine)platinum(IV) Complexes with Higher Cytotoxicity than Cisplatin. *Eur. J. Med. Chem.* **2011**, *46*, 5456–5464. [CrossRef] [PubMed]

36. Low, Y.W.; Blasco, F.; Vachaspati, P. Optimised Method to Estimate Octanol Water Distribution Coefficient (Log D) in a High Throughput Format. *Eur. J. Pharm. Sci.* **2016**, *92*, 110–116. [CrossRef] [PubMed]

37. Hall, M.D.; Mellor, H.R.; Callaghan, R.; Hambley, T.W. Basis for Design and Development of Platinum(IV) Anticancer Complexes. *J. Med. Chem.* **2007**, *50*, 3403–3411. [CrossRef]

38. Schoenmakers, P.J.; Billiet, H.A.H.; Tussen, R.; De Galan, L. Gradient Selection in Reversed-Phase Liquid Chromatography. *J. Chromatogr. A* **1978**, *149*, 519–537. [CrossRef]

39. Baczek, T.; Markuszewski, M.; Kaliszan, R.; van Straten, M.A.; Claessens, H.A. Linear and Quadratic Relationships between Retention and Organic Modifier Content in Eluent in Reversed Phase High-Performance Liquid Chromatography: A Systematic Comparative Statistical Study. *J. High Resolut. Chromatogr.* **2000**, *23*, 667–676. [CrossRef]

40. Goschl, S.; Varbanov, H.P.; Theiner, S.; Jakupec, M.A.; Galanski, M.; Keppler, B.K. The Role of the Equatorial Ligands for the Redox Behavior, Mode of Cellular Accumulation and Cytotoxicity of Platinum(IV) Prodrugs. *J. Inorg. Biochem.* **2016**, *160*, 264–274. [CrossRef]

41. Hofer, D.; Varbanov, H.P.; Hejl, M.; Jakupec, M.A.; Roller, A.; Galanski, M.; Keppler, B.K. Impact of the Equatorial Coordination Sphere on the Rate of Reduction, Lipophilicity and Cytotoxic Activity of Platinum(IV) Complexes. *J. Inorg. Biochem.* **2017**, *174*, 119–129. [CrossRef]

42. Hofer, D.; Varbanov, H.P.; Legin, A.; Jakupec, M.A.; Roller, A.; Galanski, M.; Keppler, B.K. Tetracarboxylato-platinum(IV) Complexes Featuring Monodentate Leaving Groups—A Rational Approach Towards Exploiting the Platinum(IV) Prodrug Strategy. *J. Inorg. Biochem.* **2015**, *153*, 259–271. [CrossRef]

43. Reithofer, M.; Galanski, M.; Roller, A.; Keppler, B.K. An Entry to Novel Platinum Complexes: Carboxylation of Dihydroxoplatinum(IV) Complexes with Succinic Anhydride and Subsequent Derivatization. *Eur. J. Inorg. Chem.* **2006**, *2006*, 2612–2617. [CrossRef]

44. Reithofer, M.R.; Valiahdi, S.M.; Jakupec, M.A.; Arion, V.B.; Egger, A.; Galanski, M.; Keppler, B.K. Novel Di- and Tetracarboxylatoplatinum(IV) Complexes. Synthesis, Characterization, Cytotoxic Activity and DNA Platination. *J. Med. Chem.* **2007**, *50*, 6692–6699. [CrossRef] [PubMed]

45. Reithofer, M.R.; Schwarzinger, A.; Valiahdi, S.M.; Galanski, M.; Jakupec, M.A.; Keppler, B.K. Novel Bis(carboxylato)dichlorido(ethane-1,2-diamine)platinum(IV) Complexes with Exceptionally High Cytotoxicity. *J. Inorg. Biochem.* **2008**, *102*, 2072–2077. [CrossRef] [PubMed]

46. Pichler, V.; Valiahdi, S.M.; Jakupec, M.A.; Arion, V.B.; Galanski, M.; Keppler, B.K. Mono-Carboxylated Diaminedichloridoplatinum(IV) Complexes—Selective Synthesis, Characterization and Cytotoxicity. *Dalton Trans.* **2011**, *40*, 8187–8192. [CrossRef] [PubMed]

47. Pichler, V.; Heffeter, P.; Valiandi, S.M.; Kowol, C.R.; Egger, A.; Berger, W.; Jakupec, M.A.; Galanski, M.; Keppler, B.K. Unsymmetric Mono- and Dinuclear Platinum(IV) Complexes Featuring an Ethylene Glycol Moiety: Synthesis, Characterization, and Biological Activity. *J. Med. Chem.* **2012**, *55*, 11052–11061. [CrossRef] [PubMed]

 *inorganics*

Article

Anti-Proliferative and Anti-Migration Activity of Arene–Ruthenium(II) Complexes with Azole Therapeutic Agents

Legna Colina-Vegas [1], Katia M. Oliveira [1], Beatriz N. Cunha [1,2], Marcia Regina Cominetti [3], Maribel Navarro [4,5,*] and Alzir Azevedo Batista [1,*]

[1] Departamento de Química, Universidade Federal de São Carlos, CEP 13565-905 São Carlos, SP, Brazil; esleg_24@hotmail.com (L.C.-V.); kmoliveiraq@gmail.com (K.M.O.); beatriz.cunha@ifgoiano.edu.br (B.N.C.)
[2] Instituto Federal Goiano—IFG, Campus Ceres, CEP 76300-00 Ceres, GO, Brazil
[3] Departamento de Gerontologia, Universidade Federal de São Carlos, CEP 13565-905 São Carlos, SP, Brazil; mcominetti@ufscar.br
[4] Instituto Nacional de Metrologia, Qualidade e Tecnologia, INMETRO, CEP 25250-020 Xerem, RJ, Brazil
[5] Departamento de Química, ICE, Universidade Federal de Juiz de Fora, CEP 36036-900 Juiz de Fora, MG, Brazil
*   Correspondence: maribel.navarro@ufjf.edu.br (M.N.); daab@ufscar.br (A.A.B.)

Received: 24 October 2018; Accepted: 5 December 2018; Published: 11 December 2018

Abstract: The efficacy of organoruthenium complexes containing ergosterol biosynthesis inhibitors (CTZ: clotrimazole, KTZ: ketoconazole and FCZ: fluconazole) against tumor cells, and their interaction with important macro-biomolecules such as human serum albumin and DNA have been investigated here. Our experimental results indicated that these ruthenium(II) complexes present spontaneous electrostatic interactions with albumin, and act as minor groove binders with the DNA. The ability of these Ru(II)–azole complexes to inhibit the proliferation of selected human tumor and non-tumor cell lines was determined by MTT assay. Complexes $[RuCl(CTZ)(\eta^6\text{-}p\text{-cymene})(PPh_3)]PF_6$ (**3**) and $[RuCl(KTZ)(\eta^6\text{-}p\text{-cymene})(PPh_3)]PF_6$ (**4**) were shown to be between 3- and 40-fold more cytotoxic than the free ligands and the positive control cisplatin. Complex **3** was selected to continue studies on the triple negative breast tumor cell line MDA-MB-231, inducing morphological changes, loss of adhesion, inhibition of colony formation, and migration through Boyden chambers, cell cycle arrest in the sub-G1 phase, and a mechanism of cell death by apoptosis. All these interesting results show the potential of this class of organometallic Ru(II) complexes as an antiproliferative agent.

Keywords: ruthenium complexes; antiproliferative; antimigration; DNA interaction; HSA binding

1. Introduction

According to the WHO, cancer is a generic term for a group of diseases involving abnormal cell growth with the potential to invade or spread to other parts of the body. There are over 100 different known cancer types that affect humans, causing about 8.2 million deaths, estimated as 13% of all deaths worldwide. In 2012, about 14.1 million new cases of cancer occurred globally (not including skin cancer other than melanoma) [1]. One of the principal causes of cancer deaths is the developing of tumor metastasis. Metastasis is defined as the capacity of tumor cells to move from the original tumor to adjacent or distant tissues and spread to other organs [2].

Transition metals, particularly multiple platinum derivatives, have been tested in clinical trials against several types of cancers and cisplatin is one of the most potent chemotherapy drugs approved for clinical practice worldwide. However, its use is limited due to severe side effects. The other two FDA-approved agents are carboplatin and oxaliplatin, while nedaplatin, lobaplatin and heptaplatin received restricted approval for clinical use [3]. Furthermore, many non-platinum compounds have been evaluated against tumor cells. The clinical phase I trials of three ruthenium compounds: indazolium *trans*-[tetrachlorobis(1*H*-indazole)ruthenate(III)] KP1019, sodium *trans*-[tetrachlorobis(1*H*-indazole)ruthenate(III)] NKP1339 and imidazolium *trans*-[tetrachloro-(*S*-dimethylsulfoxide)(1*H*-imidazole)ruthenate(III)] NAMI-A has led to considerable interest in anticancer drugs based on this metal center [4,5]. Over the last two decades, Ru–arene complexes have become a focus of interest due to their anticancer properties [6–9]. These kind of metal complexes with monodentate or bidentate ligands showing different modes of action [10] such as apoptosis induction via DNA damage and anti-angiogenic properties [11], protein kinase inhibitors [12] protein RNase A [13] or a multi-target concept inhibit human topoisomerase IIα and covalently bind to DNA [14].

Organometallics and coordination ruthenium complexes containing well-known antifungal compounds like clotrimazole (CTZ) and ketoconazole (KTZ) have shown promising biological properties against *Leishmania major*, *Trypanosoma cruzi* [15], *Mycobacterium tuberculosis* [16] and also display antiproliferative activities on human tumor cells lines [17]. Some of these complexes present a mechanism of cell death preferentially through apoptosis. The triazole compound Fluconazole (FCZ) together with other azole compounds were evaluated against human breast adenocarcinomas MCF-7 and MDA-MB-231 and the compounds CTZ and KTZ showed induction of apoptosis, cell cycle arrest in the G1 phase, anti-migration, and anti-invasion properties [18].

We recently reported four organoruthenium complexes incorporated in the coordination sphere of ligands CTZ, KTZ or FCZ, triphenylphosphine and chloride, presenting the formulas [RuCl(η^6-*p*-cymene)(μ-FCZ)]$_2$[Cl]$_2$ (**1**), [RuCl(FCZ)(η^6-*p*-cymene)(PPh$_3$)]PF$_6$ (**2**), [RuCl(CTZ)(η^6-*p*-cymene)(PPh$_3$)]PF$_6$ (**3**) and [RuCl(KTZ)(η^6-*p*-cymene)(PPh$_3$)]PF$_6$ (**4**), which display high antiparasitic activity and ultrastructural alterations against *Leishmania amazonensis* [19]. In this work, we evaluated the ability of complexes **1–4** to interact with important biological targets such as human serum albumin (HSA) and DNA, as well as, their cytotoxic activity against three human tumor cell lines: Prostate (DU-145), breast (MDA-MB-231), lung (A549) and non-tumor (MRC-5 and L929). The most active ruthenium complex was selected to study the mechanism of action in the breast tumor cell line through analysis of morphology, clonogenic, migration, cell cycle and cell death assays.

2. Results and Discussion

2.1. Interaction Studies with Macro-Biological Targets: Blood Human Serum Albumin and DNA

The organoruthenium complexes were previously synthetized from [RuCl$_2$(η^6-*p*-cymene)]$_2$ and [RuCl$_2$(η^6-*p*-cymene)(PPh$_3$)] by reaction with FCZ, CTZ or KTZ in methanol at different molar ratios. All complexes were characterized by usual techniques including X-ray; their chemical structures are shown in Scheme 1 [19].

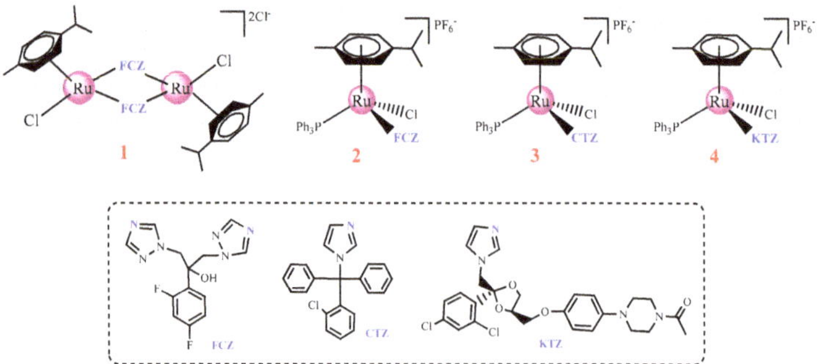

Scheme 1. Structure of organoruthenium complexes 1–4.

In order to study the interaction of complexes **1–4** with blood human serum albumin (HSA), fluorescence quenching was used, which is considered the earliest and simplest method for measuring binding affinities between this protein and an organic or inorganic compound. The advantage of this technique is that it has shown a decrease in the quantum yield of fluorescence from a fluorophore induced by a variety of molecular interactions with a quencher molecule. Albumin proteins are considered to have intrinsic fluorescence due to the presence of three fluorophores: tryptophan, tyrosine and phenylalanine, the latter two contributing to its fluorescence to only a minor extent [20]. The results obtained by exciting HSA at 280 nm and recording its emission (λ_{max} = 305 nm) as a function of the concentration of complexes **2–4** are shown in Figure S1. Compound **1** does not decrease protein fluorescence intensity in the experimental conditions used, and therefore it was not possible to calculate its interaction with the protein. The Stern–Volmer quenching constant (Ksv), binding constants (K_b), the number of binding sites (n) and thermodynamic parameters ΔH, ΔS and ΔG were calculated accordingly with the equation descripted in the supplementary material. The results show that Ksv is inversely correlated with temperature (Table 1), indicating a quenching static mechanism caused by a specific interaction between the HSA and Ru–azole complexes (**2–4**).

The values of n approximate to 1 suggest that only one reactive site exists in HSA for these Ru(II)–azole complexes. The magnitude of K_b values calculated for complexes **2–4** were between $(1.78–7.90) \times 10^5$ M^{-1}, which suggested a moderate interaction with the HSA molecule, when compared with other metal complexes [21]. Ross et al. [22] have used the signal and magnitude of the thermodynamic parameters to interpret the nature of the interaction in a variety of host-guest systems. The negative values of ΔG support the assertion that the binding process is spontaneous. The positive ΔH and ΔS values of the interaction of all Ru(II)–azole complexes to HSA indicate that the electrostatic interactions played a major role in the binding reaction, which is allowed by the positive charge of the complexes with negative regions of the protein. Indeed, the fluorescence spectral results of two ruthenium(II) arene complexes of curcumin (O–O) analogs with the general formula [RuCl(η^6-*p*-cymene)(O–O)] indicated that they quenched the intrinsic fluorescence of HSA through static quenching mode and the thermodynamic parameters showed that Van der Waals and hydrogen bond interactions played major roles in complex stabilization [23].

Table 1. Stern–Volmer quenching constant (K_{sv}, L·mol^{-1}), bimolecular quenching rate constant (K_q, L·mol^{-1}·s^{-1}), binding constant (K_b, M^{-1}), the number of binding sites (n), ΔG (KJ·mol^{-1}), ΔH (KJ·mol^{-1}) and ΔS (J·mol^{-1}·K) values for the complex-HSA system.

Compound	T (K)	K_{sv} (10^4)	K_b (10^5)	n	ΔG	ΔH	ΔS
2	295	1.32 ± 0.06	2.35 ± 0.04	1.30	−30.35	15.73	156.12
	310	1.24 ± 0.10	1.72 ± 0.02	1.25	−31.10		
3	295	1.93 ± 0.08	5.35 ± 0.70	1.25	−32.40	38.90	241.32
	310	1.83 ± 0.03	2.50 ± 0.20	1.25	−32.04		
4	295	2.71 ± 0.04	7.90 ± 0.14	1.35	−34.30	17.70	172.90
	310	2.45 ± 0.03	5.57 ± 0.22	1.30	34.10		

The interaction of compounds **1–4** with calf thymus DNA (ct-DNA) were studied by UV–vis titration, viscosity, circular dichroism (CD) and agarose gel electrophoresis assay using procedures previous reported by us [24]. From the UV–vis titration experiments, changes in the UV–vis spectrum were clearly observed, showing an intense absorption band around 280 nm for all Ru(II)–azole complexes. After adding increasing amounts of ct-DNA to the compound, as previously reported, they showed a decrease in the intensity of the absorbance and also bathochromic shifts; however, it was not possible to calculate the binding constant because the data presented a non-linear correlation. Viscosity measurements have often been used to evaluate structural changes in the DNA helix and to determine intercalation or non-intercalation binding modes of metal complexes to DNA in solution. The ct-DNA viscosity was constant in different complex/DNA ratios. This behavior in DNA viscosity ruled out the intercalative binding mode of the complexes to DNA.

To investigate in more detail the interaction of complexes **1–4** with ct-DNA, CD assessments were performed. The CD spectrum of ct-DNA free consists of two bands due to base stacking and to the helicity, which is a characteristic of DNA in the right-handed B form [25]. All Ru(II)–azole complexes did not induce changes in the ct-DNA CD spectrum, neither in the helicity of the negative or positive bands. Considering that, the interaction between pBR322 DNA with compound **1–4** was studied by agarose gel electrophoresis assays, observing a lower intensity of the bands corresponding to supercoiled (SC), open circular (OC) and linear (L) forms with increasing concentration of these compounds. An explanation for this result is the replacing of ethidium bromide (EB), used as staining in this assay, resulting in no fluorescent bands, even though this behavior has been reported mainly by DNA intercalation binding compounds, and it is also reported that several compounds are able to replace EB through DNA groove interaction [26,27]. For example, Hoechst 33258 (H33258), a known benzimidazole dye (not intercalator compound) binds to the minor groove of ds-DNA with a preference for adenine and thymine-rich, and its fluorescence intensity greatly increases when it is bound to DNA [28,29]. For this reason, a competitive Hoechst 33258 (H33258) displacement assay was performed by fluorescence spectroscopy. The emission spectra of the H33258-DNA adduct in the presence of increasing amounts of **1** is shown in Figure 1. The emission spectra for complexes **2–4** is shown in Figure S2.

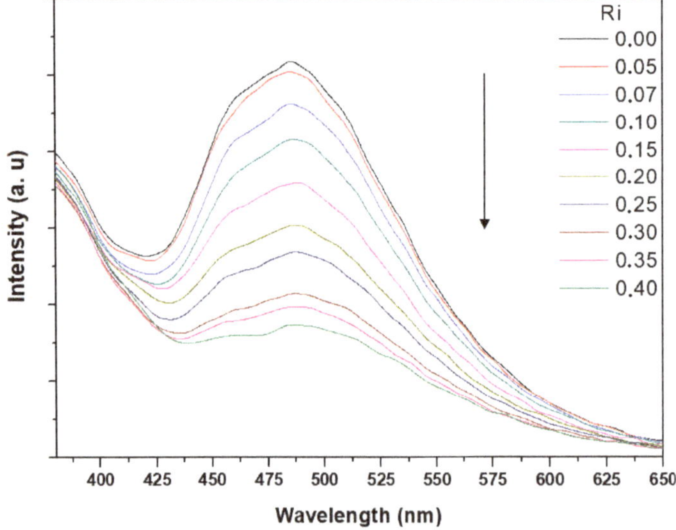

Figure 1. Hoechst 33258 (H33258) displacement assay for metal complex **1**.

A decrease in the fluorescence intensity occurred and fluorescence quenching was observed due to the release of free H33258 molecules from the H33258-DNA adduct, which suggested that **1–4** acted as minor groove binders. The K_{SV} values calculated and shown in Table 2 presented an inverse correlation with the temperature, which indicates a static quenching mechanism, initiated by an adduct formation in the ground state. The quenching efficiency shows the following trend **1** > **2** > **3** > **4**. Thus, the interaction between new complexes **1–4** and ct-DNA is probably by groove binding that is based upon intermolecular interactions, such as electrostatic and/or van der Waals attractions resulting in relatively minor changes to the structure of the double helix [30].

Table 2. Stern–Volmer quenching constant (K_{sv}, L·mol^{-1}) values for the complex-H33528 system.

	T (K)	1	2	3	4
Ksv (10^4)	298	56.71 ± 0.49	39.12 ± 0.01	11.99 ± 0.38	2.76 ± 0.29
	310	32.04 ± 0.14	24.42 ± 2.57	9.71 ± 0.81	2.30 ± 0.41

In order to confirm that no covalent binding occurs between these Ru(II)–azole complexes and DNA, due to the possible lability of the chlorido ligand present in these Ru(II) complexes, two experiments were included. A reaction between complex **3** and guanosine (DNA nitrogen base) was carried out at different times. These reactions were followed by ^{1}H (Figure S3) and ^{31}P{^{1}H} NMR. No changes were observed in the NMR spectra, indicating that non-covalent binding occurred between them. The quantification of ruthenium by inductively coupled plasma optical emission spectrometry (ICP OES) of the final solution of complex **4** with ct-DNA after 24 h of incubation (see experimental section) indicated that the amount of metal bounded to DNA was 75 ± 3 nmol of Ru/mg of DNA (corresponding to 1 atom of Ru per 20 DNA base), which can be taken as evidence that no metal–DNA covalent binding is taking place. These binding levels are lower than those observed for other metal complexes with DNA covalent binding such as cisplatin [31] and other ruthenium complexes [32]. Another family of potential ruthenium–arene complexes containing dipyrrinato ligands showed electrostatic/intercalative interaction with ct-DNA and protein affinity toward bovine serum albumin (BSA) [33].

2.2. Biological Evaluations against Tumor Cell Lines

Antiproliferative assay: The ability of complexes **1–4** and their respective ligands to inhibit the proliferation of selected human tumor and non-tumor cell lines were determined by MTT assay (Table 3). It is worth mentioning that before performing the biological screening, the stability of the complexes was tested using the ^{31}P{^{1}H} NMR technique in DMSO or Tris–HCl solution containing 70% DMSO. After seven days, the spectra of these complexes were the same, when compared with those recorded using fresh solutions (Figure S4). The Ru(II)–azole complexes presented good activities in vitro and were also considerably more active than their free ligand in all the cell lines tested, except for compound binuclear **1,** which was not active against the tumor cells studied, and nor was the free ligand. The present organoruthenium complexes **2–4** showed even higher activity than some of the other reported Ru-*p*-cymene complexes containing KTZ and CTZ. The cytotoxicity of [RuCl$_2$(η^6-*p*-cymene)(L)] or [Ru(η^6-*p*-cymene)(L)(N–N)]$^{2+}$ (L = KTZ, CTZ and N–N = bipyridine, ethylenediamine) type complexes showed IC$_{50}$ values ranging from 4 to 74 μM for KTZ and from 5 to 546 μM for CTZ, against prostate and other tumor cells [17]. It is clear from the comparison that the presence of triphenylphosphine ligand in the complexes showed a better activity profile. Additionally, it is possible to observe that imidazole (KTZ and CTZ) ruthenium derivatives are more active against tumor cells (A549 and MDA-MB-231) than the triazole (FLZ) ruthenium derivative, showing them to be more selective to these tumor cells. To continue the biological studies and investigate the mechanism of cell death in MDA-MB-231 cells, compound **3** was selected.

Cell morphology and colony formation: To investigate the effects on cell morphology, complex **3** was incubated with the MDA-MB-231 cells at different concentrations. Figure 2A depicts the effect of the treatment on the morphology. The untreated control cells appeared phenotypically as spindle-shaped, whereas the cells treated with complex **3**, especially in 0.60 and 1.20 μM, were found to be mostly spherically shaped, demonstrating damaged cell bodies, a loss of adhesion and confluence, where there was a clear concentration-response tendency. This result suggests a clear change in cell morphology induced by complex **3**. The clonogenic cell survival assay determines the ability of a cell to proliferate, thereby retaining its reproductive ability to form a large colony (>50 cells) or a clone after treatment with a cytotoxic agent. Complex **3** at a concentration of 0.06 μM inhibited the number of colonies of MDA-MB-231 and A549 cells, when compared to the control (Figure 2B,C); however, in the

MRC-5 cells, the compound did not significantly inhibit colony formation until concentration 0.6 µM, indicating the selectivity of the compound by tumor cells. The highest concentrations (0.6 µM and 6.0 µM) completely abolished the capacity of breast and lung tumor cells to form colonies. Barr et al. described the ability to inhibit the colony number in JWA-overexpressing BGC823 cells by treatment with 2.7 µM of cisplatin [34]. In the case of treatment with 1 µM and 10 µM of NAMI-A in B16F1 cells, a decrease of cell survival was reported, however no colonies were observed with cells exposed to 100 µM NAMI-A [35]. Thus, it can be concluded that compound **3** is capable of inhibiting the cell colonies more efficiently for tumor cells than non-tumor ones.

Table 3. In vitro cytotoxicity in micromolar concentrations of the complexes in tumor cells A549, DU-145, MDA-MB-231 and non-tumor cell MRC-5, L929 by 48 h.

Compound	A549	DU-145	MDA-MB-231	MRC-5	L929
1	>100	>100	>100	>100	>100
2	2.94 ± 0.73	3.90 ± 0.85	2.35 ± 0.42	2.02 ± 0.10	2.00 ± 0.16
3	0.61 ± 0.07	5.13 ± 0.98	0.63 ± 0.03	1.16 ± 0.01	1.15 ± 0.03
4	0.64 ± 0.04	4.45 ± 0.75	0.62 ± 0.02	1.09 ± 0.06	1.80 ± 0.13
Fluconazole	>100	>100	>100	>100	>100
Clotrimazole	14.47 ± 0.95	15.82 ± 0.23	10.11 ± 2.43	12.70 ± 0.65	9.74 ± 2.05
Ketoconazole	41.85 ± 2.54	47.54 ± 2.53	10.26 ± 1.04	37.50 ± 2.25	16.35 ± 0.47
Cisplatin	14.42 ± 1.45	2.33 ± 0.40	2.44 ± 0.20	23.90 ± 0.70	16.53 ± 2.38

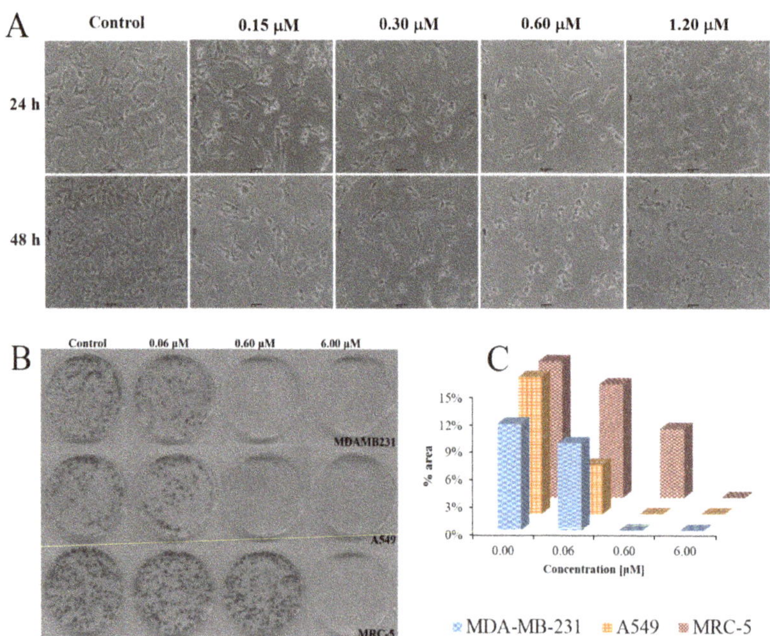

Figure 2. (**A**) Cellular Morphology of MDA-MB-231 treated with complex **3**. (**B**) Effects of complex **3** on MDA-MB-231, A549 and MRC-5 colony formation. Clonogenic assay of untreated cells (control) or treated with complex **3**. (**C**) Graph of colony area quantification.

Cell migration: Cellular migration occurs during physiological and pathological processes that play an important role in the progression of various diseases, including cancer. The Boyden chamber assay is commonly used to measure cell mobility through the PET (Polyester) transwell membrane. In vitro migration assays are necessary to understand mechanism cell migration and identify inhibitory or stimulatory compounds [36]. To investigate the inhibition of cell migration, the invasive breast MDA-MB-231 cells were treated with different concentrations of compound **3** (lower concentrations that IC_{50} value 1.22 ± 0.07 µM, in 24 h). Moderate inhibition of the migration was observed in the MDA-MB-231 cells, as shown in Figure 3, treated with 0.15 µM and 0.30 µM of compound **3**, an average of 30% and 52% respectively, while treatment with 0.60 µM resulted in significant inhibition of migration cells (an average of 97%). Similar results described by Chen et al. demonstrated the inhibition of the migration of MDA-MB-231 cells treated with concentrations of 1.0 to 4.0 µM of [Ru(phen)$_2$-*p*-MOPIP](PF$_6$)$_2$·2H$_2$O (PIP = 2-phenylimidazo[4,5-*f*][1,10]phenanthroline), with –OCH$_3$ on the *p*-site substitution, a type of Ru polypyridyl complex that has been identified as a potent antimetastatic agent [37].

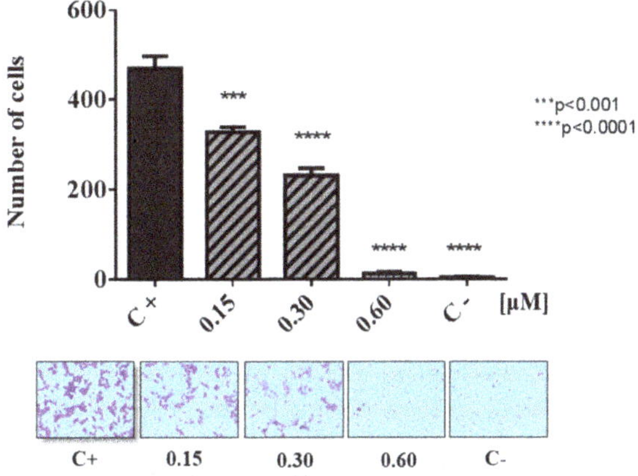

Figure 3. Effects of complex **3** on MDA-MB-231 cell migration in Boyden chambers. Up: Graph of number of cells. Down: The positive control (C+) represents migrating cells without any treatment and the negative control (C−) was cells migrating toward an FBS-free medium.

Cell cycle analysis: The ability of a compound to interfere with the distribution of cell cycle phases can provide information about their mechanism of action. To investigate the mechanism of cell division and cell cycle analysis, cytometry was performed using propidium iodide DNA staining of MDA-MB-231 cells following 24 h of treatment with 0.15, 0.30 and 0.60 µM of compound **3**. Figure 4 shows the percentage of the sub G_1, G_0/G_1, S and G_2/M phases of breast tumor cells treated with compound **3**, indicating a decrease in the number of cells in G_0/G_1 and inducing an accumulation in the number of cells in the sub-G_1 phase, similar to other ruthenium-arene complexes [33]. A small decrease of cells in the S and G_2/M phases was also observed. Identification of a sub-G1 cell population is usually related to apoptotic cells [38], which is a marker of cell death caused by apoptosis, consistent with the morphological observations described above. A large number of studies have shown that anticancer drugs can induce tumor cell apoptosis, which is the main objective of malignant tumor treatment [39].

Cell death analysis: Apoptosis occurs normally during development and aging as a homeostatic mechanism to maintain cell populations in tissues; in apoptotic cells, in addition to the morphological changes discussed above, the membrane phospholipid phosphatidylserine (PS) is translocated from the inner to the outer part of the plasma membrane, thereby exposing PS to the external cellular environment. Annexin V is a recombinant phosphatidylserine-binding protein that interacts strongly and specifically with PS residues and can be used to detect apoptosis [40]. Using a vital dye such as 7-Amino-Actinomycin (7-AAD) can identify early apoptotic cells. The extent of apoptosis for MDA-MB-231 cells caused by different concentrations of complex **3** was investigated by flow cytometry. Figure 5A shows the percentage of live cells (PE AnnexinV$^-$, 7-ADD$^-$, PI$^-$), cells in early apoptosis (PE AnnexinV$^+$, 7-ADD$^-$,

PI$^-$), cells in late apoptosis (PE AnnexinV$^+$, 7-ADD$^+$, PI$^+$) and necrotic cells (PE AnnexinV$^-$, 7-ADD$^-$, PI$^+$) after 24 h of exposure. The total of apoptotic cell populations in Figure 5B is expressed as the sum of percentages of early and late stages of apoptosis. These results indicate that cell death in MDA-MB-231 cells induced by compound **3** is mainly caused by apoptosis in a concentration-dependent way, in agreement with what was observed for the investigations of the cell death pathway of clotrimazole ligand, through apoptosis triggered by the displacement of key glycolytic enzymes in breast cancer cell proliferation [41], as well as for other organometallic Ru–CTZ complexes [17].

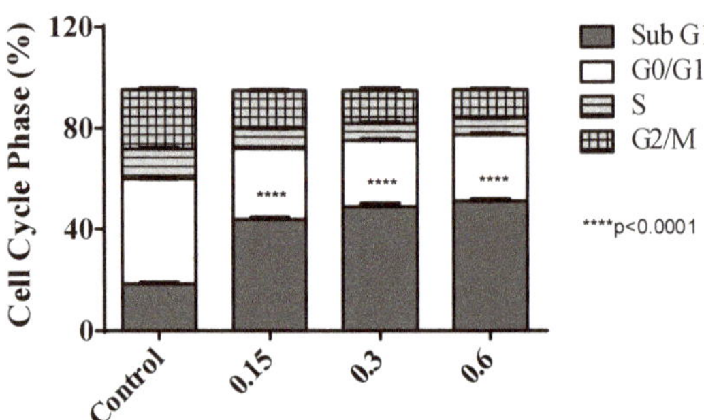

Figure 4. Effects of compound **3** on cell cycle distribution of MDA-MB-231 cells. Cell population percentages of sub-G1, G0/G1, S and G2/M phases were indicated and statistical analyses are shown as averages with indicated standard errors (n = 3).

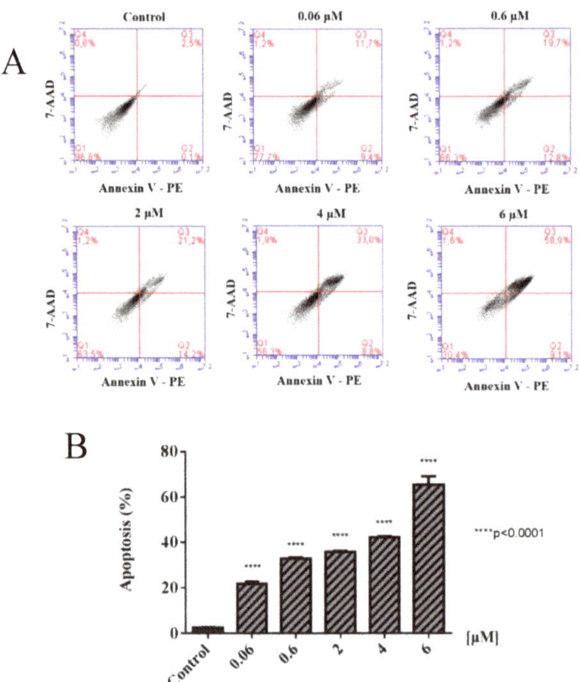

Figure 5. Effect of compound **3** on apoptosis in MDA-MB-231 tumor cells. Camptothecin was used as a positive control for apoptosis. (**A**) Cytometry analysis, the fluorescence of 7AAD is detected in the FL3-A channel and the fluorescence of PE Annexin V is detected in the FL2-A channel. (**B**) Bar chart showing percentages of apoptotic cells.

3. Materials and Methods

All the syntheses of the complexes were performed under an argon atmosphere. The ct-DNA, Hoechst 33258 and HSA were purchased from Sigma-Aldrich (St. Louis, MO, USA). Starting materials [RuCl$_2$(η^6-*p*-cymene)]$_2$ and [RuCl$_2$(η^6-*p*-cymene)(PPh$_3$)] were prepared following the method described in the literature [42,43]. The organoruthenium complexes **1–4** were synthetized and characterized previously by us [19].

3.1. Biomolecules Interaction

Fluorescence measurements with HSA: HSA solutions were prepared in a Tris–HCl buffer (5 mM Tris–HCl and 50 mM NaCl, pH 7.4). A quantitative analysis of the potential interaction between the complexes and HSA was performed by fluorimetric titration (excitation at 280 nm and emission at 305 nm) monitored at different temperatures (295 and 310 K) in 96-well plates used for fluorescence assays. The HSA concentration was kept constant in all samples, while the complex concentration was increased from 100 to 0.78 µM. The experiments were carried out in triplicate on a Synergy H1 Multi-Mode Reader (BioTek, Winooski, WI, USA). The equations used to determine the constant interaction between the albumin and the studied Ru(II)–azole complexes are in the supplementary material.

Interaction studies with DNA: Hoechst 33258 displacement assays were recorded from 385 to 650 nm with an excitation wavelength of 343 nm for solutions of ct-DNA (50.0 µM), Hoechst 33258 (5.0 µM), and complexes dissolved in DMSO (0–100 µM).

DNA interaction by spectroscopic titration, viscosity, gel electrophorese, circular dichroism and reaction with guanosine were performed following the same procedure as reported previously [16].

For the determination of total metal content per mg of DNA, 10 mL of complex **4** (0.9 mM) was mixed with 10 mL of ct-DNA (4.45 mM) obtained a molar ratio of 0.2. After 24 h of incubation, the DNA was precipitated by adding 30 mL of EtOH and 1 mL of NaCl (2 M) for 10 min. At this point, the tubes were centrifuged (30 min. at 5000 rpm), the supernatant decanted, and the DNA was resuspended in water. This precipitation–resuspension cycle was repeated twice and the final DNA was solubilized in water [31]. The DNA quantification was determinate by electronic absorption measurements (Hewlett-Packard diode array-8452A, HP, Palo Alto, CA, USA). The Ru content was analyzed by IPC emission spectrometry (icap 6000 ICP OES, Thermo Fisher Scientific, Waltham, MA, USA), assisting the digestion of the sample by microwave (Berghof MW, speedwave® DIRC, Eningen, Germany). Instrument settings, data acquisition parameters and Microwave-assisted acid digestion specifications are in the supplementary material.

3.2. Biological Evaluations

Cell culture, viability and morphological observations: The cell lines were obtained from the American Type Culture Collection (ATCC), human tumor cells: A549 (lung, ATCC CCL-185), MDA-MB-231 (breast, ATCC HTB-26), DU-145 (prostate, ATCC HTB-81), non-tumor cell lines: MRC-5 (lung, ATCC CCL-171) and L929 (mouse fibroblast, ATCC CCL-1) and were cultivated under sterile conditions in Dulbecco's Modified Eagle's medium (DMEM, A549, MDA-MB-231, L929 and MRC-5) or RPMI 1640 (DU-145) supplemented with 10% of fetal bovine serum (FBS), at 37 °C in a humidified 5% CO$_2$ atmosphere. The growth, confluence and the morphology of the cells were observed using an inverted microscope (Nikon, T5100, Minato, Japan) and microscopic images were captured using a camera (Motic Moticam 1000, Kowloon Bay, Hong Kong, China). For antiproliferative assays, cells (1.5 × 10^4 cells) were seeded in flat-bottom 96-well plates, allowed to attach for 24 h prior to adding the compounds dissolved in DMSO in eight different concentrations in noneplicate. Cells were incubated at the time required, and then cell proliferation was determined by the MTT-reduction method. Briefly, 30 µL/well of MTT (1 mg/mL) was added and plates were incubated for 1–4 h at 37 °C. Finally, the formazan crystal was dissolved by adding 100 µL/well of isopropanol and quantified at 540 nm using a multi scanner microplate absorbance reader (Labtech, LT 4000, Heathfield, England). For the morphological study, 0.8×10^5 cells/well were seeded into 12-well plates. After allowing 24 h to adhere, images of cells treated with or without compounds were taken at 0, 24 and 48 h.

Clonogenic assay: Growing tumor and non-tumor cells were harvested, counted and seeded (300 cells) into Petri dishes. Cells were allowed to grow at 37 °C in a humidified 5% CO_2 atmosphere overnight and then treated with different concentrations of the compounds **3** for 48 h. At this time point, the medium was changed without any compound. After incubation for an additional 10 days, the cells were rinsed with PBS, fixed with a solution 3:1 of methanol: acetic acid for 5 min and stained with 0.5% crystal violet for 25 min. Relative survival was calculated from the number of single cells that formed colonies of >50 cells on the 10th day. The covert area by the colonies was calculated with the Image J software.

Cell migration: The migration assay was performed using a 24-well chamber (BD Biosciences, Franklin Lakes, NJ, USA). MDA-MB-231 cells (0.5×10^5/well) were resuspended in FBS free medium and were added to the upper chamber, together with different concentrations of compound **3**. In the lower chamber, only the medium with FBS was added. The negative control contained FBS free medium. The chamber incubated for 22 h at 37 °C and 5% CO_2. Then, the cells that did not migrate to the upper surface were removed and the cells attached to the lower section were fixed with methanol, stained with toluidine blue, and washed with distilled water. Images of migrated cells were captured and counted using Image J software.

Cell cycle analysis: Cell cycle arrest was evaluated by flow cytometry. Briefly, 1.5×10^5 cells were seeded into 12-well plates and incubated for 24 h. Compound **3** (0.15, 0.30 and 0.60 μM) was added to the wells and incubated for 24 h. Then, the cells were trypsinized, collected and harvested in cold PBS and fixed in ethanol (70%) at -20 °C overnight. After this period, the cells were centrifuged; the supernatant was discarded and RNase A was added (0.2 mg·mL^{-1}) for 30 min at 37 °C. Then, the cells were stained with hypotonic fluorochrome solution (PI 5 μg·mL^{-1}, sodium citrate 0.1% and Triton-X-100 0.1%) for 1 h. Finally, the cells were analyzed by flow cytometry (Accuri C6 BD Biosciences).

Cell death analysis: The apoptosis-mediated cell death of MDA-MB-231 cells was examined by using the Annexin V-FITC Apoptosis Detection Kit (BD Biosciences), according to the manufacturer's instructions. MDA-MB-231 cells were treated with compound **3** (6.0, 4.0, 2.0, 0.60, 0.06 μM) for 24 h. In brief, 1.0×10^5 cells were harvested and washed with PBS and resuspended in 200 μL binding buffer. Next, 2.5 μL of Annexin V-FITC and 2.5 μL of PI were added. Flow cytometric analysis was performed immediately after supravital staining. The reading was performed in Accuri C6 flow cytometer (BD Biosciences) and fluorescence emitted by each dye was quantified using CellQuest software (BD Biosciences). The criteria for positivity in cells in the early stages of apoptosis were Annexin V positive and PI negative, whereas the criteria for cells in the late stages of apoptosis were Annexin V positive and PI positive.

4. Conclusions

Complexes **2–4** bind to HSA with moderate affinity based on the magnitude of their K_b (1.78–$7.90 \times 10^5 \text{ M}^{-1}$). The thermodynamic parameters helped to understand the interaction between complexes **2–4** and the HSA molecule. The negative values of ΔG support the assertion that the binding process between complex/HSA is spontaneous. The positive ΔH and ΔS values of the interaction of all complexes with HSA indicate that the electrostatic interactions played a major role in the binding interaction of complex/HSA. A minor groove binding is the interaction suggested between these ruthenium(II)-azole complexes and DNA. The biological results showed that complex **2** displayed a very good activity against the tumor cells studied here. Moreover, it was demonstrated that ruthenium(II)-azole complexes **3–4** are significantly active in the inhibition of the proliferation of selected human tumor cells at a very low level of concentration. The phosphine used as an auxiliary ligand seems to have a significant effect on the remarkable biological activity observed for complexes

3 and **4**. Further studies will be conducted to establish a better structure–activity correlation and to identify the main target of these promising metallodrugs.

Supplementary Materials: The following are available online at http://www.mdpi.com/2304-6740/6/4/132/s1, Figure S1: Spectrofluorometric titration spectra of HSA with ruthenium compounds **2–4**. Figure S2: ^{1}H NMR spectrum of guanosine and complex **3** at different times (only resonances of H8, NH and NH2 are assignments). Figure S3: Hoechst 33258 (H33258) displacement assay for metal complexes **2–4**. Figure S4: ^{31}P{^{1}H} NMR spectrum in the mixture 70:30 DMSO: Culture medium (DMEM) of complex **3** at different times.

Author Contributions: L.C-V., K.M.O. and B.N.C. acquired and analyzed all data. L.C-V., M.R.C., M.N. and A.A.B. writing, review and editing.

Funding: Legna Colina-Vegas thanks FAPESP for her postdoctoral fellowship (grant #2016/23130-5 and 2017/23254-9).

Acknowledgments: The authors are grateful to the São Paulo Research Foundation (FAPESP), the National Counsel of Technological and Scientific Development (CNPq) and the Coordination for the Improvement of Higher Level-or Education-Personnel (CAPES) for their financial support.

Conflicts of Interest: The authors declare no conflict of interest.

References

1. *World Cancer Report 2014*; World Health Organization: Geneva, Switzerland, 2014; ISBN 9283204298.
2. Steeg, P.S. Targeting metastasis. *Nat. Rev. Cancer* **2016**, *16*, 201–218. [CrossRef] [PubMed]
3. Farrell, N.P. Multi-platinum anti-cancer agents. Substitution-inert compounds for tumor selectivity and new targets. *Chem. Soc. Rev.* **2015**, *44*, 8773–8785. [CrossRef] [PubMed]
4. Bergamo, A.; Sava, G. Ruthenium anticancer compounds: Myths and realities of the emerging metal-based drugs. *Dalton Trans.* **2011**, *40*, 7817–7823. [CrossRef] [PubMed]
5. Meier-Menches, S.M.; Gerner, C.; Berger, W.; Hartinger, C.G.; Keppler, B.K. Structure–activity relationships for ruthenium and osmium anticancer agents—Towards clinical development. *Chem. Soc. Rev.* **2018**, *47*, 909–928. [CrossRef] [PubMed]
6. Su, W.; Tang, Z.; Li, P. Development of arene ruthenium antitumor complexes. *Mini Rev. Med. Chem.* **2016**, *16*, 787–795. [CrossRef] [PubMed]
7. Zeng, L.; Gupta, P.; Chen, Y.; Wang, E.; Ji, L.; Chao, H.; Chen, Z.S. The development of anticancer ruthenium(II) complexes: From single molecule compounds to nanomaterials. *Chem. Soc. Rev.* **2017**, *46*, 5771–5804. [CrossRef] [PubMed]
8. Su, W.; Li, Y.; Li, P. Design of Ru-arene Complexes for Antitumor Drugs. *Mini Rev. Med. Chem.* **2018**, *18*, 184–193. [CrossRef]
9. Babak, M.B.; Ang, W.H. *Metallo-Drugs: Development and Action of Anticancer Agents*; Sigel, A., Sigel, H., Freisinger, E., Sigel, R.K.O., Eds.; Walter de Gruyter: Berlin, Germany, 2018; Volume 18, Chapter 6, p. 161.
10. Merlino, A. Interactions between proteins and Ru compounds of medicinal interest: A structural perspective. *Coord. Chem. Rev.* **2016**, *26*, 111–134. [CrossRef]
11. Nowak-Sliwinska, P.; van Beijnum, J.R.; Casini, A.; Nazarov, A.A.; Wagnieres, G.; van den Bergh, H.; Dyson, P.J.; Griffioen, A.W. Organometallic Ruthenium(II) Arene Compounds with Antiangiogenic Activity. *J. Med. Chem.* **2011**, *54*, 3895–3902. [CrossRef]
12. Pagano, N.; Maksimoska, J.; Bregman, H.; Williams, D.S.; Webster, R.D.; Xue, F.; Meggers, E. Ruthenium half-sandwich complexes as protein kinase inhibitors: Derivatization of the pyridocarbazole pharmacophore ligand. *Org. Biomol. Chem.* **2007**, *5*, 1218–1227. [CrossRef]
13. Hildebrandt, J.; Görls, H.; Häfner, N.; Ferraro, G.; Dürst, M.; Runnebaum, I.B.; Weigand, W.; Merlino, A. Unusual mode of protein binding by a cytotoxic π-arene ruthenium(II) piano-stool compound containing an O,S-chelating ligand. *Dalton Trans.* **2016**, *45*, 12283–12287. [CrossRef] [PubMed]
14. Kurzwernhart, A.; Kandioller, W.; Bartel, C.; Bächler, S.; Trondl, R.; Mühlgassner, G.; Jakupec, M.A.; Arion, V.B.; Marko, D.; Keppler, B.K.; et al. Targeting the DNA-topoisomerase complex in a double-strike approach with a topoisomerase inhibiting moiety and covalent DNA binder. *Chem. Commun.* **2012**, *48*, 4839–4841. [CrossRef] [PubMed]

15. Martinez, A.; Carreon, T.; Iniguez, E.; Anzellotti, A.; Sanchez, A.; Tyan, M.; Sattler, A.; Herrera, L.; Maldonado, R.A.; Sánchez-Delgado, R. Searching for new chemotherapies for tropical diseases: Ruthenium-clotrimazole complexes display high *in vitro* activity against Leishmania major and Trypanosoma cruzi and low toxicity toward normal mammalian cells. *J. Med. Chem.* **2012**, *55*, 3867–3877. [CrossRef] [PubMed]

16. Colina-Vegas, L.; Lucena Dutra, J.; Villarreal, W.; Neto, J.H.; Cominetti, M.R.; Pavan, F.; Navarro, M.; Batista, A.A. Ru(II)/clotrimazole/diphenylphosphine/bipyridine complexes: Interaction with DNA, BSA and biological potential against tumor cell lines and Mycobacterium tuberculosis. *J. Inorg. Biochem.* **2016**, *162*, 135–145. [CrossRef] [PubMed]

17. Robles-Escajeda, E.; Martínez, A.; Varela-Ramirez, A.; Sánchez-Delgado, R.A.; Aguilera, R.J. Analysis of the cytotoxic effects of ruthenium–ketoconazole and ruthenium–clotrimazole complexes on cancer cells. *Cell Biol. Toxicol.* **2013**, *29*, 431–443. [CrossRef] [PubMed]

18. Bae, S.H.; Park, J.H.; Choi, H.G.; Kim, H.; Kim, S.H. Imidazole Antifungal Drugs Inhibit the Cell Proliferation and Invasion of Human Breast Cancer Cells. *Biomol. Ther.* **2018**, *26*, 494–502. [CrossRef]

19. Colina-Vegas, L.; Coutinho, T.; Correa, R.S.; de Souza, W.; Rodrigues, J.C.F.; Batista, A.A.; Navarro, M. Antiparasitic activity and ultrastructural alterations provoked by organoruthenium complexes against Leishmania amazonensis. *New J. Chem.*, under review.

20. Lakowicz, J.R. *Principles of Fluorescence Spectroscopy*, 3rd ed.; Springer: New York, NY, USA, 2006; ISBN 978-0-38-746312-4.

21. Sun, J.; Huang, Y.; Zheng, C.; Zhou, Y.; Liu, Y.; Liu, J. Ruthenium (II) complexes interact with human serum albumin and induce apoptosis of tumor cells. *Biol. Trace Elem. Res.* **2015**, *163*, 266–274. [CrossRef]

22. Ross, D.; Subramanian, S. Thermodynamics of protein association reactions: Forces contributing to stability. *Biochemistry* **1981**, *20*, 3096–3102. [CrossRef]

23. Huang, S.; Peng, S.; Zhu, F.; Lei, X.; Xiao, Q.; Su, W.; Liu, Y.; Huang, C.; Zhang, L. Multispectroscopic investigation of the interaction between two ruthenium(II) arene complexes of curcumin analogs and human serum albumin. *Biol. Trace Elem. Res.* **2016**, *169*, 189–203. [CrossRef]

24. Colina-Vegas, L.; Luna-Dulcey, L.; Plutín, A.M.; Castellano, E.E.; Cominetti, M.R.; Batista, A.A. Half sandwich Ru(II)-acylthiourea complexes: DNA/HSA-binding, anti-migration and cell death in a human breast tumor cell line. *Dalton Trans.* **2017**, *46*, 12865–12875. [CrossRef] [PubMed]

25. Bishop, G.R.; Chaires, J.B. Characterization of DNA structures by circular dichroism. *Curr. Protoc. Nucleic Acid Chem.* **2003**, *11*, 7–11. [CrossRef]

26. Han, M.J.; Duan, Z.M.; Hao, Q.; Zheng, S.Z.; Wang, K.Z. Molecular Light Switches for Calf Thymus DNA Based on Three Ru(II) Bipyridyl Complexes with Variations of Heteroatoms. *J. Phys. Chem. C* **2007**, *111*, 16577–16585. [CrossRef]

27. Fu, X.B.; Liu, D.D.; Lin, Y.; Hu, W.; Mao, Z.W.; Le, X.Y. Water-soluble DNA minor groove binders as potential chemotherapeutic agents: Synthesis, characterization, DNA binding and cleavage, antioxidation, cytotoxicity and HSA interactions. *Dalton Trans.* **2014**, *43*, 8721–8737. [CrossRef] [PubMed]

28. Sabnis, R.W. *Handbook of Biological Dyes and Stains: Synthesis and Industrial Applications*, 1st ed.; Wiley: New York, NY, USA, 2010; ISBN 978-0-47-040753-0.

29. Fornander, L.H.; Wu, L.; Billeter, M.; Lincoln, P.; Nordén, B. Minor-Groove binding drugs: Where is the second Hoechst 33258 molecule? *J. Phys. Chem. B* **2013**, *117*, 5820–5830. [CrossRef] [PubMed]

30. Pages, B.J.; Ang, D.L.; Wrightb, E.P.; Aldrich-Wright, A.R. Metal complex interactions with DNA. *Dalton Trans.* **2015**, *44*, 3505–3526. [CrossRef] [PubMed]

31. Navarro, M.; Castro, W.; Higuera-Padilla, A.R.; Sierraalta, A.; Abad, M.J.; Taylor, P.; Sánchez-Delgado, R.A. Synthesis, characterization and biological activity of *trans*-platinum(II) complexes with chloroquine. *J. Inorg. Biochem.* **2011**, *105*, 1684–1691. [CrossRef]

32. Frik, M.; Martínez, A.; Elie, B.T.; Gonzalo, O.; de Mingo, D.R.; Sanau, M.; Sanchez-Delgado, R.A.; Sadhukha, T.; Prabha, S.; Ramos, J.W.; et al. *In vitro* and *in vivo* Evaluation of water-soluble iminophosphorane ruthenium(II) compounds. A potential chemotherapeutic agent for triple negative breast cancer. *J. Med. Chem.* **2014**, *57*, 9995–10012. [CrossRef]

33. Gupta, K.R.; Kumar, A.; Paitandi, R.P.; Singh, R.S.; Mukhopadhyay, S.; Verma, S.P.; Dasb, P.; Pandey, P.S. Heteroleptic arene Ru(II) dipyrrinato complexes: DNA, protein binding and anti-cancer activity against the ACHN cancer cell line. *Dalton Trans.* **2016**, *45*, 7163–7177. [CrossRef]

34. Barr, M.P.; Gray, S.G.; Hoffmann, A.C.; Hilger, R.A.; Thomale, J.; O'Flaherty, J.D.; Fennell, D.A.; Richard, D.; O'Leary, J.J.; O'Byrne, K.J. Generation and characterization of cisplatin-resistant non-small cell lung cancer cell lines displaying a stem-like signature. *PLoS ONE* **2013**, *8*, e54193. [CrossRef]

35. Bicek, A.; Turel, I.; Kanduser, M.; Miklavcic, D. Combined therapy of the antimetastatic compound NAMI-A and electroporation on B16F1 tumour cells *in vitro*. *Bioelectrochemistry* **2007**, *71*, 113–117. [CrossRef] [PubMed]

36. Chen, H.C. Cell Migration. In *Developmental Methods and Protocols*, 1st ed.; Guan, J.L., Ed.; Humana Press: Totowa, NJ, USA, 2005; pp. 15–22. ISBN 978-1-59-259860-1.

37. Cao, W.; Zheng, W.; Chen, T. Ruthenium Polypyridyl Complex Inhibits Growth and Metastasis of Breast Cancer Cells by Suppressing FAK signaling with Enhancement of TRAIL-induced Apoptosis. *Sci. Rep.* **2015**, *5*, 9157. [CrossRef] [PubMed]

38. Huang, X.; Halicka, H.D.; Traganos, F.; Tanaka, T.; Kurose, A.; Darzynkiewicz, Z. Cytometric assessment of DNA damage in relation to cell cycle phase and apoptosis. *Cell Prolif.* **2005**, *38*, 223–243. [CrossRef] [PubMed]

39. Elmore, S. Apoptosis: A review of programmed cell death. *Toxicol Pathol.* **2007**, *35*, 495–516. [CrossRef] [PubMed]

40. Schutte, B.; Nuydens, R.; Geerts, H.; Ramaekers, F. Annexin V binding assay as a tool to measure apoptosis in differentiated neuronal cells. *J. Neurosci. Methods* **1998**, *86*, 63–69. [CrossRef]

41. Glass-Marmor, L.; Morgenstern, H.; Beitner, R. Calmodulin antagonists decrease glucose 6-bisphosphate, fructose 6-bisphosphate, ATP and viability of melanoma cells. *Eur. J. Pharmacol.* **1996**, *313*, 265–271. [CrossRef]

42. Bennett, M.A.; Huang, T.N.; Matheson, T.W.; Smith, A.K. (η^6-Hexamethylbenzene) ruthenium Complexes. *Inorg. Synth.* **1982**, *21*, 74–78. [CrossRef]

43. Hodson, E.; Simpson, S.J. Synthesis and characterization of [(η^6-cymene)Ru(L)X2] compounds: Single crystal X-ray structure of [(η^6-cymene)Ru(P{OPh}$_3$)Cl$_2$] at 203 K. *Polyhedron* **2004**, *23*, 2695–2707. [CrossRef]

 *inorganics*

Article

Investigation of 1-Methylcytosine as a Ligand in Gold(III) Complexes: Synthesis and Protein Interactions

James Beaton and Nicholas P. Farrell *

Department of Chemistry, Virginia Commonwealth University, Richmond, VA 23284-2006, USA; beatonjf@vcu.edu
* Correspondence: npfarrell@vcu.edu; Tel.: +1-804-828-6320

Received: 5 November 2018; Accepted: 15 December 2018; Published: 20 December 2018

Abstract: The HIV nucleocapsid protein NCp7 was previously shown to play a number of roles in the viral life cycle and was previously identified as a potential target for small molecule intervention. In this work, the synthesis of the previously unreported complexes $[Au(dien)(1MeCyt)]^{3+}$, $[Au(N\text{-}Medien)(1MeCyt)]^{3+}$, and $[Au(dien)(Cyt)]^{3+}$ is detailed, and the interactions of these complexes with the models for NCp7 are described. The affinity for these complexes with the target interaction site, the "essential" tryptophan of the C-terminal zinc finger motif of NCp7, was investigated through the use of a fluorescence quenching assay and by ^{1}H-NMR spectroscopy. The association of $[Au(dien)(1MeCyt)]^{3+}$ as determined through fluorescence quenching is intermediate between the previously reported DMAP and 9-EtGua analogs, while the associations of $[Au(N\text{-}Medien)(1MeCyt)]^{3+}$ and $[Au(dien)(Cyt)]^{3+}$ are lower than the previously reported complexes. Additionally, NMR investigation shows that the self-association of relevant compounds is negligible. The specifics of the interaction with the C-terminal zinc finger were investigated by circular dichroism spectroscopy and electrospray-ionization mass spectrometry. The interaction is complete nearly immediately upon mixing, and the formation of Au_xF^{n+} ($x = 1$, 2, or 4; F = apopeptide) concomitant with the loss of all ligands is observed. Additionally, oxidized dimerized peptide was observed for the first time as a product, indicating a reaction via a charge transfer mechanism.

Keywords: gold; zinc finger proteins; 1-methylcytosine; gold fingers; fluorescence quenching; π–π stacking; protein-DNA recognition

---

1. Introduction

Gold compounds have a long history in medicinal chemistry. The thiophilic nature of gold suggests sulfur-rich proteins, such as thioredoxin, as reasonable cellular targets for gold action [1]. Au(III) is quite labile in its chemistry and is susceptible to reduction to Au(I) [1]. Apart from the inherent difficulty in understanding speciation of an Au(III) complex in a biological medium, nonselective interactions are also highly likely. This is especially true when highly nucleophilic sulfur sites in proteins are considered as viable targets. Formally, substitution-inert coordination spheres, such as MN_4 or MN_3L (where M = Pt, Au and N_4 may be an N_3 chelate such as diethylenetriamine (dien) and L a purine or pyrimidine ligand), have been explored by our group for greater specificity with biomolecules [2]. In principle, we suggested that highly nucleophilic cysteines of biomolecules can be targeted using weak nucleophiles, such as platinum-nucleobase PtN_4 complexes for selective peptide reactions [2]. Although Au(III) is more chemically reactive than Pt(II), the same principle applies [3]. A complementary approach to enhance selectivity in metal ion-biomolecule interactions involves using an inherent property of the protein along with appropriate design of the small molecule. A two-stage approach to selectivity can then be envisaged: Molecular recognition followed by "fixation", where the inorganic moiety forms covalent linkages to the protein of interest. In this contribution, we summarize how this approach might work using formally substitution-inert platinum-metal nucleobase complexes and a tryptophan-containing zinc finger peptide from the HIV nucleocapsid protein HIVNCp7. Specifically, we describe the synthesis and properties of Au(III)-1-MeCytosine complexes and their interactions with the C-terminal NCp7-F2 targeting the essential tryptophan of the protein.

Zinc binding proteins are one of the most common types of metalloprotein and make up as much as 10% of proteins found in humans [4]. In these proteins, zinc can serve a catalytic role, a structural role, and even a recently discovered inhibitory role [5,6]. In zinc-finger proteins, the zinc serves a structural role, binding in a tetrahedral geometry to four surrounding amino acids [7]. These amino acids are generally a combination of cysteines and histidines, and take the forms Cys_2His_2, $Cys_2HisCys$ (Cys_3His), or Cys_4. The modification of the zinc-coordinating residues can have a dramatic effect on protein function [8]. The HIV nucleocapsid protein (NCp7), a 55 amino acid protein that contains two Cys_3His zinc fingers (Figure 1), has been identified as a potential target for chemical intervention through the modification of these structural elements. It was previously shown that alteration of the zinc coordination sphere significantly hinders viral replication [5]. NCp7 interacts with its polynucleotide substrate, viral RNA, through guanine stacking with both W37 on the second C-terminal zinc finger motif, as well as through the phenylalanine (F16) located on the first zinc finger [5,9].

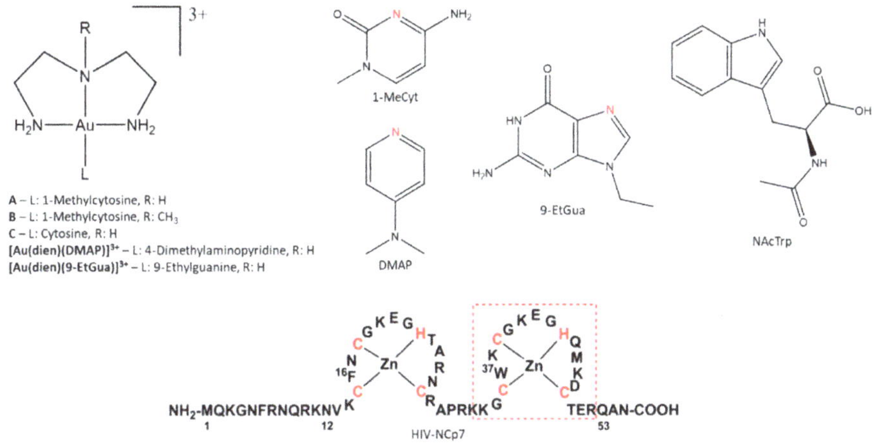

Figure 1. Structures of the complexes studied. Metal binding sites are highlighted in red.

NCp7 recognition by small molecules can be optimized through a noncovalent π–π stacking interaction between the "essential" tryptophan residue and metallated nucleobases, analogous to the RNA(DNA) interaction. Previously reported gold (III) complexes based on [Au(dien)L]$^{3+}$ (dien = diethylenetriamine; L = *N*-heterocycle) were investigated for their ability to stabilize the Au(III) oxidation state, as well as their ability to eject zinc from zinc-finger proteins [10–13]. In the case of coordination complexes, this interaction brings the metal center in close contact with the highly nucleophilic zinc-bound cysteine residues, allowing for a secondary electrophilic attack by the metal center, resulting in zinc ejection [2,10]. Changing the *N*-heterocyclic ligand can be used to fine-tune the electronic properties of the metal center in order to impart higher selectivity for the protein of interest. To this end, the [Au(dien)L]$^{3+}$ coordination sphere was studied with the previously uninvestigated ligand 1-methylcytosine (1-MeCyt). Compared to the *N*-heterocycles 4-Dimethylaminopyridine (DMAP) and 9-Etylguanine (9-EtGua), the free ligand 1-MeCyt has a higher association constant with tryptophan [14]. Additionally, 1-MeCyt has an intermediate basicity (pK_a = 4.45) relative to DMAP (pK_a = 9.1) and 9-EtGua ($pK_{a(N7)}$ = 2.7) [15]. Previously, basicity of the heterocycle was thought to be proportional to association constant in the [Au(dien)(*N*-heterocycle)]$^{3+}$ motif [10]. The [Au(dien)(1-MeCyt)]$^{3+}$ is, to our knowledge, the first 1-MeCytosine compound of Au(III) to be synthesized, and it completes the series of Pt(II) and Pd(II) complexes [M(dien)L]$^{n+}$ (L = 9-EtGua, 1-MeCyt) studied for their isostructural and isoelectronic similarities [14,16,17]. We report the synthesis of the [Au(dien)(1-MeCyt)]$^{3+}$ (**A**), [Au(*N*-Medien)(1-MeCyt)]$^{3+}$ (**B**), and [Au(dien)(Cyt)]$^{3+}$ (**C**) (Figure 1). The associative interaction with *N*-Acetyl-L-tryptophan (NAcTrp) as a model for the recognition site of NCp7 (W37) was probed through the use of a fluorescence quenching assay. Additionally, the interactions with the C-terminal zinc finger of NCp7 (Figure 1) were investigated by mass spectrometry and circular dichroism spectroscopy.

2. Results

2.1. Synthesis and Characterization

The previously unreported complexes **A**, **B**, and **C** (Figure 1) were successfully synthesized in accordance with previously published methods [10]. The ^{1}H-NMR spectra of the complexes **A** and **B** show multiple protonation states with varying pH* (Figure S1, see Supplementary Materials), analogous to previously reported 9-EtGua analogs [10]. As 1-MeCyt does not have a labile proton, the changes in the shifts observed with changing pH may be due to the previously reported deprotonation of the dien-chelate, likely at the central nitrogen in the case of **A**, and at the terminal nitrogen in the case of **B** (pK$_a$ = 3.3 for 9-EtGua analog) [10]. By analogy with the Pd and Pt complexes, [14], the N3 is assigned as the binding site to Au(III). Unique to the 1-MeCyt complexes, the signals for the aromatic H6 protons split. This splitting of the aromatic protons is not observed in the complex of **C** nor in the platinated analog, and highlights the increased electronegativity of Au(III) relative to Pt(II) [3,17].

The mass spectra of **A** and **B** are typical of many coordination complexes because the parent ion peak is nearly indistinguishable from the background. However, the Au(III)-diethylenetriamine backbone bound to fragments of 1-MeCyt is observed, indicating that the desired product had been obtained (Figure S2). Synthesis of the ribosylated analog of [Au(Dien)(Cyd)]$^{3+}$ (Cyd = cytidine) was attempted, but the recovered yellow product quickly turned purple, indicating the formation of colloidal gold (reduction of the metal center to Au(0)).

2.2. Determination of Tryptophan Affinity by Fluorescence Quenching

Complexes **A**, **B**, and **C** were titrated into a solution of NAcTrp, and fluorescence quenching was used to determine the affinity (Figure 2). The data was found to be statistically significant (*p*-values = 4×10^{-4}, 5×10^{-5}, and 9×10^{-10} for **A**, **B**, and **C**, respectively). Complex **B** shows a lower affinity for NAcTrp (K$_a$ = 13.0 $\times$ 10^3 M^{-1}) than the DMAP and 9-EtGua analogs. The association constant for **A**, as determined by tryptophan quenching (24.2 $\times$ 10^3 M^{-1}), is intermediate between that of the DMAP and 9-EtGua analogs, while C has a lower association constant (9.31 $\times$ 10^3 M^{-1}). The synthesis and study completes the isoelectronic and isostructural series [M(dien)L]$^{n+}$, where M = Pd(II), Pt(II), and Au(III), and L = purine, 9-Ethylguanine, or pyrimidine, 1-meCytosine [10,17]. Figure 2 shows how, for this closely related series, the electronic factors from individual metal ions can be modulated to control the tryptophan affinity to some extent [14].

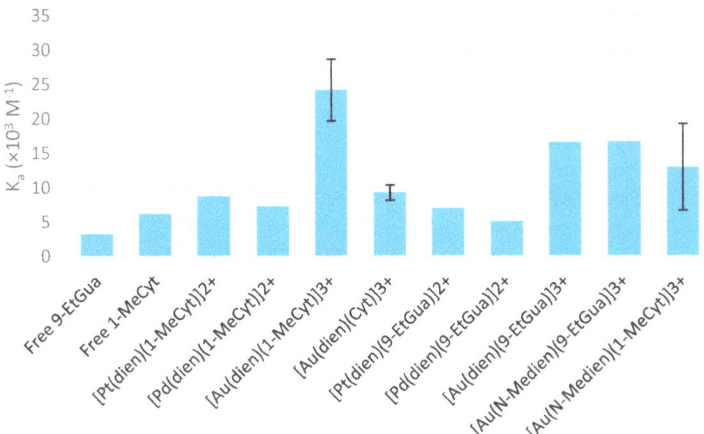

Figure 2. Association constants for [M(dien)(nucleobase)]$^{n+}$ determined by fluorescence quenching of NAcTrp [10,16,17].

2.3. Interactions With Zinc Finger Proteins

Circular dichroism (CD) can be used to gain qualitative information about the structure of the zinc-finger protein. As previously described, the zinc-finger structural motif, shown in the second finger of HIV-NCp7, has a characteristic positive band at 220 nm and a negative band at 195–200 nm [18]. The CD spectrum of the C-terminal zinc-finger motif of NCp7 (NCp7-F2) was recorded before and after incubation with one equivalent of **A**, and the interaction was monitored for 4 h (Figure 3).

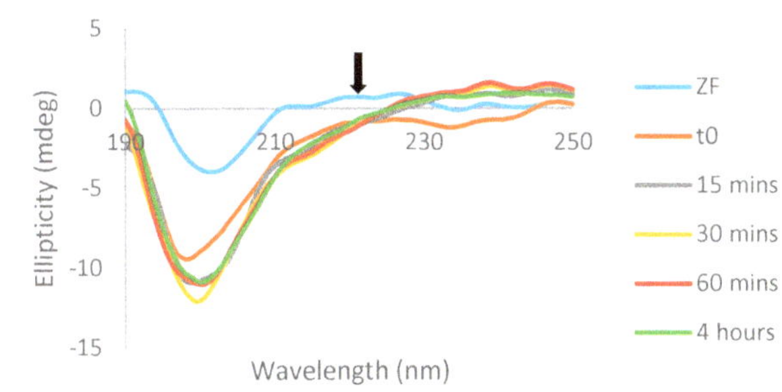

Figure 3. Circular dichroism (CD) spectrum of a 1:1 mixture of the C-terminal zinc-finger of NCp7 and **A**, monitored over 4 h. The characteristic positive band is indicated by an arrow.

Similarly to all previously studied gold(III)-dien complexes, the interaction with **A** results in rapid zinc ejection. The decrease seen in the positive band at 220 nm and the increase seen in the negative band at 200 nm upon interaction are indicative of random coil, and are similar to the previously reported CD spectrum of the apopeptide [11]. Thus, it is assumed that the zinc was ejected very rapidly. Additionally, the spectrum does not change over the remainder of the next 4 h, indicating that the reaction is complete after 15 min. After a 15 min incubation with **B**, the result was also zinc ejection (data not shown).

The mass spectra of a 1:1 mixture of NCp7-F2 and **A** were recorded immediately after mixing, after 6 h, and after 24 h (Figure 4a, Figures S3 and S4). The spectra did not change substantially from the initial timepoint to the final measurement, supporting the results obtained in the CD spectra and in the previous literature that the interaction between Au(III) complexes and zinc-finger peptides occurs rapidly. At the time of the initial measurement, the zinc-coordinated peptide is not observed. Additionally, all Au-bound ligands are lost, and only the well-described "gold fingers" are observed [11]. Replacement of zinc by one, two, and four atoms of gold was observed: 605.4805 *m/z* (4+), 438.1595 *m/z* (6+), and 1003.2671 *m/z* (3+), respectively. This is the first instance that incorporation of four gold atoms has been observed upon interaction with a single zinc-finger motif. Oxidized apopeptide (without zinc) is the most readily observable species: 741.3176 *m/z* (3+) and 1111.4739 *m/z* (2+). An unusual feature, previously unseen in our studies, is the presence of oxidized dimerized apopeptide assigned to the mass observed at 889.5802 *m/z* for the *n* = 5 charge state.

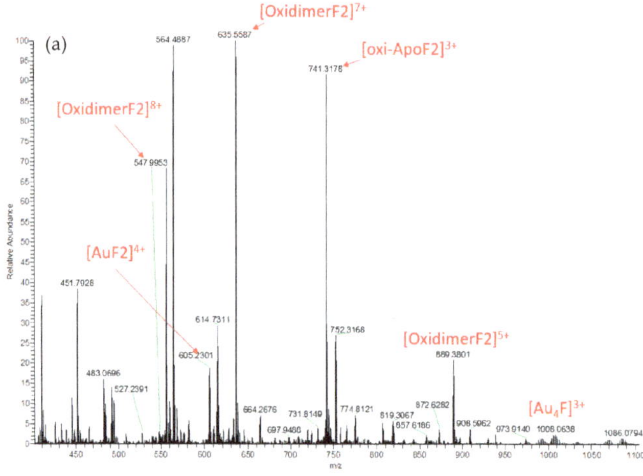

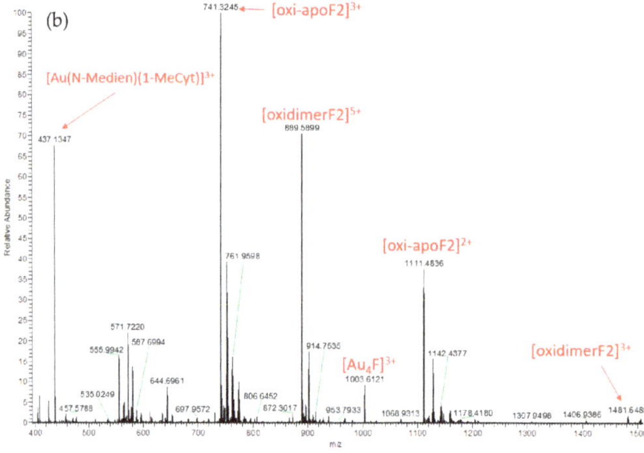

Figure 4. Mass spectra of a 1:1 mixture of **A** (**a**) and **B** (**b**) with the C-terminal zinc-finger of NCp7.

The mass spectra of a 1:1 mixture of NCp7-F2 and **B** taken immediately upon mixing and after one hour of incubation show essentially the same products as the spectra observed upon interaction with **A** (Figure 4b, Figure S5). Additionally, there is little difference in the initial spectrum and spectra obtained at later time points, again indicating that the reaction between NCp7-F2 and **B** is very rapid.

Oxidized peptide and oxidized dimerized peptide were seen again, as well as the incorporation of one, two, and four gold ions. Generally, since the ligands are not observed attached to gold after interaction with zinc-finger proteins, the species observed in the mass spectra of the reaction with **A** and **B** are identical.

2.4. Cyclic Voltammetry

Cyclic voltammetry (CV) studies of complexes **A** and **B** showed that the new complexes follow the trend of resistance toward reduction, seen in the previously studied DMAP and 9-EtGua analogs (Table 1) [10]. The reduction is irreversible in both complexes, and no other redox events were observed. The reduction potentials were determined relative to $[AuCl(dien)]^{2+}$. The reduction potential of **A** is intermediate between the previously studied DMAP and 9-EtGua analogs, and the Au(III) oxidation state is stabilized relative to the $AuClN_3$ coordination sphere. The reduction potential of **B** is slightly higher than the previously studied DMAP analog, but is still slightly lower than the chloride analog.

Table 1. Peak potential values (vs Ag/AgCl) for the reduction of selected Au(III) complexes at a platinum disc electrode relative to $[AuCl(dien)]^{2+}$. Measurements were obtained in 50 mM phosphate buffer (4 mM NaCl, pH = 7.4).

Complex	ΔE_p (V)	Ref.
$[Au(dien)(9\text{-}EtGua)]^{3+}$	−0.069	[10]
$[Au(dien)(1\text{-}MeCyt)]^{3+}$ (**A**)	−0.052	-
$[Au(dien)(DMAP)]^{3+}$	−0.048	[10]
$[AuCl(N\text{-}Medien)]^{2+}$	+0.048	[10]
$[Au(N\text{-}Medien)(1\text{-}MeCyt)]^{3+}$ (**B**)	+0.045	-
$[Au(N\text{-}Medien)(DMAP)]^{3+}$	+0.042	[10]
$[Au(N\text{-}Medien)(9\text{-}EtGua)]^{3+}$	+0.017	[10]

3. Discussion

The synthesis of complexes **A** and **B** serves to complete the series of complexes of the form $[M(dien)(1\text{-}MeCyt)]^{n+}$, in which the Pt(II) and Pd(II) complexes, both isoelectronic and isostructural to the Au(III) complexes investigated in this work, had been previously investigated. However, the Au(III) analog had remained unreported up to this point.

Tryptophan is present on the C-terminal zinc-finger of NCp7 and plays an essential role in protein-DNA/RNA recognition, which occurs through a π–π stacking interaction between the aromatic tryptophan (or in the case of the full peptide, phenylalanine) residues and the nucleobases in DNA/RNA [14,19]. Previously, it was shown that association with free *N*-Acetyl-L-tryptophan (NAcTrp) is a good indicator of a compound's affinity for the full protein. The fluorescent emission of NAcTrp can be quenched by a complex that interacts with NAcTrp in a π–π stacking interaction, and the degree to which the fluorescent emission is quenched is directly related to the strength of the interaction [3,10–12,14,15].

Compared to the Pt(II) and Pd(II) analogs, the Au(III) complexes show a higher affinity for NAcTrp, consistent with the previous literature [5,20]. The lower affinity of **B** for NAcTrp than the DMAP analog and similar affinity compared to the 9-EtGua analog is surprising given the intermediate basicity of 1-MeCyt. The result for **A** is as expected, as the basicity of free 1-MeCyt is intermediate between that of DMAP and 9-EtGua, but the results obtained for **B** and **C** show that there are factors other than the basicity that influence the associative interaction. Therefore, it is likely that the *N*-methyl group present on the chelating ligand in **B** causes a lower association constant for the π–π stacking interaction with NAcTrp than in **A**. This shows that the modulation of the electronic environment of the metal center can influence the interactions between the complex and the protein recognition element. However, this reduction upon methylation of the chelating ligand is only observed to a lesser extent in the 9-EtGua analog.

Mass spectrometry studies with the previously reported Au(III) analogs showed that the formation of random coil observed by CD was accompanied by the formation of "gold fingers," in which one or more atoms of gold replaced the zinc in the zinc-finger structure upon zinc ejection [10]. The observation of oxidized apopeptide is indicative of disulfide bond formation between cysteine molecules of the same peptide, while the observation of oxidized dimerized apopeptide is indicative of disulfide bond formation between cysteine residues of different protein molecules. These interactions would likely be facilitated by the reduction of the metal center from Au(III) to Au(I)/Au(0). This interaction is in contrast to the interactions observed in the Pt(II) analogs, in which reduction and incorporation of the Pt(II) metal center is not seen. Instead, the metal center is observed with the ligands intact and with the zinc still incorporated [3,5,20]. Like previously reported aurated DMAP and 9-EtGua analogs, but unlike the platinated analogs, peaks representing the unreacted zinc-finger protein are not seen at the initial time point, highlighting the increased reactivity of the Au(III) metal center. The incorporation of four equivalents of gold was thought to be due to a difference in the redox properties of **A** and **B** when compared to the previously studied Au(III) analogs. Therefore, CV studies were carried out in order to assess these differences.

CV showed that it is significantly easier to reduce cytosine and thymine than it is to reduce adenine and guanine [21]. In the context of interaction with zinc-finger peptides, this reduction would be accompanied by disulfide bond formation. In this case, the greater ease of reduction may cause the occurrence of a charge transfer mechanism, in which the disulfide bond formation occurs across two equivalents of the peptide, forming the oxidized dimerized peptide uniquely observed in the case of **A** and **B**, in contrast to the simple oxidation of one equivalent of zinc finger. Whereas this latter oxidation is easily understood with reduction of Au(III) to Au(I) and concomitant oxidation of two Zn-bound cysteines, the first step in peptide dimerization may be an electron transfer between the 1-MeCyt ligand to the tryptophan ligand. This suggestion would explain why similar products would not be seen in the isostructural Pt(II) complexes, as the platinum center is not reduced upon interaction with zinc-finger peptides, while a charge transfer mechanism must occur with concomitant reduction of the metal center [22]. No evidence for self-association was observed by NMR studies for **A** or for free ligands (Figure S6). Previously published work discussed the reduction potential and accessibility of the cysteine residues of the C-terminal zinc finger, and this data suggests that disulfide bond formation would likely occur between C39 residues (Figure 5), as C39 has a reduction potential nearly as low as that of C49 (−33.35 vs. −35.91 mV) but has a much greater accessibility (52.88 vs. 15.17) [22]. It is also possible that reduction of multiple molecules of **A** or **B** could also result in the formation of multiple disulfide bonds, in which case both C49 and C39 could form disulfide bonds to form the dimer.

Figure 5. Illustration of disulfide bond formation. Zinc coordinating residues are highlighted in red.

4. Materials and Methods

[AuCl(dien)]Cl$_2$ and [AuCl(*N*-Medien)]Cl$_2$ were synthesized according to previously published procedures [23]. The 1-methylcytosine was synthesized and purified according to previously published procedures [24,25]. The addition of the *N*-heterocycles were carried out analogously to previously published procedures [10]. The apoNCp7-F2 was purchased from GenScript (Piscataway, NJ, USA). NCp7-F2 was prepared according to previously published procedures [10]. All nuclear magnetic resonance (NMR) studies were carried out in deuterium oxide on a Bruker AVANCE III (400 MHz) (Bruker, Billerica, MA, USA). CD studies were carried out at 50 µM concentrations in water on a Jasco J-1500 spectrometer (Jasco, Easton, MD, USA). Mass spectrometry studies were carried out on a

Thermo Electron Corporation Orbitrap Velos mass spectrometer (Thermo Scientific, Waltham, MA, USA). Mass spectrometry samples were prepared at 10 μM concentrations in 10% methanol, and were sprayed at 2.30 kV and 0.9 μL/minute at 230 °C.

Synthesis of [Au(dien)(1-Mecyt)](NO$_3$)$_3$ (**A**): [AuCl(dien)]Cl$_2$ (101.2 mg, 0.249 mmol) was dissolved in dH$_2$O (10 mL).The 1-methylcytosine (30.2 mg, 0.242 mmol) and silver nitrate (125.4 mg, 0.738 mmol) were added, and the reaction mixture was stirred under light protection for 72 h. The precipitated silver chloride was removed by filtration through celite, and the filtrate was evaporated to dryness under reduced pressure. The orange/yellow residue was dissolved in a small amount of dH$_2$O, and acetone was added to the cloud point. The orange/yellow solid product was collected by vacuum filtration and characterized by ^{1}H-NMR, infrared spectroscopy, and elemental analysis (experimental (theoretical)): C: 14.46 (14.62) H: 2.81 (2.86) N: 17.70 (17.17). ^{1}H NMR (D$_2$O, 400 MHz): 7.83, 7.73 (d, 1H), 6.14 (d, 1H), 3.69 (broad, 6H), 3.45 (s, 3H), 3.18 (broad, 2H). UV–Vis (1 mM, H$_2$O): 237 nm (Abs = 3.98), 357 nm (Abs = 0.408).

Synthesis of [Au(*N*-Medien)(1-MeCyt)](NO$_3$)$_3$ (**B**): [AuCl(*N*-MeDien)]Cl$_2$ (100.12 mg, 0.238 mmol) was dissolved in dH$_2$O (10 mL). Silver nitrate (121.4 mg, 0.715 mmol) and 1-methylcytosine (29.8 mg, 0.238 mmol) were added and the reaction mixture was stirred under light protection for 4 days. The reaction mixture was filtered through celite and the solvent was evaporated under reduced pressure. The resulting crude product was dissolved in a small amount of water and excess acetone was added to precipitate the yellow product, characterized by ^{1}H-NMR, Far-IR, and elemental analysis (experimental (theoretical)): C: 16.77 (17.20) H: 3.33 (3.46) N: 17.69 (18.05). ^{1}H-NMR (D$_2$O, 400 MHz): 7.82, 7.72 (d,1H), 6.13(d, 1H), 4.21, 4.09 (m, 4H), 3.65 (m, 3H), 3.46 (d, 2H), 3.38 (s, 3H), 3.19 (m, 2H). UV–Vis (1 mM, H$_2$O): 237 nm (Abs = 3.94), 352 nm (Abs = 0.921).

Synthesis of [Au(dien)(Cyt)](NO$_3$)$_3$ (**C**): [AuCl(dien)]Cl$_2$ (100.44 mg, 0.168 mmol) was dissolved in dH$_2$O (10 mL). Cytosine (28.2 mg, 0.164 mmol) and silver nitrate (85.61 mg, 0.506 mmol) were added, and the reaction mixture was stirred under light protection overnight. The precipitated silver chloride was removed by filtration through celite, and the filtrate was evaporated to dryness under reduced pressure. The orange residue was dissolved in a small amount of dH$_2$O, and acetone was added to the cloud point. The orange solid product was collected by vacuum filtration. ^{1}H-NMR (D$_2$O, 400 MHz): 7.64 (d, 1H), 6.06 (d, 1H), 3.52 (t, 4H), 2.99 (broad, 4H).

Fluorescence quenching of tryptophan: Tryptophan quenching methods were adapted from those previously published [10]. A solution of *N*-acetyl-L-tryptophan (5 μM) was titrated with the Au(III) complexes (7.5 mM) in amounts ranging from 1-10 equivalents of quencher. The decrease in fluorescence intensity of *N*-acetyl-L-tryptophan was plotted using the Eadie-Hofstee method in order to determine the association constant.

Determination of self-association by NMR spectroscopy: Varying concentrations (0.1–2.5 mM) of *N*-heterocycle and gold(III) complex were analyzed by ^{1}H-NMR. The degree of self-association was determined by previously published methods [26]. The change in chemical shift with increasing concentration was fitted to the equation $\delta_{obs} = \{[-K_d + (K_d^2 + 8CK_d)^{1/2}]/4C\}(\delta_M - \delta_D) + \delta_D$. Microsoft Excel Solver was used to solve the three-variable equation.

Cyclic voltammetry: CV studies were carried out following the previously published procedures [10]. A three-electrode cell was employed, containing a platinum disk working electrode, a platinum mesh counter electrode, and a silver/silver chloride reference electrode (0.1 M KCl). Solutions were degassed with nitrogen for 15 minutes before analysis. Measurements were obtained in 50 mM phosphate buffer (4 mM NaCl, pH = 7.4).

5. Conclusions

In conclusion, the previously unreported 1-methylcytosine complexes based on the [Au(dien)(*N*-heterocycle)]$^{3+}$ motif, **A**, and **B**, have been synthesized and investigated for their ability to interact with the C-terminal finger of the HIV nucleocapsid protein NCp7 as a model for a protein with highly nucleophilic cysteines. Gold compounds are considered to primarily interfere with protein, rather than DNA, function. In considering the series $[M(dien)L]^{n+}$, the electronic properties can be modulated systematically for optimal protein interaction. A key feature leading to the rational design of complexes of the form $[Au(dien)L]^{3+}$ is the enhancement of the redox stability of the complexes relative to the precursor $[AuCl(dien)]^{2+}$, due to the incorporation of a more substitution inert nitrogen donor in the form of an *N*-heterocycle [9]. In the case of the more sterically hindered complex **B**, the reduction potential is greater than the chloride analog $[AuCl(N\text{-Medien})]^{2+}$, and is in line with the reduction in stabilization of the gold(III) oxidation state seen in the complex $[Au(N\text{-Medien})(DMAP)]^{3+}$. The aurated nucleobase may interact preferentially with an RNA(DNA) recognition motif, such as tryptophan. Formation of oxidized dimerized peptide was observed in the mass spectrometry studies conducted with NCp7-F2, and indicated a difference in the reactivity profile of **A** and **B** compared to the previously studied DMAP and 9-EtGua analogs. These results further confirm that tuning the electronic and steric properties of the ligands on the Au(III) coordination sphere can modulate the strength of interaction between complexes of the $[Au(dien)(N\text{-heterocycle})]^{3+}$ motif and a zinc finger a systematic manner.

Supplementary Materials: The following are available online at http://www.mdpi.com/2304-6740/7/1/1/s1: Figure S1: The ^{1}H-NMR spectra of **A** (left) and **B** (right) at varying pH* values, where pH* is the reading of the pH meter; Figure S2: Mass spectra of **A** (left) and **B** (right); Figure S3: Mass spectrum of **A** with NCp7-F2 immediately after mixing; Figure S4: Mass spectrum of **A** with NCp7-F2 6 hours after mixing; Figure S5: Mass spectrum of **B** with NCp7-F2 1 hour after mixing; Figure S6: Self-association determination of DMAP, NAcTrp, 9-EtGua, and **A**.

Author Contributions: Conceptualization: N.P.F. Investigation and draft writing: J.B. Writing, reviewing, and editing: N.P.F.

Funding: This research was funded by National Science Foundation CHE CHE1413189.

Acknowledgments: We gratefully acknowledge Christopher J. Freeman and Maryanne M. Collinson for assistance in carrying out the CV experiments.

Conflicts of Interest: The authors declare no conflict of interest.

References

1. Berners-Price, S.J.; Filipovska, A. Gold Compounds as Therapeutic Agents for Human Diseases. *Metallomics* **2011**, *3*, 863–873. [CrossRef] [PubMed]

2. De Paula, Q.A.; Tsotsoros, S.D.; Qu, Y.; Bayse, C.A.; Farrell, N.P. Platinum-nucleobase PtN4 Complexes as Chemotypes for Selective Peptide Reactions with Biomolecules. *Inorg. Chim. Acta* **2012**, *393*, 222–229. [CrossRef]

3. Bernardes, V.H.F.; Qu, Y.; Du, Z.; Beaton, J.; Vargas, M.D.; Farrell, N.P. Interaction of the HIV NCp7 Protein with Platinum(II) and Gold(III) Complexes Containing Tridentate Ligands. *Inorg. Chem.* **2016**, *55*, 11396–11407. [CrossRef] [PubMed]

4. Andreini, C.; Banci, L.; Bertini, I.; Rosato, A. Counting the Zinc-Proteins Encoded in the Human Genome. *J. Proteome Res.* **2006**, *5*, 196–201. [CrossRef] [PubMed]

5. Morellet, N.; Jullian, N.; De Rocquigny, H.; Maigret, B.; Darlix, J.; Roques, B.P. Determination of the Structure of the Nucleocapsid Protein NCp7 from the Human Immunodeficiency Virus Type 1 by ^{1}H NMR. *EMBO J.* **1992**, *11*, 3059–3065. [CrossRef]

6. Daniel, A.G.; Peterson, E.J.; Farrell, N.P. The Bioinorganic Chemistry of Apoptosis: Potential Inhibitory Zinc Binding Sites in Caspase-3. *Angew. Chem.* **2014**, *126*, 4182–4185. [CrossRef]

7. Klug, A.; Rhodes, D. Zinc Fingers: A Novel Protein Fold for Nucleic Acid Recognition. *Cold Spring Harb. Symp. Quant. Biol.* **1987**, *52*, 473–482. [CrossRef]

8. Krishna, S.S.; Majumdar, I.; Grishin, N.V. Structural Classification of Zinc Fingers. *Nucleic Acids Res.* **2003**, *31*, 532–550. [CrossRef]

9. South, T.L.; Blake, P.R.; Sowder, R.C., III; Arthur, L.O.; Henderson, L.E.; Summers, M.F. The Nucleocapsid Protein Isolated from HIV-1 Particles Binds Zinc and Forms Retroviral-Type Zinc Fingers. *Biochemistry* **1990**, *29*, 7786–7789. [CrossRef]

10. Spell, S.R.; Farrell, N.P. Synthesis and Properties of the First $[Au(dien)(N\text{-heterocycle})]^{3+}$ Compounds. *Inorg. Chem.* **2014**, *53*, 30–32. [CrossRef]

11. Spell, S.R.; Farrell, N.P. [Au(dien)(*N*-heterocycle)]³⁺: Reactivity with Biomolecules and Zinc Finger Peptides. *Inorg. Chem.* **2015**, *54*, 79–86. [CrossRef] [PubMed]

12. Spell, S.R.; Mangrum, J.B.; Peterson, E.J.; Fabris, D.; Ptak, R.; Farrell, N.P. Au(III) Compounds as HIV Nucleocapsid Protein (NCp7)-Nucleic Acid Antagonists. *Chem. Commun.* **2017**, *53*, 91–94. [CrossRef] [PubMed]

13. Anzellotti, A.I.; Farrell, N.P. Zinc Metalloproteins as Medicinal Targets. *Chem. Soc. Rev.* **2008**, *37*, 1629–1651. [CrossRef] [PubMed]

14. Anzellotti, A.I.; Bayse, C.A.; Farrell, N.P. Effects of Nucleobase Metalation on Frontier Molecular Orbitals: Potential Implications for π-Stacking Interactions with Tryptophan. *Inorg. Chem.* **2008**, *47*, 10425–10431. [CrossRef]

15. Verdolino, V.; Cammi, R.; Munk, B.H.; Schlegel, H.B. Calculation of pK$_a$ Values of Nucleobases and Guanine Oxidation Products Guanidinohydantoin and Spiroiminodihydantoin using Density Functional Theory and a Polarizable Continuum Model. *J. Phys. Chem. B* **2008**, *112*, 16860–16873. [CrossRef]

16. Anzellotti, A.I.; Ma, E.S.; Farrell, N. Platination of Nucleobases to Enhance Noncovalent Recognition in Protein-DNA/RNA Complexes. *Inorg. Chem.* **2005**, *44*, 483–485. [CrossRef]

17. Anzellotti, A.I.; Sabat, M.; Farrell, N. Covalent and Noncovalent Interactions for [Metal(dien)nucleobase]²⁺ Complexes with L-Tryptophan Derivatives: Formation of Palladium-Tryptophan Species by Nucleobase Substitution under Biologically Relevant Conditions. *Inorg. Chem.* **2006**, *45*, 1638–1645. [CrossRef]

18. dePaula, Q.A.; Mangrum, J.B.; Farrell, N.P. Zinc Finger Proteins as Templates for Metal Ion Exchange: Substitution Effects on the C-Finger of HIV Nucleocapsid NCp7 Using M(chelate) Species (M = Pt, Pd, Au). *J. Inorg. Biochem.* **2009**, *103*, 1347–1354. [CrossRef]

19. Ma, E.S.F.; Daniel, A.G.; Farrell, N.P. Dinuclear Platinum Complexes Containing Planar Aromatic Ligands to Enhance Stacking Interactions with Proteins. *ChemMedChem* **2014**, *9*, 1155–1160. [CrossRef]

20. Tsotsoros, S.D.; Lutz, P.B.; Daniel, A.G.; Peterson, E.J.; de Paiva, R.E.F.; Rivera, E.; Qu, Y.; Bayse, C.A.; Farrell, N.P. Enhancement of the Physiochemical Properties of [Pt(dien)(nucleobase)]²⁺ for HIVNCp7 Targeting. *Chem. Sci.* **2017**, *8*, 1269–1281. [CrossRef]

21. Manoj, P.; Mohan, H.; Mittal, J.P.; Manoj, V.M.; Aravindakumar, C.T. Charge Transfer from 2-Aminopurine Radical Cation and Radical Anion to Nucleobases: A Pulse Radiolysis Study. *Chem. Phys.* **2007**, *331*, 351–358. [CrossRef]

22. Abbehausen, C.; de Paive, R.E.F.; Bjornsson, R.; Gomes, S.Q.; Du, Z.; Corbi, P.P.; Lima, F.A.; Farrell, N. X-ray Absorption Spectroscopy Combined with Time-Dependent Density Functional Theory Elucidates Differential Substitution Pathways of Au(I) and Au(III) with Zinc Fingers. *Inorg. Chem.* **2018**, *57*, 218–230. [CrossRef] [PubMed]

23. Nardin, G.; Randaccio, L.; Annibale, G.; Natile, G.; Pitteri, B. Comparison of Structure and Reactivity of bis(2-aminoethyl)amine- and bis(2-aminoethyl)amido-chlorogold(III) Complexes. *J. Chem. Soc. Dalton Trans.* **1980**, *2*, 220–223. [CrossRef]

24. Greco, E.; Aliev, A.E.; Lafitte, V.G.H.; Bala, K.; Duncan, D.; Pilon, L.; Golding, P.; Hailes, H.C. Cytosine Modules in Quadruple Hydrogen Bonded Arrays. *New J. Chem.* **2010**, *34*, 2634–2642. [CrossRef]

25. Helfer, D.L., II; Hosmane, R.S.; Leonard, N.L. Selective Alkylation and Aralkylation of Cytosine at the 1-Position. *J. Org. Chem.* **1981**, *46*, 4803–4804. [CrossRef]

26. Waterhous, D.V.; Muccio, D.D. ¹H and ¹³C Studies on the Self-Association of Retinoic Acid. *Magn. Reson. Chem.* **1990**, *28*, 223–226. [CrossRef]

 *inorganics*

Review

Nano-Based Systems and Biomacromolecules as Carriers for Metallodrugs in Anticancer Therapy

Mina Poursharifi [1,2], **Marek T. Wlodarczyk** [1,3] and **Aneta J. Mieszawska** [1,2,3,*]

1   Department of Chemistry, Brooklyn College, The City University of New York, 2900 Bedford Avenue, Brooklyn, NY 11210, USA; mina.poursharifi@brooklyn.cuny.edu (M.P.); marekw@brooklyn.cuny.edu (M.T.W.)
2   Ph.D. Program in Biochemistry, The Graduate Center of the City University of New York, New York, NY 10016, USA
3   Ph.D. Program in Chemistry, The Graduate Center of the City University of New York, New York, NY 10016, USA
*   Correspondence: aneta.mieszawska@brooklyn.cuny.edu; Tel.: +1-718-951-5000 (ext. 2839)

Received: 1 November 2018; Accepted: 11 December 2018; Published: 20 December 2018

Abstract: Since the discovery of cisplatin and its potency in anticancer therapy, the development of metallodrugs has been an active area of research. The large choice of transition metals, oxidation states, coordinating ligands, and different geometries, allows for the design of metal-based agents with unique mechanisms of action. Many metallodrugs, such as titanium, ruthenium, gallium, tin, gold, and copper-based complexes have been found to have anticancer activities. However, biological application of these agents necessitates aqueous solubility and low systemic toxicity. This minireview highlights the emerging strategies to facilitate the in vivo application of metallodrugs, aimed at enhancing their solubility and bioavailability, as well as improving their delivery to tumor tissues. The focus is on encapsulating the metal-based complexes into nanocarriers or coupling to biomacromolecules, generating efficacious anticancer therapies. The delivery systems for complexes of platinum, ruthenium, copper, and iron are discussed with most recent examples.

Keywords: metallodrugs; nanoparticles; dendrimers; nanotubes; liposomes; biomacromolecules; encapsulation; targeting; micelles

---

1. Introduction

Cisplatin was the first Food and Drug Administration (FDA)-approved metallodrug for the treatment of solid tumors. Cisplatin, or *cis*-diamminedichloroplatinum (II), is a square planar coordination complex with two chlorine and two amine ligands in *cis* configuration. The cytotoxic effects of cisplatin originate from the formation of active *cis*-[Pt(NH$_3$)$_2$(H$_2$O)$_2$]$^{2+}$ species that bind to nuclear DNA distorting the helical structure, and interfere with DNA replication and transcription inducing cancer cell apoptosis [1,2]. However, the major limitation of cisplatin is the formation of active species in the systemic circulation. The fast exchange of chloride ligands to water leads to high systemic toxicity and limits the clinical dose of the drug [3]. Therefore, often sub-therapeutic concentrations of cisplatin reaching cancer cells leads to minor DNA damage, DNA repair, and eventually might lead to the growth of resistant cells. Developing platinum-resistance is a major problem in cisplatin-based therapy [4,5]. Carboplatin and oxaliplatin are second-generation platinum complexes approved by the FDA for use as anticancer drugs. They are characterized with lower toxicity profiles, broader spectrum of action, and overcoming resistance in some types of tumors. However, analogous to cisplatin mechanism of action hampers elimination of all problems associated with Pt(II) therapy. Therefore "*trans*" geometries [6–9], polynuclear [10–14], or platinum (IV) prodrugs [15,16] have been explored as possible alternatives, but none have yet entered the clinic. Importantly, high efficacy of platinum paved the way for the development of other transition metal complexes, many with outstanding cytotoxic properties against cancer cells.

Transition metals form complexes of multiple geometries based on the number of coordination sites, offering a stereoisomeric diversity higher than carbon. In addition, a synthesis of metallodrugs requires fewer steps when compared to organic compounds. This is accompanied by a myriad choice of coordinating ligands [17] facilitating fine-tuning the properties of metal-based complexes [17,18]. The octahedral titanium species *cis*-diethoxy-bis(1-phenylbutane-1,3-dionato)titanium(IV) [(bzac)$_2$ Ti(OEt)$_2$] (budotitane), was the first non-Pt(II) metal compound that reached clinical trials [19]. Following this, ruthenium-based complexes were explored showing fewer side effects when compared to platinum-based agents and activity against Pt-resistant cancers [20–22]. Two ruthenium derivates, imidazolium *trans*-DMSO-imidazole-tetrachlororuthenate (NAMI-A) and imidazolium *trans*-[tetrachlorido(DMSO)(1*H*-imidazole)ruthenate(III)] (KP-1019), have been tested in clinical trials [22,23]. KP-1019 showed efficacy towards primary tumors, such as colon cancer, without triggering systemic toxicity, while NAMI-A was not cytotoxic but demonstrated activity against metastases. Moreover, NKP-1339 (the sodium salt analogue of KP1019, sodium *trans*-[tetrachloridobis(1*H*-indazole)ruthenate(III)]) has successfully completed a phase I clinical trial [24]. NKP-1339 is Ru(III) that binds to serum proteins albumin and transferrin, which transport and deliver the metallodrug to the tumor tissue. Once inside endosomes, NKP-1339 reduces to Ru(II). Recently, new designs of ruthenium complexes are emerging that include Ru(η^5-C$_5$H$_5$) core. This class of Ru(II) compounds demonstrates enhanced selectivity and effectiveness, for example ruthenium pyridocarbazole (DW1/2) showed inhibition of protein kinase in vitro [22,25,26]. Other transition metal complexes include redox-active mono(thiosemicarbazone) copper compounds [27–29], nucleic acid-targeting gold(I) complex auranofin [30], or osmium(II) arene complexes that act on mitochondria [31]. Figure 1 shows a few examples of metallodrugs, including platinum, ruthenium, gallium, and titanium complexes that were found to have anti-cancer activities [32,33].

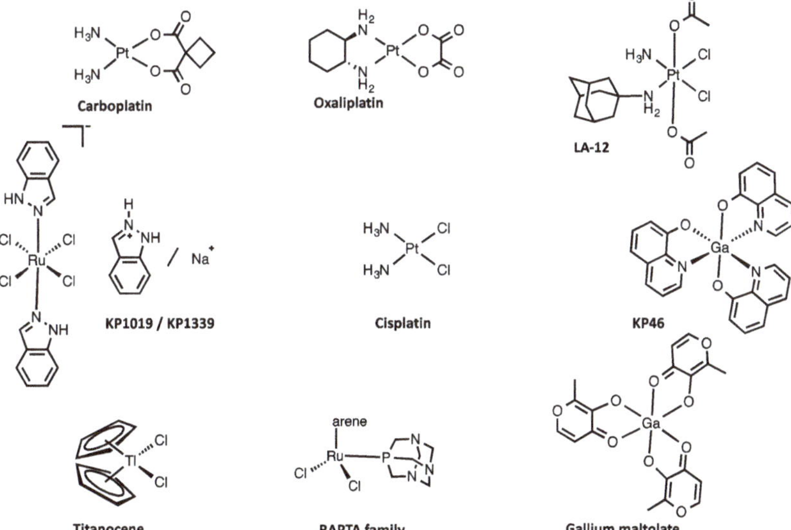

Figure 1. Chemical structures of representative Pt, Ru, Ti, and Ga anticancer drugs.

However, major challenges in the clinical application of metallodrugs, when delivered via conventional intravenous methods, include limited aqueous solubility and short in vivo half-lives, resulting in inadequate bioavailability and low accumulation in the tumors. Therefore, many approaches have been proposed to address these limitations, and the main strategies focus on exploring nanotechnology-based systems [34,35]. This minireview discusses the examples of synthetic nanoparticles as well as biological carriers used to increase the solubility, reduce the systemic toxicity, increase cancer-cell delivery, or achieve cancer-specific targeting of metallodrugs.

2. Synthetic Nanocarriers

The important advantages of nano-based systems for drug delivery applications are high stability, loading capacity, possibility to incorporate hydrophobic or highly toxic drug molecules, and, depending on the core material used, achieving controlled or sustained drug release. This can sharply improve the solubility and bioavailability of the drugs, as well as their pharmacokinetics, while reducing the side effects associated with therapy [36,37]. The nanoparticles can accumulate in the tumor through the enhanced permeability and retention (EPR) effect [38]. In EPR the nanoparticles extravasate through the impaired tumor vasculature, and the lack of appropriate lymphatic drainage leads to their prolonged retention in the tumor tissue. The degree of nanoparticle accumulation in the tumor via EPR depends on the tumor's type and size. The most pronounced EPR effect was observed in pancreatic, colon, breast, and stomach cancers. Also, a higher accumulation of nanoparticles was found in larger tumors [39].

In addition, the versatility of nanoparticle technologies and coating strategies, allows the design of nanoparticle systems capable of cancer cell recognition and selective tumor targeting [40–42]. The advantages of nanocarriers heighten the drug payload within the tumor, thereby improving the therapeutic outcomes. Table 1 summarizes the nanocarrier-based systems for metallodrug delivery that have entered clinical trials.

Table 1. Representative metallodrug delivery systems that have entered clinical trials.

Delivery System	Active Drug	Formulation	Clinical Status	Tumor Target	Identifier
Polymeric nanoparticles	oxaliplatin	AP5346	Phase II	Head and Neck cancer	NCT00415298
	oxaliplatin	NC-4016	Phase I	Colorectal cancer	NCT03168035
	cisplatin	NC-6004	Phase III	Pancreatic cancer	NCT03109158
Liposome	cisplatin	Lipoplatin	Phase I	Lung cancer	NCT02702700
	cisplatin	SPI-077	Phase I, II	Ovarian, Breast and Skin cancer	NCT01861496 NCT00004083
	oxaliplatin	Lipoxal	Phase I	Advanced gastrointestinal cancer	NCT00964080
Inorganic Nanoparticles	iron oxide	Ferumoxytol	Early Phase I	Glioblastoma	NCT00660543

Data are gathered from www.clinicaltrials.gov.

2.1. Liposomes

Liposomes, discovered in 1960s, are one of the first nanosized vehicles used in drug delivery applications. They are spherical vesicles having an aqueous core surrounded by lipid bilayers, a similar morphology to cellular membranes [43]. Liposomes are ideal to encapsulate hydrophilic drugs in an aqueous core, but the hydrophobic therapeutics can also be entrapped within a lipid bilayer. Liposomes are considered one of the most successful drug delivery vehicles to date [44], with several formulations approved by the FDA for anticancer therapy and available in the clinic [43,45].

The liposomal formulations of cisplatin have been the most studied including LiPlaCis, SPI-77, L-NDDP, or lipoplatin [46,47]. Lipoplatin has successfully completed Phase I, Phase II, and Phase III human clinical trials. It is a liposome with diameter of ~110 nm with long in vivo half-life of ~120 h, and was found to effectively accumulate into tumors, while reducing major toxicities associated with cisplatin [48,49]. Tumoral accumulations with up to ten to 200-fold increase versus normal tissue were reported [48,49]. However, the liposomal formulations did not enhance the clinical efficacy of cisplatin, possibly due to inadequate release of the drug in the tumor [50].

The liposomal formulations of Ru have also been proposed [51,52] Shen et al. demonstrated encapsulation of ruthenium polypirydine complex [Ru(phen)$_2$ dppz](ClO$_4$)$_2$ in a liposome carrier (Lipo-Ru), composed of dipalmitoylphosphatidylcholine (DPPC), cholesterol, and distearoylphosphatidylethanolamine (DSPE)-PEG (polyethylene glycol). The schematic of the Lipo-Ru is presented in Figure 2A. The presence of dipyridophenazine (dppz) ligand in the complex served as a "light switch" that emitted when nitrogen atoms of dppz were exposed to the hydrophobic environment of the lipid bilayer [51]. Lipo-Ru was an example of a theranostic system, where drug delivery, as well as imaging studies, can be accomplished using a single platform [53]. The fluorescent properties of the nanocarrier were demonstrated using confocal imaging of cells incubated with Lipo-Ru (Figure 2B). The average diameter of Lipo-Ru was 82.53 ± 2.66 nm with Ru loading of 4% [51]. The liposomes were stable under physiological conditions and biologically active. The in vitro studies using triple negative breast cancer cells demonstrated high cytotoxicity of Lipo-Ru and low IC$_{50}$ value of 4 µM.

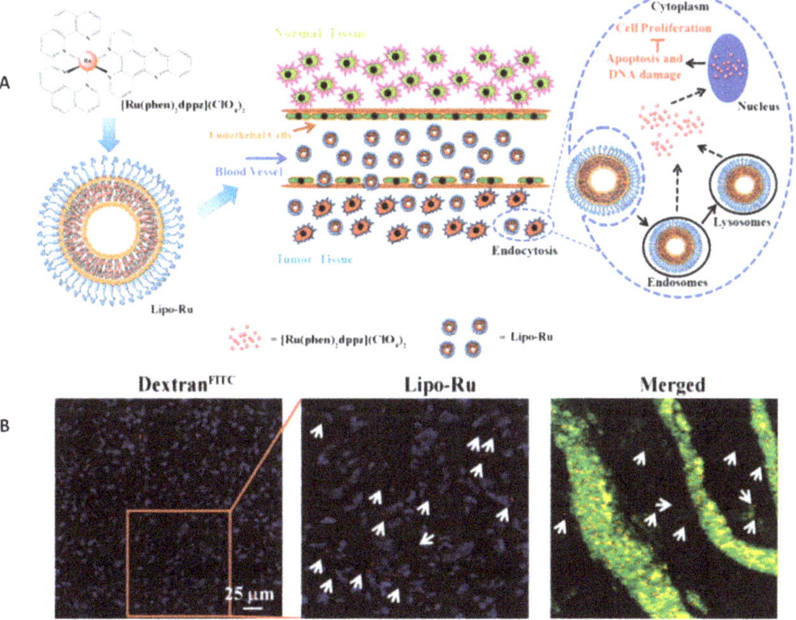

Figure 2. (**A**) Schematic Representation of Lipo-Ru liposomes, and (**B**) confocal microscopy images of tumors 2 h post injection of Lipo-Ru. Lipo-Ru particles are shown in red and highlighted with white arrows. Nuclei are stained in blue with DAPI. Reprinted with permission from reference [51].

2.2. Micelles

Micelles are simple nanostructures formed from amphiphilic molecules, they range in size between 5 and 100 nm, and are considered to be suitable for drug delivery applications [54]. Micelles are spherical ellipsoid systems that self-assemble in an aqueous environment to form a hydrophobic monolayer in the core and a polar surface. Micelles disintegrate slowly upon the interaction of polar groups with water, which leads to pore formation within the hydrophobic part and effective drug escape from the core.

Micelles delivering metallodrugs have been investigated [55–60]. One strategy involved the covalent attachment of the drug directly to the micelle to avoid its premature release in the circulation [55–59]. Sentzel's group has developed micelles to deliver Pt(II) and Ru(II) complexes to cancer cells [61–64]. They used various lengths of poly[oligo(ethylene glycol) methyl ether methacrylate] (POEGMEMA) and 1,1-di-tert-butyl 3-(2-(methacryloyloxy)ethyl) butane-1,1,3-tricarboxylate (PMAETC) block copolymers with pendant bidentate carboxylate ligands to form a carboplatin-like complex directly on the polymer [62]. The length of hydrophobic PMAETC block was varied and micelles with the shortest PMAETC block showed highest Pt(II) loading of 83%, smallest hydrodynamic diameter of ~77 nm, faster Pt(II) release, and higher cytotoxicity towards lung cancer cells than micelles with longer PMAETC blocks. A similar approach was used to prepare ruthenium micelles of RAPTA-C [$RuCl_2$ (p-cymene)-(PTA)] [63]. RAPTA-C limits growth of solid tumor metastases, and its mechanism of action relies on inducing cancer cell apoptosis by triggering G2/M phase arrest. However, RAPTA-C suffers from poor delivery efficacy [65]. The amphiphilic polymers containing the water-soluble 1,3,5-phosphaadamantane (PTA) ligand were used to attach RAPTA-C [63]. The polymers were modified with fluorescein enabling in vitro evaluation of the micelle uptake by cancer cells. Encapsulation into micelles showed increased cellular uptake of RAPTA-C and a ten-fold increase in toxicity in ovarian cancer cell lines, when compared to a free drug.

2.3. Polymeric Nanoparticles

Poly(lactic-*co*-glycolic) acid (PLGA) [66] is an FDA-approved biodegradable slow-releasing polymer exhibiting a wide range of erosion times [67]. At present, many PLGA-based formulations are at the pre-clinical stage [68,69]. Recently, PLGA nanoparticles incorporating ruthenium-based radiosensitizer, $Ru(phen)_2(tpphz)^{2+}$ (phen = 1,10-phenanthroline, tpphz = tetrapyridophenazine) (**Ru1**), were investigated as a combination therapy in oesophageal cancer cells [70]. The nanoparticles were formed via a double emulsion evaporation method from a polymer:drug mixture at the ratio of 1:6.25, respectively, resulting in drug loading of 1.14%. The surface of the nanoparticles was modified with metal ion chelator diethylenetriaminepentaacetic acid (DTPA) for ^{111}In radiolabelling, and the targeting ligand for human epidermal growth factor receptor (EGFR). The schematic of the nanoparticle is shown in Figure 3A. The **Ru1** is electron-dense and can be clearly distinguished inside the PLGA core. The in vitro studies showed higher nanoparticle uptake in EGFR-overexpressing oesophageal cancer cells, when compared to cells with normal EGFR levels. Importantly, co-delivery of ^{111}In and **Ru1** led to synergistic cytotoxic effects through ^{111}In radiotoxicity and DNA damage fortified by **Ru1** intercalating properties, resulting in enhanced therapeutic outcomes (Figure 3B). In another approach, Ru(II) was encapsulated into PLGA nanoparticles with a diameter of ~100 nm and evaluated in two-photon-excited photodynamic therapy (PDT) [71]. The nanoparticles showed cytotoxic effects in vitro on C6 glioma cells.

Yang et al. coordinated Pt (IV) to diamminedichlorodihydroxyplatinum (DHP) polymer or its dicarboxyl derivative diamminedichlorodisuccinatoplatinum (DSP) [72]. The platinum(IV)–polymer conjugates were easily reduced in acidic conditions producing active Pt(II) species. The constructs were cytotoxic in vitro towards breast (MDA-MB-468, MCF-7) and ovarian cancer (SKOV-3) cell lines, and in vivo evaluation showed accumulation of the conjugates in the tumor.

Earlier reports on Pt(IV) prodrugs delivery using polymeric carriers include the work of Dhar et al., where Pt(IV) complex was loaded into PEG-PLGA nanoparticles, and the nanoparticle's surface was decorated with aptamer targeting ligand for prostate-specific membrane antigen (PSMA) [73]. The in vitro studies demonstrated higher cytotoxicity of nanoparticle Pt(IV) formulation towards prostate cancer cells than the free drug. Furthermore, Aryal et al. conjugated Pt(IV) to PLA of PEG-PLA block copolymer, using acid labile bonds [74]. The nanoparticles formed from the construct had an average diameter of about ~100 nm, and showed a controlled acid-responsive drug release.

Also, Rieter et al. synthesized the nanoparticles by precipitation of disuccinatocisplatin (DSCP) with Tb^{3+} ions [75]. The nanoparticles were stabilized with a shell of amorphous silica and showed higher than cisplatin in vitro cytotoxicity.

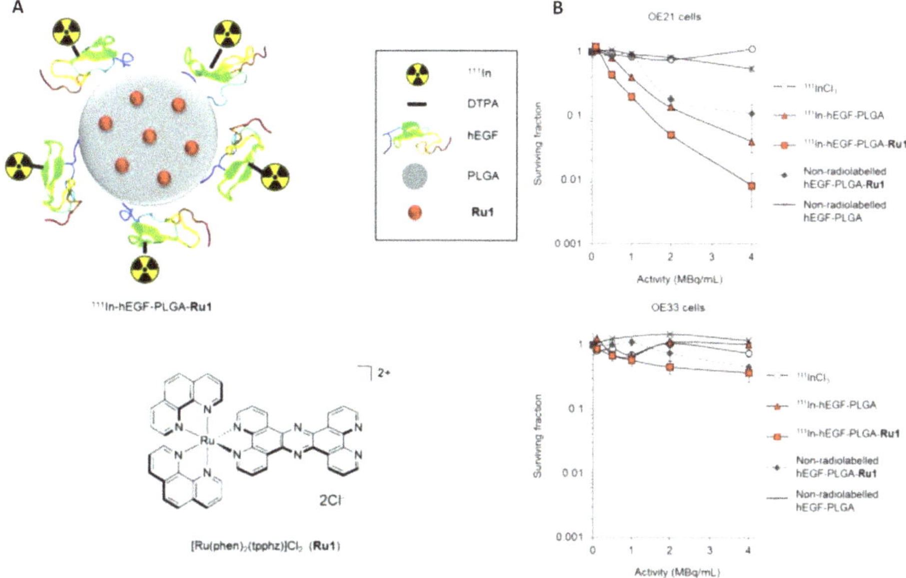

Figure 3. (**A**) The ruthenium(II) metallo-intercalator and radiosensitizer, **Ru1**, is encapsulated within Poly(lactic-*co*-glycolic) acid (PLGA) core and nanoparticles are surface labeled with ^{111}In-DTPA-hEGF, bottom—chemical structure of **Ru1**. (**B**) Clonogenic survival assays of OE21 or OE33 cells exposed to ^{111}In radiolabelled hEGF-PLGA nanoparticles with or without **Ru1**-loading (24 h incubation time). Non-radiolabelled nanoparticles (in equivalent concentrations; 0 to 1000 µg mL^{-1}) and free ^{111}InCl$_3$ (in equivalent specific activity) are included as controls. Data points are the mean of triplicates ± S.D. Reprinted with permission from reference [70].

In another example, Keppler's group used poly(lactic acid) nanoparticles stabilized with Tween 80 surfactant, as a delivery vehicle for KP1019 [76]. The mean diameter of the nanoparticles was 164 ± 10 nm with ruthenium loading of about 2.4 µmol and 95% loading efficiency. The nanoparticles were stable for at least 25 days. The in vitro testing with colon carcinoma SW480 and the hepatoma Hep3B cell lines demonstrated higher cytotoxicity of the formulation versus free KP1019. Interestingly, the interaction of Ru(III) with Tween 80 led to an exchange of one chlorido ligand and reduction of Ru(III) to Ru(II), the 20-fold higher in vitro biological activity of the construct was observed.

2.4. Carbon Nanotubes (CNTs)

Carbon nanotubes, high aspect ratio nanomaterials, have been considered as drug delivery vehicles primarily due to their excellent cellular penetration ability, and hollow interior capable of incorporating high payloads of drug molecules. The major challenge in biological use of CNTs is their toxicity. However, large surface area of CNTs and the choice of surface chemistries can potentially alleviate this problem by introducing additional functionalities to increase biocompatibility [77].

Several carbon nanotube-based carriers have been developed for Pt(IV) prodrug delivery [78–83]. In one example, two different diameters of multi-walled carbon nanotubes were examined to deliver Pt(IV) to HeLa cells [82]. The cells internalized the CNTs-Pt effectively, with the larger diameter

CNTs-Pt showing higher cytotoxicity than the smaller one or free drugs. Importantly, low toxicity was observed after the incubation of CNTs-Pt with murine macrophages.

Zhang et al. encapsulated ruthenium(II) complexes into single-walled carbon nanotubes for photothermal and photodynamic therapy with near-infrared (NIR) irradiation [84]. Ru(II) complexes, were sonicated with SWCNTs for 4 h to produce Ru@SWCNTs. The schematic of the construct and TEM image are shown in Figure 4A. Ru(II) encapsulated into SWCNTs via π–π interactions, and Ru(II) loading was 15.4 µg/mL at SWCNTs concentration of 34.6 µg/mL. The irradiation with an 808 nm diode laser led to localized photothermal effects due to light absorption by SWCNTs, while released Ru(II) complexes produced singlet oxygen species ($^1O_2$). The synergistic effect led to tumor ablation in vivo. The thermographs from the in vivo studies are shown in Figure 4B.

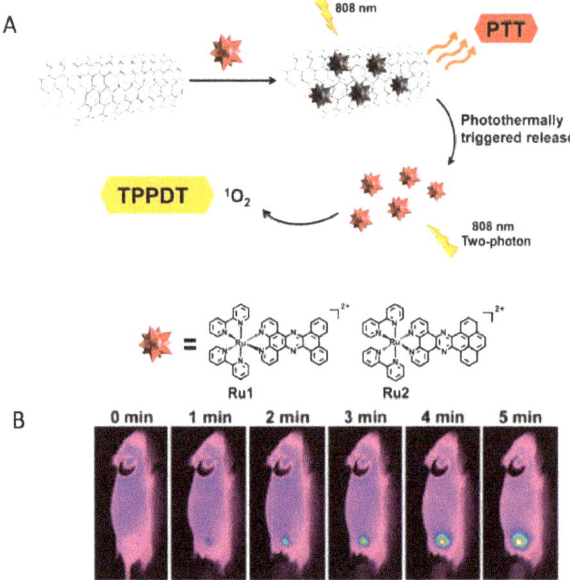

Figure 4. (**A**) Schematic design of Ru@SWCNTs. (**B**) Thermographs of tumor-bearing mice receiving photothermal treatment with **Ru1@SWCNTs** different periods of time (0 to 5 min). Reprinted with permission from reference [84].

A similar approach was used by Wang et al., where a Ru(II) polypyridyl complex was loaded in multi-walled carbon nanotubes producing RuPOP@MWCNTs delivery system [85]. The cellular uptake and efficacy of the construct was evaluated in drug-resistant R-Hep-G2 liver cancer cells. The RuPOP@MWCNTs entered cells via endocytosis, and higher uptake was observed in cancer cells than in normal cells. Also, the system was evaluated with clinical X-rays, and resulted in reactive oxygen species (ROS) overproduction, which induced cancer cell apoptosis.

2.5. Inorganic Nanoparticles

Inorganic nanoparticles have been applied in anticancer therapy as either therapeutic agents or as carrier vehicles for small drug molecules. Their advantages include unique physicochemical properties, large surface area, and versatility of surface chemistries for ligand attachment, as well as biocompatibility [86].

Several groups have studied gold nanoparticles as a delivery system for platinum-based agents [87–91]. In one example by Xiong et al., cisplatin was complexed onto a gold nanoparticle modified with tripeptide and a folic acid-PEG linker [87]. The laser irradiation led to Pt (II) release either through the ligand exchange with chloride ions, or Au–S bond cleavage and peptide–folate–Pt conjugate release from the surface of the nanoparticle. The cytotoxic effects were evaluated against folate-overexpressing human epidermoid carcinoma KB cells, where the nanoparticles showed high biological activity.

Pramanik et al. also used gold nanoparticles to deliver a copper(II) diacetyl-bis(*N*4-methylthiosemicarbazone) complex [92]. The loading strategy involved the attachment of the Cu(II) complex to the linker containing disulfide bond in the middle and carboxylic acid end. The linker-Cu(II) was loaded onto the gold nanoparticle surface modified with the amine terminated PEG. The schematic of the nanoparticle is shown in Figure 5A. In addition, the attachment of biotin on the nanoparticle's surface provided cancer specific targeting, as fast proliferating cells demand biotin and often overexpress the biotin receptors on their surface. The in vitro and in vivo studies showed activity towards HeLa cells and a HeLa cell xenograft tumor model, respectively. The 3.8-fold reduction in tumor volume was observed with the Cu(II)-nanoparticle treatment (Figure 5B).

The use of selenium nanoparticles as carrier vehicles for metallodrugs has also been reported [93,94]. Sun et al. investigated the cytotoxicity of selenium nanoparticles stabilized with thiolated Ru(II)–polypyridyl complexes [93]. The fluorescent properties of the nanoconstructs allowed the imaging studies of Ru(II)–Se nanoparticle uptake by cancer cells. The Ru(II)–Se nanoparticles inhibited tumor growth and angiogenesis in vivo in a xenograft HepG2 tumor model.

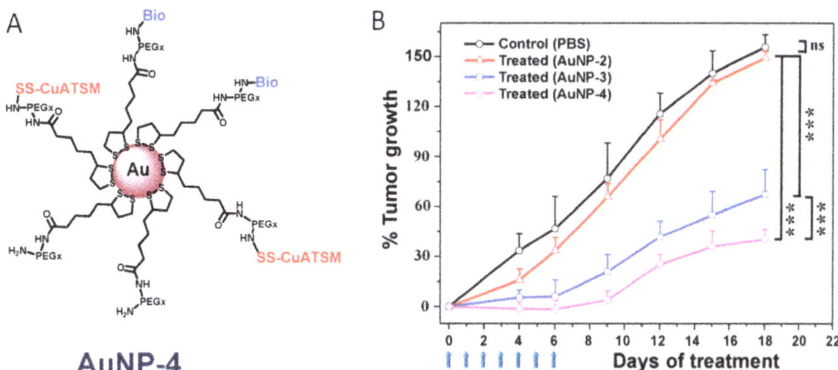

Figure 5. (**A**) Schematic illustration for the synthesis of PEG stabilized gold nanoparticles and nanoconjugates. (**B**) In vivo anticancer activity of AuNPs in mice bearing HeLa xenografts. *** indicates $p < 0.001$. Reprinted with permission from reference [92].

Liu et al. synthesized folic acid-conjugated selenium nanoparticles as a carrier for ruthenium polypyridyl [94]. Similarly to previous reports, the nanoparticle was modified with folic acid for cancer cell targeting, and the system exhibited fluorescent properties allowing the imaging of the nanoparticle trafficking inside the cell. The main goal of this study was to overcome the multidrug resistance using the proposed nanoparticle system. The in vitro studies with multidrug resistant cells R-HepG2 demonstrated high toxicity through up-regulating the ROS level inside the cells triggering mitogen-activated protein kinases (MAPKs) and protein kinase B (AKT) pathways for cancer cell death.

Mesoporous silica nanoparticles also emerged as nanocarriers for metal-based complexes. The advantages of silica carriers include chemical stability, ease of surface modification with different functional groups for subsequent drug conjugation, as well as a tunable pore diameter [95]. Gomes-Ruiz's group employed mesoporous silica nanoparticles as carriers for tin, titanium, and ruthenium complexes [34,96–103]. Interestingly, they observed that the entire nanostructure of mesoporous silica–drug conjugate was involved in triggering cancer cell apoptosis, and only a small release of the encapsulated metallodrug was contributing to cytotoxic effects. This work was pioneering as it pointed towards a non-classical mechanism of action usually observed for these types of systems. This was highlighted in a recent publication by this group, where the complexes of tin and titanium were immobilized onto silica-based material SBA-15 (Santa Barbara Amorphous-15) modified with the aminodiol ligand 3-[bis(2-hydroxyethyl)amino] propyltriethoxysilane [102]. This construct was used as a support to attach diorganotin(IV) compound, $SnPh_2Cl_2$, and two titanocene derivatives, $TiCp_2Cl_2$ ($[Ti(\eta^5-C_5H_5)_2Cl_2]$ and $TiCpCpPhNfCl_2$ ($[Ti(\eta^5-C_5H_5)$ ($\eta^5-C_5H_4CHPhNf)Cl_2]$ (Ph = C_6H_5; Nf = $C_{10}H_7$). The tin loading was determined to be 12 wt % and titanium 3.2 and 2.0 wt %, respectively. The pore diameter of mesoporous silica decreased from 5.4 nm to 3.3 nm after the drug conjugation. The apoptotic effects have been studied in three different cancer cell lines: Human DLD-1 colon carcinoma, A2780 ovarian carcinoma and A431 epidermoid carcinoma. Although the observed drug release was low, only 10% of loaded tin and 1% of titanium from the constructs, the cancer cell viability was markedly reduced. It was found that silica-tin nanostructures induce apoptosis through the Fas–FasL system while silica-titanium constructs interfere with the TNF-α pathway. Therefore, it was concluded that the factors contributing to cancer cell death were small release of metallodrug from mesoporous silica as well as the presence of the nanoconstruct itself.

Other groups have investigated biomimetic hydroxyapatite nanocrystals as the delivery vehicles for platinum (II) [104–106]. The advantage of hydroxyapatite nanostructures is their intrinsic biocompatibility with biological systems since they are mineral constituents of bone [107]. For example, Natile's group used hydroxyapatite nanocrystals to deliver two platinum derivatives of *cis*-1,4-diaminocyclohexane ($[PtX_2(cis$-1,4-DACH)], X_2 = Cl_2 and 1,1-cyclobutanedicarboxylate (CBDCA) [105]. The complexes were adsorbed on the surface of the nanocrystals and the systems were tested in different cancer cell cultures. The in vitro studies included human colon (LoVo) and lung (A549) cancer cells as well as human osteosarcoma cells sensitive (U2OS) and resistant (U2OS/Pt) to cisplatin. The release of the complexes was accelerated at lower pH associated with the tumor's microenvironment, upon dissolution of hydrohyapatite nanocrystals. The hydroxyapatite-$PtX_2(cis$-1,4-DACH) nanoconstruct has proven to be more cytotoxic than the other derivative, primarily due to its fast release from the nanocrystals.

2.6. Dendrimers

Dendrimers are nanoscale macromolecules that have a spherical three-dimensional morphology and are formed by a successive addition of layers of branching groups. The step-wise synthesis of dendrimers provides the control over the number of branches, producing different generations of the dendrimers with distinct size and molecular weight. The external layer of the dendrimer has functional groups that can provide solubility or serve as an anchoring site to attach additional ligands. The drug molecules can be encapsulated in the interior of the branches or attached to the surface groups. Dendrimers can be composed of any type of monomer, and thus a variety of dendrimers exists, with many showing low cytotoxicity profiles while applied in biological settings [108].

Dendrimers have been investigated for the delivery of metallocomplexes [109–114]. For example, Gouveia et al. produced low generation dendrimers modified with an organometallic compound of Ru(II) [110]. The nitrile poly(alkylidenimine) dendritic scaffolds of different sizes were modified with an organometallic fragment $[Ru(\eta^5-C_5H_5)(PPh_3)_2]^+$ attached to the peripheral end groups of the dendrimers. The Ru(II)-dendrimers showed high biological activity in vitro against different cancer cell lines, such as Caco-2 (colon adenocarcinoma), CAL-72 (osteocarcoma), and MCF-7 (breast cancer). In addition, the metallodendrimers showed cytotoxicity higher than cisplatin in ovarian A2780 cells and A2780-platinum resistant cells.

In another example, Smith's group studied organometallic dendrimers of diaminobutane containing ruthenium(II)–*p*-cymene, ruthenium(II)–hexamethylbenzene, rhodium(III)-cyclopentadienyl, and iridium(III)–cyclopentadienyl moieties [113]. The biological activity of the metallodendrimers was evaluated in vitro using A2780 cisplatin-sensitive and A2780cisR cisplatin-resistant human ovarian cancer cell lines as well as a non-tumorigenic HEK-293 human embryonic kidney cell line. All constructs showed cytotoxicity but the most pronounced effects were observed for ferrocenyl-derived ruthenium(II)–hexamethylbenzene metallodendrimer. Importantly, all metallodendrimers showed lower cytotoxicity towards non-malignant cells. Interestingly, it was proposed that the possible mode of action of these constructs might involve non-covalent interactions with DNA.

Li et al. formulated supramolecular dendritic system for Pt(II) delivery (Figure 6). The building blocks included the lipoic acid functionalized dendrons, Pt(II)–PEG conjugate, and a NIR fluorophore for in vivo imaging purposes [114]. The core was bridged via disulfide bonds and incorporated Cy5.5 dye, while Pt(II)–PEG conjugate was attached to the terminal carboxyl groups of lipoic acid. Pt(II) formed carboplatin-like complex with the carboxyl groups of the acid, and PEG moiety provided the shell of the nanoplatform. The in vitro studies with lung adenocarcinoma A549 cells showed internalization of the nanostructures via endocytosis and high cytotoxicity, while in vivo studies demonstrated efficacy comparable to cisplatin with avoiding the nephrotoxicity. NIR imaging revealed the localization of nanostructures in vitro and in vivo facilitating the determination of Pt(II) delivery to target sites.

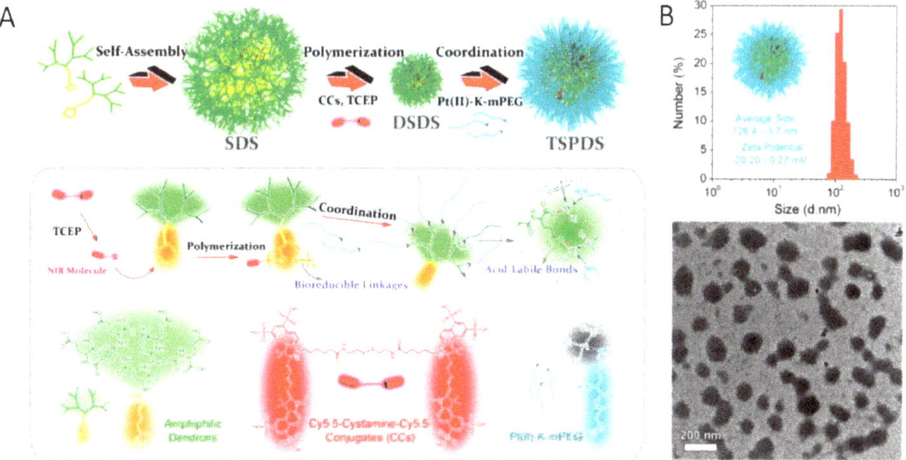

Figure 6. (**A**) Schematic illustrations for molecular and supramolecular engineering of theranostic supramolecular PEGylated dendritic systems. (**B**) Dynamic Light Scattering (DLS) results (in aqueous solutions) and TEM images for Theranostic supramolecular PEGylated dendritic systems (TSPDSs). Reprinted with permission from reference [114].

3. Biological Carriers

Natural biomacromolecules, such as proteins and peptides, are attractive carriers for small drug molecules due to their inherent biodegradability, biocompatibility, low toxicity, and high aqueous solubility. Additionally, biological drug carriers can recognize tumor markers overexpressed by cancer cells, which has been readily used in targeted therapies. Finally, peptides and proteins are easy to manufacture and stable, which is often challenging to accomplish with synthetic nanoparticles [115].

3.1. Peptides

The delivery of metallodrugs using peptides as biological carriers has been investigated. Wlodarczyk et al. used the SV40 large T antigen-derived PKKKRKV peptide known as nuclear localization sequence (NLS) peptide, as a carrier to deliver Pt(II) therapy [116]. The NLS peptide is a recognition signal on proteins facilitating the nuclear transport of biomolecules. In the study, the N-terminal of the NLS peptide was modified with malonic acid derivative during the solid-state peptide synthesis, and the carboplatin-like complex was formed directly on the peptide from activated Pt(II). The schematic of the construct is presented in Figure 7A. The solubility of the Pt-NLS hybrid was 50 mg/mL, which was much higher than that of native carboplatin (10 mg/mL). The construct was found to effectively translocate through the cellular membrane (Figure 7B) and deliver platinum to the nucleus. Importantly, the Pt-NLS hybrid showed high biological activity in ovarian cancer cell lines, overcoming platinum resistance. This can be attributed to higher Pt(II) content in the resistant cells. The quantification of Pt(II) in the nucleus from two isogenic A2780 (Pt sensitive) and CP70 (Pt resistant) cells showed that in A2780 cells one Pt(II) was attached every 20th DNA turn, while in CP70 cells every 10th turn.

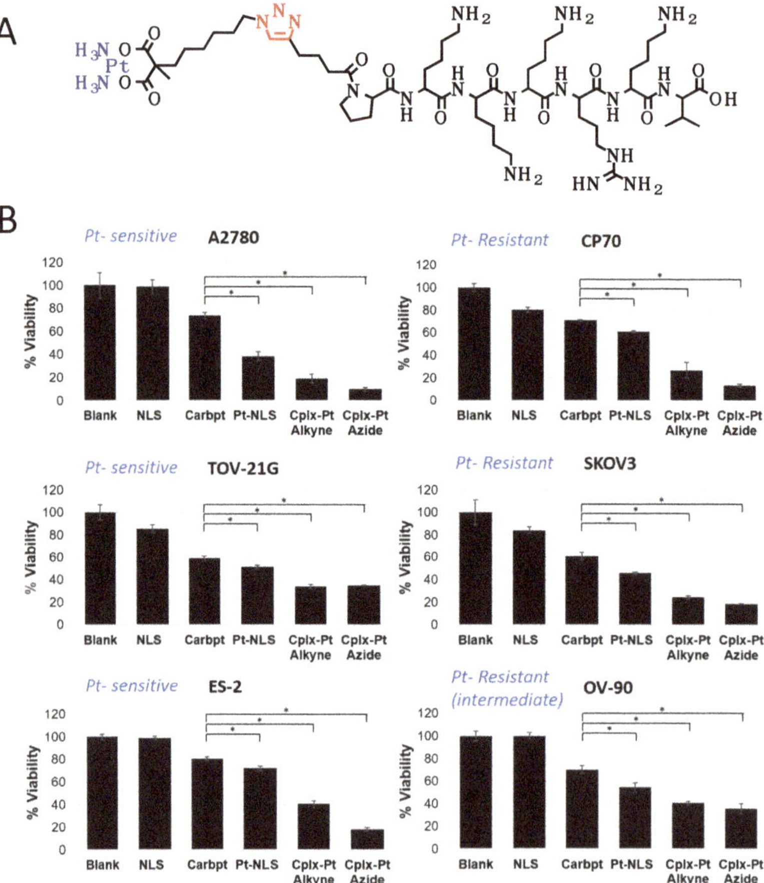

Figure 7. (**A**) Chemical structure of the Pt(II)- nuclear localization sequence (NLS) hybrid. (**B**) Viability of platinum sensitive (left) and resistant (right) cells after 72 h incubation with Pt-NLS hybrid and controls. * Each column represents the mean and standard deviation of $N = 3$ and $p < 0.005$. Reprinted with permission from reference [116].

Similarly, Noor et al. used the NLS peptide to form ferrocene–NLS conjugate [117]. The ferrocene carboxylic acid was attached to the N-terminus of the NLS peptide. The evaluation of the conjugate with Hep G2 cells did not show the cytotoxic effects. However, the presence of metallocene moiety increased the cellular uptake of the construct, which can potentially serve as a platform to add additional functionalities. The same group tested the cobaltocenium–NLS conjugates prepared using the same approach, and also observed the enhanced cellular accumulation of the construct [118].

3.2. Antibodies and Proteins

Antibody–drug conjugates (ADC) represent a highly potent class of therapeutics that combines precise cancer cell recognition by monoclonal antibodies mAb with cell killing by small molecule drugs. The ADC conjugates with metallodrugs have also been developed. Recently, Gupta et al. reported a Pt(II)-based linker that can re-bridge the inter-chain cysteines in the antibody [119]. The schematic of the construct is shown in Figure 8A. This strategy resulted in ADC-Pt(II) constructs with higher stability than conventional ADCs connected via maleimide-based chemistry, reducing the drug exchange with blood serum albumin and off-target toxicity. The ADC-Pt(II) demonstrated high toxicity in vitro, and the in vivo studies using the A549 non-small cell lung cancer model showed marked tumor growth inhibition (Figure 8B) with up to ~80% of apoptotic cells in tumor sections. This novel technology can be extended to various types of antibodies with a potential to target multiple malignancies.

Hanif et al. demonstrated the functionalization of Ru(II) (arene) complexes with maleimide. The complexes alone were characterized in vitro with human ovarian (CH1), colon (SW480), and non-small cell lung cancer (A549) cells, and showed high biological activity. The complexes were attached to thiolated biomolecules, such as Cysteine (Cys), glutathione (GSH), N-acetylcysteine (NAC), and N-acetylcysteine methyl ester (N-AcCysMe), as well as human serum albumin (HSA) [120]. The coupling of the complexes to biomolecules was fast and was completed within 1 h, while binding to HSA occurred in 72 h. These results demonstrate an effective approach to attach the maleimide-modified metallodrugs to Cys-containing carrier biomolecules.

In A different approach, Zhang et al. studied the delivery of vanadocene dichloride (Cp_2VCl_2) using human serum transferrin (h-Tf) and apotransferrin (apo-Tf) plasma proteins [121]. The transferrin protein has two Fe(III) binding sites and its primary role is to transport iron. It was demonstrated that Cp_2VCl_2 could bind with h-Tf/apo-Tf by noncovalent conjugation at pH 7.4 and 4.8. Moreover, the conjugates showed dose-dependent cytotoxicity towards human lung adenocarcinoma cell line A549. Interestingly, higher cytotoxicity was observed for the vanadocene–protein conjugates than for the free complex.

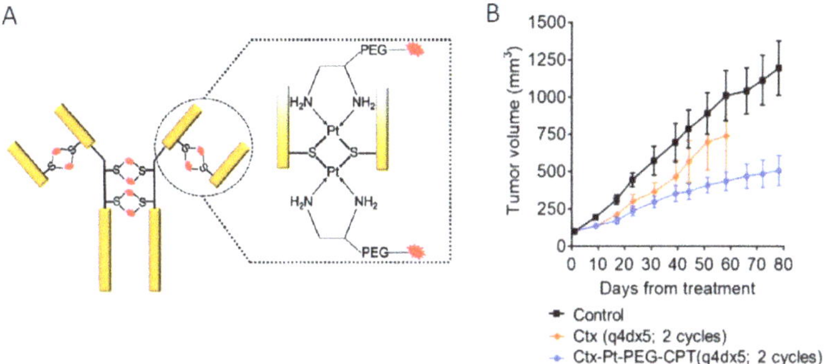

Figure 8. (**A**) Pt(II)-linker re-bridges the antibody chains with strong Pt–S interaction imparting stability, homogeneity and site-specificity. (**B**) In vivo efficacy of the Ctx–Pt–PEG–CPT treated animals showed significant differences in tumor volume compared with the control group ($p < 0.016$), with a tumor growth inhibition (TGI) of 55%. Reprinted with permission from reference [119].

4. Conclusions

In summary, nanocarriers have proven to be powerful tools in enhancing the delivery of metallodrugs to cancer cells. Nanotechnology offers versatility in nanostructure synthesis methods and choice of materials, giving the control over physicochemical properties of the final nanodelivery system. This also allows encapsulating metallodrugs at sufficient payload, increasing their solubility, limiting the release in the systemic circulation preventing toxicities, and eventually heightening their efficacy in vivo. The main advantage of long-circulating nanocarriers is their accumulation in the tumor via EPR, which leads to increased drug payloads in the tumor. Moreover, the nanodelivery vehicles can be decorated with targeting ligands that further enhance the entry into cancer cells via receptor-mediated endocytosis. Furthermore, incorporating fluorescent dyes to the nanocarriers allows unique theranostic applications. Another attractive strategy involves conjugation of metallodrugs to biomacromolecules, such as antibodies or functional peptides. They are cheap, easy to produce, stable, and, most importantlym inherently biodegradable. Also, superior cancer cell recognition can be achieved using mAb, which, coupled to metallodrugs, amplifies their intracellular delivery. The biological carriers might be extended even further to include vitamin-mediated drug targeting [122,123] or steroid-conjugates [124]. These systems have the potential to evolve into highly efficacious metallodrug-based therapies.

Funding: This research received no external funding.

Conflicts of Interest: The authors declare no conflicts of interest.

References

1. Oberoi, H.S.; Nukolova, N.V.; Kabanov, A.V.; Bronich, T.K. Nanocarriers for delivery of platinum anticancer drugs. *Adv. Drug Deliv. Rev.* **2013**, *65*, 1667–1685. [CrossRef] [PubMed]

2. Fichtinger-Schepman, A.M.; van Oosterom, A.T.; Lohman, P.H.; Berends, F. *cis*-Diamminedichloroplatinum(II)-induced DNA adducts in peripheral leukocytes from seven cancer patients: Quantitative immunochemical detection of the adduct induction and removal after a single dose of *cis*-diamminedichloroplatinum(II). *Cancer Res.* **1987**, *47*, 3000–3004. [PubMed]

3. Englander, E.W. DNA damage response in peripheral nervous system: Coping with cancer therapy-induced DNA lesions. *DNA Repair* **2013**, *12*, 685–690. [CrossRef] [PubMed]

4. Galluzzi, L.; Senovilla, L.; Vitale, I.; Michels, J.; Martins, I.; Kepp, O.; Castedo, M.; Kroemer, G. Molecular mechanisms of cisplatin resistance. *Oncogene* **2012**, *31*, 1869–1883. [CrossRef] [PubMed]

5. Boeckman, H.J.; Trego, K.S.; Turchi, J.J. Cisplatin sensitizes cancer cells to ionizing radiation via inhibition of nonhomologous end joining. *Mol. Cancer Res. MCR* **2005**, *3*, 277–285. [CrossRef] [PubMed]

6. Aris, S.M.; Farrell, N.P. Towards Antitumor Active *trans*-Platinum Compounds. *Eur. J. Inorg. Chem.* **2009**, *2009*, 1293. [CrossRef]

7. Montero, E.I.; Diaz, S.; Gonzalez-Vadillo, A.M.; Perez, J.M.; Alonso, C.; Navarro-Ranninger, C. Preparation and characterization of novel *trans*-[PtCl(2)(amine)(isopropylamine)] compounds: Cytotoxic activity and apoptosis induction in *ras*-transformed cells. *J. Med. Chem.* **1999**, *42*, 4264–4268. [CrossRef]

8. Kasparkova, J.; Novakova, O.; Marini, V.; Najajreh, Y.; Gibson, D.; Perez, J.M.; Brabec, V. Activation of *trans* geometry in bifunctional mononuclear platinum complexes by a piperidine ligand. Mechanistic studies on antitumor action. *J. Biol. Chem.* **2003**, *278*, 47516–47525. [CrossRef]

9. Novakova, O.; Kasparkova, J.; Malina, J.; Natile, G.; Brabec, V. DNA-protein cross-linking by *trans*-[PtCl$_2$(E-iminoether)$_2$]. A concept for activation of the *trans* geometry in platinum antitumor complexes. *Nucleic Acids Res.* **2003**, *31*, 6450–6460. [CrossRef]

10. Mangrum, J.B.; Farrell, N.P. Excursions in polynuclear platinum DNA binding. *Chem. Commun.* **2010**, *46*, 6640–6650. [CrossRef]

11. Kloster, M.B.; Hannis, J.C.; Muddiman, D.C.; Farrell, N. Consequences of nucleic acid conformation on the binding of a trinuclear platinum drug. *Biochemistry* **1999**, *38*, 14731–14737. [CrossRef] [PubMed]

12. Billecke, C.; Finniss, S.; Tahash, L.; Miller, C.; Mikkelsen, T.; Farrell, N.P.; Bogler, O. Polynuclear platinum anticancer drugs are more potent than cisplatin and induce cell cycle arrest in glioma. *Neuro-Oncology* **2006**, *8*, 215–226. [CrossRef] [PubMed]

13. Brabec, V.; Kasparkova, J.; Menon, V.; Farrell, N.P. Polynuclear Platinum Complexes. Structural Diversity and DNA Binding. *Met. Ions Life Sci.* **2018**, *18*. [CrossRef]

14. Malina, J.; Farrell, N.P.; Brabec, V. Substitution-Inert Polynuclear Platinum Complexes That Inhibit the Activity of DNA Polymerase in Triplex-Forming Templates. *Angew. Chem. Int. Ed. Engl.* **2018**, *57*, 8535–8539. [CrossRef] [PubMed]

15. Venkatesh, V.; Sadler, P.J. Platinum(IV) Prodrugs. *Met. Ions Life Sci.* **2018**, *18*. [CrossRef]

16. Najjar, A.; Rajabi, N.; Karaman, R. Recent Approaches to Platinum(IV) Prodrugs: A Variety of Strategies for Enhanced Delivery and Efficacy. *Curr. Pharm. Des.* **2017**, *23*, 2366–2376. [CrossRef] [PubMed]

17. Noffke, A.L.; Habtemariam, A.; Pizarro, A.M.; Sadler, P.J. Designing organometallic compounds for catalysis and therapy. *Chem. Commun.* **2012**, *48*, 5219–5246. [CrossRef] [PubMed]

18. Liang, J.X.; Zhong, H.J.; Yang, G.; Vellaisamy, K.; Ma, D.L.; Leung, C.H. Recent development of transition metal complexes with in vivo antitumor activity. *J. Inorg. Biochem.* **2017**, *177*, 276–286. [CrossRef] [PubMed]

19. Keppler, B.K.; Friesen, C.; Vongerichten, H.; Vogel, E. *Metal Complexes in Cancer Chemotherapy*; VCH: Weinheim, Germany, 1993; pp. 297–323, ISBN 1-56081-216-8.

20. Hanif, M.; Hartinger, C.G. Anticancer metallodrugs: Where is the next cisplatin? *Future Med. Chem.* **2018**, *10*, 615–617. [CrossRef]

21. Alessio, E.; Messori, L. The Deceptively Similar Ruthenium(III) Drug Candidates KP1019 and NAMI-A Have Different Actions. What Did We Learn in the Past 30 Years? *Met. Ions Life Sci.* **2018**, *18*.

22. Hartinger, C.G.; Jakupec, M.A.; Zorbas-Seifried, S.; Groessl, M.; Egger, A.; Berger, W.; Zorbas, H.; Dyson, P.J.; Keppler, B.K. KP1019, a new redox-active anticancer agent—preclinical development and results of a clinical phase I study in tumor patients. *Chem. Biodivers.* **2008**, *5*, 2140–2155. [CrossRef] [PubMed]

23. Leijen, S.; Burgers, S.A.; Baas, P.; Pluim, D.; Tibben, M.; van Werkhoven, E.; Alessio, E.; Sava, G.; Beijnen, J.H.; Schellens, J.H. Phase I/II study with ruthenium compound NAMI-A and gemcitabine in patients with non-small cell lung cancer after first line therapy. *Investig. New Drugs* **2015**, *33*, 201–214. [CrossRef] [PubMed]

24. Trondl, R.; Heffeter, P.; Kowol, C.R.; Jakupec, M.A.; Berger, A.; Keppler, B.K. NKP-1339, the first ruthenium-based anticancer drug on the edge to clinical application. *Chem. Sci.* **2014**, *5*, 2925–2932. [CrossRef]

25. Antonarakis, E.S.; Emadi, A. Ruthenium-based chemotherapeutics: Are they ready for prime time? *Cancer Chemother. Pharmacol.* **2010**, *66*, 1–9. [CrossRef] [PubMed]

26. Thota, S.; Rodrigues, D.A.; Crans, D.C.; Barreiro, E.J. Ru(II) Compounds: Next-Generation Anticancer Metallotherapeutics? *J. Med. Chem.* **2018**. [CrossRef] [PubMed]

27. Jansson, P.J.; Sharpe, P.C.; Bernhardt, P.V.; Richardson, D.R. Novel thiosemicarbazones of the ApT and DpT series and their copper complexes: Identification of pronounced redox activity and characterization of their antitumor activity. *J. Med. Chem.* **2010**, *53*, 5759–5769. [CrossRef] [PubMed]

28. Kowol, C.R.; Heffeter, P.; Miklos, W.; Gille, L.; Trondl, R.; Cappellacci, L.; Berger, W.; Keppler, B.K. Mechanisms underlying reductant-induced reactive oxygen species formation by anticancer copper(II) compounds. *J. Biol. Inorg. Chem.* **2012**, *17*, 409–423. [CrossRef]

29. Lovejoy, D.B.; Jansson, P.J.; Brunk, U.T.; Wong, J.; Ponka, P.; Richardson, D.R. Antitumor activity of metal-chelating compound Dp44mT is mediated by formation of a redox-active copper complex that accumulates in lysosomes. *Cancer Res.* **2011**, *71*, 5871–5880. [CrossRef]

30. Mirabelli, C.K.; Johnson, R.K.; Sung, C.M.; Faucette, L.; Muirhead, K.; Crooke, S.T. Evaluation of the in vivo antitumor activity and in vitro cytotoxic properties of auranofin, a coordinated gold compound, in murine tumor models. *Cancer Res.* **1985**, *45*, 32–39.

31. Van Rijt, S.H.; Romero-Canelon, I.; Fu, Y.; Shnyder, S.D.; Sadler, P.J. Potent organometallic osmium compounds induce mitochondria-mediated apoptosis and S-phase cell cycle arrest in A549 non-small cell lung cancer cells. *Metallomics* **2014**, *6*, 1014–1022. [CrossRef]

32. Kopf-Maier, P. Complexes of metals other than platinum as antitumour agents. *Eur. J. Clin. Pharmacol.* **1994**, *47*, 1–16. [CrossRef] [PubMed]

33. Lazarevic, T.; Rilak, A.; Bugarcic, Z.D. Platinum, palladium, gold and ruthenium complexes as anticancer agents: Current clinical uses, cytotoxicity studies and future perspectives. *Eur. J. Med. Chem.* **2017**, *142*, 8–31. [CrossRef] [PubMed]

34. Wani, W.A.; Prashar, S.; Shreaz, S.; Gomez-Ruiz, S. Nanostructured materials functionalized with metal complexes: In search of alternatives for administering anticancer metallodrugs. *Coord. Chem. Rev.* **2016**, *312*, 67–98. [CrossRef]

35. Sarkar, A. Novel platinum compounds and nanoparticles as anticancer agents. *Pharm. Pat. Anal.* **2018**, *7*, 33–46. [CrossRef] [PubMed]

36. Din, F.U.; Aman, W.; Ullah, I.; Qureshi, O.S.; Mustapha, O.; Shafique, S.; Zeb, A. Effective use of nanocarriers as drug delivery systems for the treatment of selected tumors. *Int. J. Nanomed.* **2017**, *12*, 7291–7309. [CrossRef] [PubMed]

37. Malam, Y.; Loizidou, M.; Seifalian, A.M. Liposomes and nanoparticles: Nanosized vehicles for drug delivery in cancer. *Trends Pharmacol. Sci.* **2009**, *30*, 592–599. [CrossRef] [PubMed]

38. Iyer, A.K.; Khaled, G.; Fang, J.; Maeda, H. Exploiting the enhanced permeability and retention effect for tumor targeting. *Drug Discov. Today* **2006**, *11*, 812–818. [CrossRef] [PubMed]

39. Natfji, A.A.; Ravishankar, D.; Osborn, H.M.I.; Greco, F. Parameters Affecting the Enhanced Permeability and Retention Effect: The Need for Patient Selection. *J. Pharm. Sci.* **2017**, *106*, 3179–3187. [CrossRef] [PubMed]

40. Mishra, B.; Patel, B.B.; Tiwari, S. Colloidal nanocarriers: A review on formulation technology, types and applications toward targeted drug delivery. *Nanomedicine* **2010**, *6*, 9–24. [CrossRef] [PubMed]

41. Sun, T.; Zhang, Y.S.; Pang, B.; Hyun, D.C.; Yang, M.; Xia, Y. Engineered nanoparticles for drug delivery in cancer therapy. *Angew. Chem. Int. Ed. Engl.* **2014**, *53*, 12320–12364. [CrossRef] [PubMed]

42. Dragulska, S.A.; Chen, Y.; Wlodarczyk, M.T.; Poursharifi, M.; Dottino, P.; Ulijn, R.V.; Martignetti, J.A.; Mieszawska, A.J. Tripeptide-Stabilized Oil-in-Water Nanoemulsion of an Oleic Acids-Platinum(II) Conjugate as an Anticancer Nanomedicine. *Bioconjugate Chem.* **2018**, *29*, 2514–2519. [CrossRef] [PubMed]

43. Bozzuto, G.; Molinari, A. Liposomes as nanomedical devices. *Int. J. Nanomed.* **2015**, *10*, 975–999. [CrossRef] [PubMed]

44. Felice, B.; Prabhakaran, M.P.; Rodriguez, A.P.; Ramakrishna, S. Drug delivery vehicles on a nano-engineering perspective. *Mater. Sci. Eng. C Mater. Biol. Appl.* **2014**, *41*, 178–195. [CrossRef] [PubMed]

45. Fanciullino, R.; Ciccolini, J. Liposome-encapsulated anticancer drugs: Still waiting for the magic bullet? *Curr. Med. Chem.* **2009**, *16*, 4361–4371. [CrossRef] [PubMed]

46. Caster, J.M.; Patel, A.N.; Zhang, T.; Wang, A. Investigational nanomedicines in 2016: A review of nanotherapeutics currently undergoing clinical trials. *Wiley Interdiscip. Rev. Nanomed. Nanobiotechnol.* **2017**, *9*. [CrossRef] [PubMed]

47. Ventola, C.L. Progress in Nanomedicine: Approved and Investigational Nanodrugs. *Pharm. Ther.* **2017**, *42*, 742–755.

48. Boulikas, T. Clinical overview on Lipoplatin: A successful liposomal formulation of cisplatin. *Expert Opin. Investig. Drugs* **2009**, *18*, 1197–1218. [CrossRef]

49. Stathopoulos, G.P.; Boulikas, T. Lipoplatin formulation review article. *J. Drug Deliv.* **2012**, *2012*, 581363. [CrossRef]

50. Zisman, N.; Dos Santos, N.; Johnstone, S.; Tsang, A.; Bermudes, D.; Mayer, L.; Tardi, P. Optimizing Liposomal Cisplatin Efficacy through Membrane Composition Manipulations. *Chemother. Res. Pract.* **2011**, *2011*, 213848. [CrossRef]

51. Shen, J.; Kim, H.C.; Wolfram, J.; Mu, C.; Zhang, W.; Liu, H.; Xie, Y.; Mai, J.; Zhang, H.; Li, Z.; et al. A Liposome Encapsulated Ruthenium Polypyridine Complex as a Theranostic Platform for Triple-Negative Breast Cancer. *Nano Lett.* **2017**, *17*, 2913–2920. [CrossRef]

52. Rodrigues, F.P.; Carneiro, Z.A.; Mascharak, P.K.; Curti, C.; da Silva, R.S. Incorporation of a Ruthenium Nitrosyl Complex into Liposomes, the Nitric Oxide Released from these Liposomes and HepG2 Cell Death Mechanism. *Coord. Chem. Rev.* **2016**, *306*, 701–707. [CrossRef]

53. Tang, W.L.; Tang, W.H.; Li, S.D. Cancer theranostic applications of lipid-based nanoparticles. *Drug Discov. Today* **2018**, *23*, 1159–1166. [CrossRef] [PubMed]

54. Hanafy, N.A.N.; El-Kemary, M.; Leporatti, S. Micelles Structure Development as a Strategy to Improve Smart Cancer Therapy. *Cancers* **2018**, *10*, 238. [CrossRef] [PubMed]

55. Cabral, H.; Nishiyama, N.; Okazaki, S.; Koyama, H.; Kataoka, K. Preparation and biological properties of dichloro(1,2-diaminocyclohexane)platinum(II) (DACHPt)-loaded polymeric micelles. *J. Control. Release* **2005**, *101*, 223–232. [CrossRef] [PubMed]

56. Nishiyama, N.; Yokoyama, M.; Aoyagi, T.; Okano, T.; Sakurai, Y.; Kataoka, K. Preparation and Characterization of Self-Assembled Polymer−Metal Complex Micelle from *cis*-Dichlorodiammineplatinum(II) and Poly(ethylene glycol)−Poly(α,β-aspartic acid) Block Copolymer in an Aqueous Medium. *Langmuir* **1999**, *15*, 377–383. [CrossRef]

57. Nishiyama, N.; Kataoka, K. Preparation and characterization of size-controlled polymeric micelle containing *cis*-dichlorodiammineplatinum(II) in the core. *J. Control. Release* **2001**, *74*, 83–94. [CrossRef]

58. Plummer, R.; Wilson, R.H.; Calvert, H.; Boddy, A.V.; Griffin, M.; Sludden, J.; Tilby, M.J.; Eatock, M.; Pearson, D.G.; Ottley, C.J.; et al. A Phase I clinical study of cisplatin-incorporated polymeric micelles (NC-6004) in patients with solid tumours. *Br. J. Cancer* **2011**, *104*, 593–598. [CrossRef] [PubMed]

59. Bontha, S.; Kabanov, A.V.; Bronich, T.K. Polymer micelles with cross-linked ionic cores for delivery of anticancer drugs. *J. Control. Release* **2006**, *114*, 163–174. [CrossRef]

60. Heffeter, P.; Riabtseva, A.; Senkiv, Y.; Kowol, C.R.; Korner, W.; Jungwith, U.; Mitina, N.; Keppler, B.K.; Konstantinova, T.; Yanchuk, I.; et al. Nanoformulation improves activity of the (pre)clinical anticancer ruthenium complex KP1019. *J. Biomed. Nanotechnol.* **2014**, *10*, 877–884. [CrossRef]

61. Huynh, V.T.; Quek, J.Y.; de Souza, P.L.; Stenzel, M.H. Block copolymer micelles with pendant bifunctional chelator for platinum drugs: Effect of spacer length on the viability of tumor cells. *Biomacromolecules* **2012**, *13*, 1010–1023. [CrossRef]

62. Huynh, V.T.; Souza, P.; Stenzel, M.H. Polymeric Micelles with Pendant Dicarboxylato Chelating Ligands Prepared via a Michael Addition for *cis*-Platinum Drug Delivery. *Macromolecules* **2011**, *44*, 7888–7900. [CrossRef]

63. Blunden, B.M.; Lu, H.; Stenzel, M.H. Enhanced delivery of the RAPTA-C macromolecular chemotherapeutic by conjugation to degradable polymeric micelles. *Biomacromolecules* **2013**, *14*, 4177–4188. [CrossRef] [PubMed]

64. Lu, M.; Chen, F.; Noy, J.M.; Lu, H.; Stenzel, M.H. Enhanced Antimetastatic Activity of the Ruthenium Anticancer Drug RAPTA-C Delivered in Fructose-Coated Micelles. *Macromol. Biosci.* **2017**, *17*. [CrossRef]

65. Chatterjee, S.; Kundu, S.; Bhattacharyya, A.; Hartinger, C.G.; Dyson, P.J. The ruthenium(II)–arene compound RAPTA-C induces apoptosis in EAC cells through mitochondrial and p53-JNK pathways. *J. Biol. Inorg. Chem.* **2008**, *13*, 1149–1155. [CrossRef]

66. Zhang, L.; Chan, J.M.; Gu, F.X.; Rhee, J.W.; Wang, A.Z.; Radovic-Moreno, A.F.; Alexis, F.; Langer, R.; Farokhzad, O.C. Self-assembled lipid–polymer hybrid nanoparticles: A robust drug delivery platform. *ACS Nano* **2008**, *2*, 1696–1702. [CrossRef] [PubMed]

67. Jagur-Grodzinski, J. Biomedical application of functional polymers. *React. Funct. Polym.* **1999**, *39*, 99–138. [CrossRef]

68. Cryan, S.A. Carrier-based strategies for targeting protein and peptide drugs to the lungs. *AAPS J.* **2005**, *7*, E20–E41. [CrossRef] [PubMed]

69. Danhier, F.; Ansorena, E.; Silva, J.M.; Coco, R.; Le Breton, A.; Preat, V. PLGA-based nanoparticles: An overview of biomedical applications. *J. Control. Release* **2012**, *161*, 505–522. [CrossRef]

70. Gill, M.R.; Menon, J.U.; Jarman, P.J.; Owen, J.; Skaripa-Koukelli, I.; Able, S.; Thomas, J.A.; Carlisle, R.; Vallis, K.A. ^{111}In-labelled polymeric nanoparticles incorporating a ruthenium-based radiosensitizer for EGFR-targeted combination therapy in oesophageal cancer cells. *Nanoscale* **2018**, *10*, 10596–10608. [CrossRef]

71. Bœuf, G.; Roullin, G.V.; Moreau, J.; Gulick, L.V.; Pineda, N.Z.; Terryn, C.; Ploton, D.; Andry, M.C.; Chuburu, F.; Dukic, S.; et al. Encapsulated Ruthenium(II) Complexes in Biocompatible Poly(D,L-lactide-*co*-glycolide) Nanoparticles for Application in Photodynamic Therapy. *ChemPlusChem* **2014**, *79*, 171–180. [CrossRef]

72. Yang, J.; Liu, W.; Sui, M.; Tang, J.; Shen, Y. Platinum (IV)-coordinate polymers as intracellular reduction-responsive backbone-type conjugates for cancer drug delivery. *Biomaterials* **2011**, *32*, 9136–9143. [CrossRef] [PubMed]

73. Dhar, S.; Gu, F.X.; Langer, R.; Farokhzad, O.C.; Lippard, S.J. Targeted delivery of cisplatin to prostate cancer cells by aptamer functionalized Pt(IV) prodrug-PLGA-PEG nanoparticles. *Proc. Natl. Acad. Sci. USA* **2008**, *105*, 17356–17361. [CrossRef] [PubMed]

74. Aryal, S.; Hu, C.M.; Zhang, L. Polymer–cisplatin conjugate nanoparticles for acid-responsive drug delivery. *ACS Nano* **2010**, *4*, 251–258. [CrossRef] [PubMed]

75. Rieter, W.J.; Pott, K.M.; Taylor, K.M.; Lin, W. Nanoscale coordination polymers for platinum-based anticancer drug delivery. *J. Am. Chem. Soc.* **2008**, *130*, 11584–11585. [CrossRef] [PubMed]

76. Fischer, B.; Heffeter, P.; Kryeziu, K.; Gille, L.; Meier, S.M.; Berger, W.; Kowol, C.R.; Keppler, B.K. Poly(lactic acid) nanoparticles of the lead anticancer ruthenium compound KP1019 and its surfactant-mediated activation. *Dalton Trans.* **2014**, *43*, 1096–1104. [CrossRef] [PubMed]

77. Mahajan, S.; Patharkar, A.; Kuche, K.; Maheshwari, R.; Deb, P.K.; Kalia, K.; Tekade, R.K. Functionalized carbon nanotubes as emerging delivery system for the treatment of cancer. *Int. J. Pharm.* **2018**, *548*, 540–558. [CrossRef] [PubMed]

78. Feazell, R.P.; Nakayama-Ratchford, N.; Dai, H.; Lippard, S.J. Soluble single-walled carbon nanotubes as longboat delivery systems for platinum(IV) anticancer drug design. *J. Am. Chem. Soc.* **2007**, *129*, 8438–8439. [CrossRef] [PubMed]

79. Dhar, S.; Liu, Z.; Thomale, J.; Dai, H.; Lippard, S.J. Targeted single-wall carbon nanotube-mediated Pt(IV) prodrug delivery using folate as a homing device. *J. Am. Chem. Soc.* **2008**, *130*, 11467–11476. [CrossRef] [PubMed]

80. Yoong, S.L.; Wong, B.S.; Zhou, Q.L.; Chin, C.F.; Li, J.; Venkatesan, T.; Ho, H.K.; Yu, V.; Ang, W.H.; Pastorin, G. Enhanced cytotoxicity to cancer cells by mitochondria-targeting MWCNTs containing platinum(IV) prodrug of cisplatin. *Biomaterials* **2014**, *35*, 748–759. [CrossRef] [PubMed]

81. Li, J.; Pant, A.; Chin, C.F.; Ang, W.H.; Menard-Moyon, C.; Nayak, T.R.; Gibson, D.; Ramaprabhu, S.; Panczyk, T.; Bianco, A.; Pastorin, G. In vivo biodistribution of platinum-based drugs encapsulated into multi-walled carbon nanotubes. *Nanomedicine* **2014**, *10*, 1465–1475. [CrossRef] [PubMed]

82. Muzi, L.; Menard-Moyon, C.; Russier, J.; Li, J.; Chin, C.F.; Ang, W.H.; Pastorin, G.; Risuleo, G.; Bianco, A. Diameter-dependent release of a cisplatin pro-drug from small and large functionalized carbon nanotubes. *Nanoscale* **2015**, *7*, 5383–5394. [CrossRef]

83. Johnstone, T.C.; Suntharalingam, K.; Lippard, S.J. The Next Generation of Platinum Drugs: Targeted Pt(II) Agents, Nanoparticle Delivery, and Pt(IV) Prodrugs. *Chem. Rev.* **2016**, *116*, 3436–3486. [CrossRef] [PubMed]

84. Zhang, P.; Huang, H.; Huang, J.; Chen, H.; Wang, J.; Qiu, K.; Zhao, D.; Ji, L.; Chao, H. Noncovalent Ruthenium(II) Complexes-Single-Walled Carbon Nanotube Composites for Bimodal Photothermal and Photodynamic Therapy with Near-Infrared Irradiation. *ACS Appl. Mater. Interfaces* **2015**, *7*, 23278–23290. [CrossRef] [PubMed]

85. Wang, N.; Feng, Y.; Zeng, L.; Zhao, Z.; Chen, T. Functionalized Multiwalled Carbon Nanotubes as Carriers of Ruthenium Complexes to Antagonize Cancer Multidrug Resistance and Radioresistance. *ACS Appl. Mater. Interfaces* **2015**, *7*, 14933–14945. [CrossRef] [PubMed]

86. Wang, F.; Li, C.; Cheng, J.; Yuan, Z. Recent Advances on Inorganic Nanoparticle-Based Cancer Therapeutic Agents. *Int. J. Environ. Res. Public Health* **2016**, *13*, 1182. [CrossRef]

87. Xiong, C.; Lu, W.; Zhou, M.; Wen, X.; Li, C. Cisplatin-loaded hollow gold nanoparticles for laser-triggered release. *Cancer Nanotechnol.* **2018**, *9*, 6. [CrossRef]

88. Tan, J.; Cho, T.J.; Tsai, D.H.; Liu, J.; Pettibone, J.M.; You, R.; Hackley, V.A.; Zachariah, M.R. Surface Modification of Cisplatin-Complexed Gold Nanoparticles and Its Influence on Colloidal Stability, Drug Loading, and Drug Release. *Langmuir* **2018**, *34*, 154–163. [CrossRef] [PubMed]

89. Shi, S.; Chen, X.; Wei, J.; Huang, Y.; Weng, J.; Zheng, N. Platinum(IV) prodrug conjugated Pd@Au nanoplates for chemotherapy and photothermal therapy. *Nanoscale* **2016**, *8*, 5706–5713. [CrossRef]

90. Sánchez-Paradinas, S.; Pérez-Andrés, M.; Almendral-Parra, M.J.; Rodríguez-Fernández, E.; Millán, A.; Palacio, F.; Orfao, A.; Criado, J.J.; Fuentes, M. Enhanced cytotoxic activity of bile acid cisplatin derivatives by conjugation with gold nanoparticles. *J. Inorg. Biochem.* **2014**, *131*, 8–11. [CrossRef]

91. Shi, Y.; Goodisman, J.; Dabrowiak, J.C. Cyclodextrin capped gold nanoparticles as a delivery vehicle for a prodrug of cisplatin. *Inorg. Chem.* **2013**, *52*, 9418–9426. [CrossRef]

92. Pramanik, A.K.; Palanimuthu, D.; Somasundaram, K.; Samuelson, A.G. Biotin Decorated Gold Nanoparticles for Targeted Delivery of a Smart-Linked Anticancer Active Copper Complex: In Vitro and In Vivo Studies. *Bioconjugate Chem.* **2016**, *27*, 2874–2885. [CrossRef] [PubMed]

93. Sun, D.; Liu, Y.; Yu, Q.; Qin, X.; Yang, L.; Zhou, Y.; Chen, L.; Liu, J. Inhibition of tumor growth and vasculature and fluorescence imagingusing functionalized ruthenium–thiol protected seleniumnanoparticles. *Biomaterials* **2014**, *35*, 1572–1583. [CrossRef] [PubMed]

94. Liu, T.; Zeng, L.; Jiang, W.; Fu, Y.; Zheng, W.; Chen, T. Rational design of cancer-targeted selenium nanoparticles to antagonize multidrug resistance in cancer cells. *Nanomedicine* **2015**, *11*, 947–958. [CrossRef] [PubMed]

95. Bharti, C.; Nagaich, U.; Pal, A.K.; Gulati, N. Mesoporous silica nanoparticles in target drug delivery system: A review. *Int. J. Pharm. Investig.* **2015**, *5*, 124–133. [CrossRef] [PubMed]

96. Ellahioui, Y.; Prashar, S.; Gomez-Ruiz, S. A Short Overview on the Biomedical Applications of Silica, Alumina and Calcium Phosphate-based Nanostructured Materials. *Curr. Med. Chem.* **2016**, *23*, 4450–4467. [CrossRef] [PubMed]

97. Kaluderovic, G.N.; Perez-Quintanilla, D.; Sierra, I.; Prashar, S.; Hierro, I.; Zizak, Z.; Juranic, Z.D.; Fajardo, M.; Gomez-Ruiz, S. Study of the influence of the metal complex on the cytotoxic activity of titanocene-fucntionalized mesoporous materials. *J. Mater. Chem.* **2010**, *20*, 806–814. [CrossRef]

98. Kaluderovic, G.N.; Perez-Quintanilla, D.; Zizak, Z.; Juranic, Z.D.; Gomez-Ruiz, S. Improvement of cytotoxicity of titanocene-functionalized mesoporous materials by the increase of the titanium content. *Dalton Trans.* **2010**, *39*, 2597–2608. [CrossRef]

99. Garcia-Penas, A.; Gomez-Ruiz, S.; Perez-Quintanilla, D.; Paschke, R.; Sierra, I.; Prashar, S.; del Hierro, I.; Kaluderovic, G.N. Study of the cytotoxicity and particle action in human cancer cells of titanocene-functionalized materials with potential application against tumors. *J. Inorg. Biochem.* **2012**, *106*, 100–110. [CrossRef]

100. Ceballos-Torres, J.; Virag, P.; Cenariu, M.; Prashar, S.; Fajardo, M.; Fischer-Fodor, E.; Gomez-Ruiz, S. Anti-cancer applications of titanocene-functionalised nanostructured systems: An insight into cell death mechanisms. *Chemistry* **2014**, *20*, 10811–10828. [CrossRef]

101. Gomez-Ruiz, S.; Garcia-Penas, A.; Prashar, S.; Rodriguez-Dieguez, A.; Fischer-Fodor, E. Anticancer applications of nanostructured silica-based materials functionalized with titanocene derivatives: Induction of cell death mechanism through TNFR1 modulation. *Materials* **2018**, *11*, 224. [CrossRef]

102. Diaz-Garcia, D.; Cenariu, D.; Perez, Y.; Cruz, P.; Del Hierro, I.; Prashar, S.; Fischer-Fodor, E.; Gomez-Ruiz, S. Modulation of the mechanism of apoptosis in cancer cell lines by treatment with silica-based nanostructured materials functionalized with different metallodrugs. *Dalton Trans.* **2018**, *47*, 12284–12299. [CrossRef] [PubMed]

103. Ellahioui, Y.; Patra, M.; Mari, C.; Kaabi, R.; Karges, J.; Gasser, G.; Gomez-Ruiz, S. Mesoporous silica nanoparticles functionalised with a photoactive ruthenium(II) complex: Exploring the formulation of a metal-based photodynamic therapy photosensitiser. *Dalton Trans.* **2018**. [CrossRef] [PubMed]

104. Benedetti, M.; De Castro, F.; Romano, A.; Migoni, D.; Piccinni, B.; Verri, T.; Lelli, M.; Roveri, N.; Fanizzi, F.P. Adsorption of the *cis*-[Pt(NH$_3$)$_2$(P$_2$O$_7$)]($^{2-}$) (*phosphaplatin*) on hydroxyapatite nanocrystals as a smart way to selectively release activated *cis*-[Pt(NH3)$_2$Cl$_2$] (*cisplatin*) in tumor tissues. *J. Inorg. Biochem.* **2016**, *157*, 73–79. [CrossRef] [PubMed]

105. Lelli, M.; Roveri, N.; Marzano, C.; Hoeschele, J.D.; Curci, A.; Margiotta, N.; Gandin, V.; Natile, G. Hydroxyapatite nanocrystals as a smart, pH sensitive, delivery system for kiteplatin. *Dalton Trans.* **2016**, *45*, 13187–13195. [CrossRef] [PubMed]

106. Iafisco, M.; Palazzo, B.; Marchetti, M.; Margiotta, N.; Ostuni, R.; Natile, G.; Morpurgo, M.; Gandin, V.; Marzano, C.; Roveri, N. Smart delivery of antitumoral platinum complexes from biomimetic hydroxyapatite nanocrystals. *J. Mater. Chem.* **2009**, *19*, 8385–8392. [CrossRef]

107. Roveri, N.; Iafisco, M. Evolving application of biomimetic nanostructured hydroxyapatite. *Nanotechnol. Sci. Appl.* **2010**, *3*, 107–125. [CrossRef] [PubMed]

108. Abbasi, E.; Aval, S.F.; Akbarzadeh, A.; Milani, M.; Nasrabadi, H.T.; Joo, S.W.; Hanifehpour, Y.; Nejati-Koshki, K.; Pashaei-Asl, R. Dendrimers: Synthesis, applications, and properties. *Nanoscale Res. Lett.* **2014**, *9*, 247. [CrossRef] [PubMed]

109. Govender, P.; Renfrew, A.K.; Clavel, C.M.; Dyson, P.J.; Therrien, B.; Smith, G.S. Antiproliferative activity of chelating *N,O*- and *N,N*-ruthenium(II) arene functionalised poly(propyleneimine) dendrimer scaffolds. *Dalton Trans.* **2011**, *40*, 1158–1167. [CrossRef] [PubMed]

110. Gouveia, M.; Figueira, J.; Jardim, M.G.; Castro, R.; Tomas, H.; Rissanen, K.; Rodrigues, J. Poly(alkylidenimine) Dendrimers Functionalized with the Organometallic Moiety [Ru(η^5-C$_5$H$_5$)(PPh$_3$)$_2$]$^+$ as Promising Drugs Against *Cisplatin*-Resistant Cancer Cells and Human Mesenchymal Stem Cells. *Molecules* **2018**, *23*, 1471. [CrossRef]

111. Govender, P.; Sudding, L.C.; Clavel, C.M.; Dyson, P.J.; Therrien, B.; Smith, G.S. The influence of RAPTA moieties on the antiproliferative activity of peripheral-functionalised poly(salicylaldiminato) metallodendrimers. *Dalton Trans.* **2013**, *42*, 1267–1277. [CrossRef]

112. Govender, P.; Pai, S.; Schatzschneider, U.; Smith, G.S. Next generation PhotoCORMs: Polynuclear tricarbonylmanganese(I)-functionalized polypyridyl metallodendrimers. *Inorg. Chem.* **2013**, *52*, 5470–5478. [CrossRef] [PubMed]

113. Govender, P.; Riedel, T.; Dyson, P.J.; Smith, G.S. Regulating the anticancer properties of organometallic dendrimers using pyridylferrocene entities: Synthesis, cytotoxicity and DNA binding studies. *Dalton Trans.* **2016**, *45*, 9529–9539. [CrossRef] [PubMed]

114. Li, Y.; Li, Y.; Zhang, X.; Xu, X.; Zhang, Z.; Hu, C.; He, Y.; Gu, Z. Supramolecular PEGylated Dendritic Systems as pH/Redox Dual-Responsive Theranostic Nanoplatforms for Platinum Drug Delivery and NIR Imaging. *Theranostics* **2016**, *6*, 1293–1305. [CrossRef] [PubMed]

115. Zhang, Y.; Sun, T.; Jiang, C. Biomacromolecules as carriers in drug delivery and tissue engineering. *Acta Pharm. Sin. B* **2018**, *8*, 34–50. [CrossRef] [PubMed]

116. Wlodarczyk, M.T.; Camacho-Vanegas, O.; Dragulska, S.A.; Jarzecki, A.A.; Dottino, P.R.; Martignetti, J.A.; Mieszawska, A.J. Platinum (II) complex-nuclear localization sequence peptide hybrid for overcoming platinum resistance in cancer therapy. *ACS Biomater. Sci. Eng.* **2018**, *4*, 463–467. [CrossRef]

117. Noor, F.; Kinscherf, R.; Bonaterra, G.A.; Walczak, S.; Wolfl, S.; Metzler-Nolte, N. Enhanced cellular uptake and cytotoxicity studies of organometallic bioconjugates of the NLS peptide in Hep G2 cells. *ChemBioChem* **2009**, *10*, 493–502. [CrossRef] [PubMed]

118. Noor, F.; Wustholz, A.; Kinscherf, R.; Metzler-Nolte, N. A cobaltocenium-peptide bioconjugate shows enhanced cellular uptake and directed nuclear delivery. *Angew. Chem.* **2005**, *44*, 2429–2432. [CrossRef]

119. Gupta, N.; Kancharla, J.; Kaushik, S.; Ansari, A.; Hossain, S.; Goyal, R.; Pandey, M.; Sivaccumar, J.; Hussain, S.; Sarkar, A.; et al. Development of a facile antibody-drug conjugate platform for increased stability and homogeneity. *Chem. Sci.* **2017**, *8*, 2387–2395. [CrossRef]

120. Hanif, M.; Nazarov, A.A.; Legin, A.; Groessl, M.; Arion, V.B.; Jakupec, M.A.; Tsybin, Y.O.; Dyson, P.J.; Keppler, B.K.; Hartinger, C.G. Maleimide-functionalised organoruthenium anticancer agents and their binding to thiol-containing biomolecules. *Chem. Commun.* **2012**, *48*, 1475–1477. [CrossRef]

121. Zhang, Y.; Xiang, J.; Liu, Y.; Zhang, X.; Tang, Y. Constructing transferrin receptor targeted drug delivery system by using doxorubicin hydrochloride and vanadocene dichloride. *Bioorg. Med. Chem. Lett.* **2011**, *21*, 5982–5986. [CrossRef]

122. Babak, M.V.; Plazuk, D.; Meier, S.M.; Arabshahi, H.J.; Reynisson, J.; Rychlik, B.; Blauz, A.; Szulc, K.; Hanif, M.; Strobl, S.; et al. Half-sandwich ruthenium(II) biotin conjugates as biological vectors to cancer cells. *Chemistry* **2015**, *21*, 5110–5117. [CrossRef] [PubMed]

123. Scrase, T.G.; Page, S.M.; Barker, P.D.; Boss, S.R. Folates are potential ligands for ruthenium compounds in vivo. *Dalton Trans.* **2014**, *43*, 8158–8161. [CrossRef] [PubMed]

124. Ruiz, J.; Rodriguez, V.; Cutillas, N.; Espinosa, A.; Hannon, M.J. A potent ruthenium(II) antitumor complex bearing a lipophilic levonorgestrel group. *Inorg. Chem.* **2011**, *50*, 9164–9171. [CrossRef] [PubMed]

Article

In Vitro Cytotoxicity and In Vivo Antitumor Efficacy of Tetrazolato-Bridged Dinuclear Platinum(II) Complexes with a Bulky Substituent at Tetrazole C5

Seiji Komeda [1,*], Masako Uemura [1], Hiroki Yoneyama [2], Shinya Harusawa [2] and Keiichi Hiramoto [1]

[1] Faculty of Pharmaceutical Sciences, Suzuka University of Medical Science, Mie 513-8670, Japan; masako-u@suzuka-u.ac.jp (M.U.); hiramoto@suzuka-u.ac.jp (K.H.)

[2] Department of Pharmaceutical Organic Chemistry, Osaka University of Pharmaceutical Sciences, Osaka 569-1094, Japan; yoneyama@gly.oups.ac.jp (H.Y.); harusawa@gly.oups.ac.jp (S.H.)

* Correspondence: komedas@suzuka-u.ac.jp; Tel.: +81-59-340-0581

Received: 7 November 2018; Accepted: 24 December 2018; Published: 8 January 2019

Abstract: Tetrazolato-bridged dinuclear platinum(II) complexes ([{cis-Pt(NH$_3$)$_2$}$_2$(μ-OH)(μ-5-R-tetrazolato-$N2,N3$)]$^{2+}$; tetrazolato-bridged complexes) are a promising source of next-generation platinum-based drugs. β-Cyclodextrin (β-CD) forms inclusion complexes with bulky organic compounds or substituents, changing their polarity and molecular dimensions. Here, we determined by ^{1}H-NMR spectroscopy, the stability constants for inclusion complexes formed between β-CD and tetrazolato-bridged complexes with a bulky, lipophilic substituent at tetrazole C5 (complexes 1–3, phenyl, n-nonyl, and adamantyl substitution, respectively). We then determined the in vitro cytotoxicity and in vivo antitumor efficacy of complexes 1–3 against the Colon-26 colorectal cancer cell line in the absence or presence of equimolar β-CD. Compared with the platinum-based anticancer drug oxaliplatin (1R,2R-diaminocyclohexane)oxalatoplatinum(II)), complex 2 had similar cytotoxicity, complex 3 was moderately cytotoxic, and complex 1 was the least cytotoxic. The cytotoxicity of the complexes decreased in the presence of β-CD. When we examined the in vivo antitumor efficacy of complexes 1–3 (10 mg/kg) against homografted Colon-26 colorectal tumors in male BALB/c mice, they showed a relatively low tumor growth inhibition compared with oxaliplatin. However, in the presence of β-CD, complex 3 had higher in vivo antitumor efficacy than oxaliplatin, suggesting a new direction for future research into tetrazolato-bridged complexes with high in vivo antitumor activity.

Keywords: anticancer drug; cancer; cyclodextrin; drug discovery; platinum

---

1. Introduction

Platinum(II) coordination compounds are an important group of pharmacophores in cancer chemotherapy. The first Pt(II) coordination compound approved for clinical use was cisplatin (cis-diamminedichloridoplatinum(II)) in the 1970s [1–3], and since then, other related compounds have been developed—such as carboplatin (cis-diammine(1,1-cyclobutanedicarboxylato)platinum(II)) [4,5] and oxaliplatin (1R,2R-diaminocyclohexane)oxalatoplatinum(II)) [6,7]—which have fewer side effects than cisplatin and have been approved for different clinical applications (Figure 1). These Pt-based drugs remain some of the most utilized agents in current cancer chemotherapy.

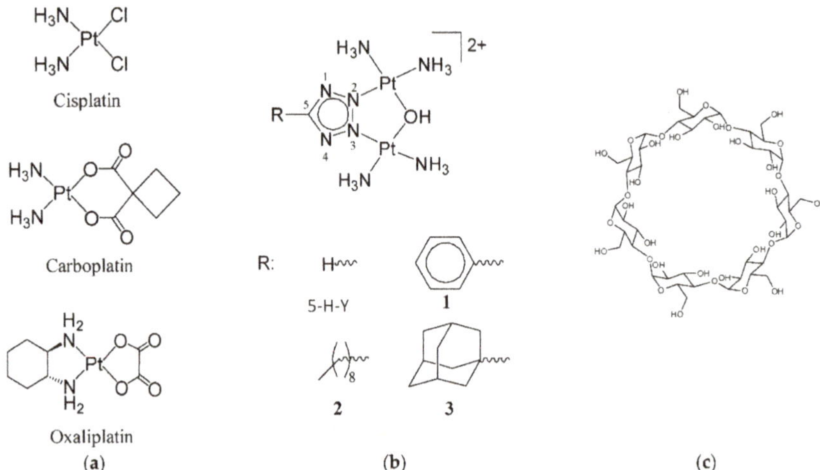

Figure 1. Chemical structures of (**a**) the platinum-based anticancer drugs cisplatin, carboplatin, and oxaliplatin, (**b**) the tetrazolato-bridged complexes 5-H-Y and **1–3**, and (**c**) β-cyclodextrin (β-CD).

Platinum(II) complexes with the general formula *cis*-[PtL$_2$X$_2$] (where L = an ammine or amine, and X = a leaving group, such as a halide or dicarboxylate) and antitumor efficacy are mostly much less effective against cisplatin-resistant cancer cells than against its parent cancer cells, due to their similar DNA-binding modes [8–10]. Therefore, we have been systematically modifying these general Pt(II) complexes [11–13] to provide structurally unconventional platinum complexes with antitumor spectra distinct from those of current platinum-based drugs [14–18].

The importance of the platinum–DNA interaction for antitumor efficacy has been shown for cisplatin, which forms covalent DNA adducts, such as 1,2-intrastrand and interstrand crosslinks [19–24], that induce local conformational changes in the DNA structure. Although these conformational changes are major determinants of the cytotoxicity of cisplatin, it remains unknown whether the most important factor in cytotoxicity is the formation of the DNA adducts themselves or the resulting conformational changes. We hypothesized that DNA adduct formation is the most important factor and thus designed azolato-bridged dinuclear Pt(II) complexes with the general formula [{*cis*-Pt(NH$_3$)$_2$}$_2$-(μ-OH)(μ-azolato)]$^{2+}$ (azolato = pyrazolato, 1,2,3-triazolato, or tetrazolato) [11–13] that can crosslink two adjacent nucleobases with minimal kinking of the double helix [25,26] and escape from the DNA repair systems of tumor cells [27,28]. These complexes consist of two Pt(II) coordination spheres bridged by azolato and hydroxo anions, the latter of which acts as a leaving group, enabling bifunctional covalent binding to DNA. Due to their positive charges, these complexes have multimodal DNA binding modes [29,30], a characteristic that makes these series of complexes cytotoxic in many human tumor cell lines and circumvents the cross-resistance to cisplatin [31–34].

Recently, we reported the structure–activity relationships of a series of tetrazolato-bridged dinuclear platinum(II) complexes ([{*cis*-Pt(NH$_3$)$_2$}$_2$-(μ-OH)(μ-5-R-tetrazolato-*N2,N3*)]$^{2+}$; tetrazolato-bridged complexes) with a diverse range of substituents introduced at tetrazole C5 of [{*cis*-Pt(NH$_3$)$_2$}$_2$-(μ-OH)(μ-tetrazolato-*N2,N3*)]$^{2+}$ (5-H-Y) and concluded that this series was a promising source of next-generation platinum-based drugs. For instance, against the mouse homografted Colon-26 colorectal tumor, two of the derivatives exhibited much higher in vivo antitumor efficacy than oxaliplatin [35], which is currently used for the treatment of colorectal cancer.

To build on our previous research, here we report the in vitro cytotoxicity and in vivo antitumor efficacy of three tetrazolato-bridged complexes, each with a bulky, lipophilic substituent at tetrazole C5 ([{*cis*-Pt(NH$_3$)$_2$}$_2$(μ-OH)(μ-5-phenyltetrazolato-*N2,N3*](NO$_3$)$_2$ (**1**), [{*cis*-Pt(NH$_3$)$_2$}$_2$(μ-OH)(μ-5-nonyltetrazolato-*N2,N3*] (NO$_3$)$_2$ (**2**), and [{*cis*-Pt(NH$_3$)$_2$}$_2$(μ-OH)(μ-5-

adamantyltetrazolato-*N2,N3*] $(NO_3)_2$ (**3**)), against the Colon-26 colorectal cancer cell line. We hypothesized that the addition of the bulky, lipophilic substituents would increase the membrane permeability of the complexes and thereby increase antitumor efficacy. We also examined the effects of the presence of β-cyclodextrin (β-CD) on the actions of the compounds because β-CD can form an inclusion complex with bulky organic compounds or substituents and change their polarity or molecular dimensions, thereby altering their efficacy compared with the non-complexed compound.

2. Results

2.1. Determination of the Stability Constant of Inclusion Complexes with β-CD

β-CD is a cyclic oligosaccharide comprising seven (α-1,4)-linked D-glucopyranose units arranged in a doughnut shape, with a hydrophilic outer surface and a somewhat lipophilic central cavity (Figure 1). Generally, β-CD increases drug permeability through biological membranes and improves drug bioavailability. To find out if this is also true for tetrazolato-bridged complexes, we determined the stability constants (K_s) for inclusion complexes formed between complexes **1–3**, oxaliplatin, or 5-H-Y and β-CD. The K_s values were determined from the ^{1}H-NMR chemical shift of the Pt(II) complexes (0.2 mM) in different concentrations of β-CD (0.2–2 mM) in D_2O at 293 K. The K_s values shown in Table 1 were obtained using the Benesi–Hildebrand equation [36–38], assuming a 1:1 guest/host interaction (Pt(II) complex/β-CD). For complexes **1–3**, the observed linear correlation (Figure 2) confirmed that they formed a 1:1 inclusion complex with β-CD. No definite directional downfield/upfield shift or chemical shift change in protons originating from the guest compounds was observed for oxaliplatin or 5-H-Y, in the presence of β-CD, within the concentration range examined. This implied that they weakly associate with β-CD and that the 1:1 inclusion complexes for complexes **1–3** were formed via interactions between the substituent at tetrazole C5 and the lipophilic surface inside the β-CD cavity. The highest stability constant was obtained for complex **3** (adamantyl group at tetrazole C5) and decreased by approximately one order of magnitude in the following order of complexes: **3** > **2** (*n*-nonyl group at tetrazole C5) > **1** (phenyl group at tetrazole C5). The order of the stability constants was somewhat consistent with previously reported data: The stability constants for complexes containing adamantane moieties are between 10^4 and 10^5 M^{-1} in water [39,40], those for a series based on the cationic surfactant cetyltrimethylammonium bromide (CTAB) are mostly within the range of 10^3 to 10^4 M^{-1} [41], and those for benzene [42] or phenylalanine [43] are less than 10^3 M^{-1}. The K_s value for the inclusion complex between complex **3** and β-CD indicates that they form a tight inclusion complex, in which the adamantyl group is located within the β-CD cavity. It is generally considered that β-CD affects drug biodistribution and elimination only when K_s is greater than 10^5 M^{-1} [44]. Therefore, the in vivo antitumor efficacy study described in later sections was performed with a 1:1 (molar ratio) mixture of complex **3** with β-CD, in addition to complex **3** alone, to examine how the formation of the β-CD inclusion complex affects tumor growth inhibition.

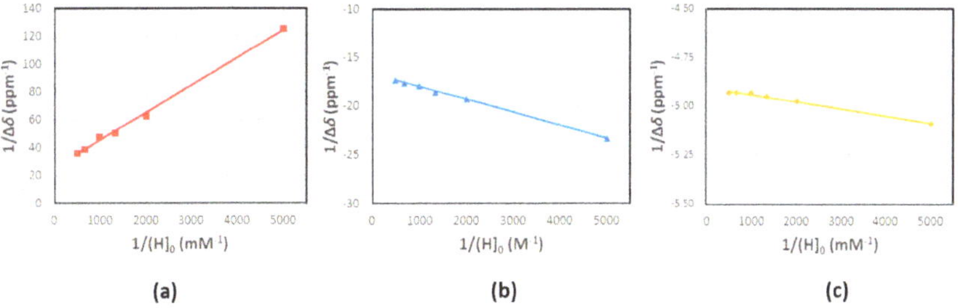

Figure 2. Benesi–Hildebrand plots of 0.2 mM of complexes **1** ((a) R^2 = 0.997), **2** ((b), R^2 = 0.999) and **3** ((c) R^2 = 0.990) for various concentrations of β-CD (0.2–2 mM) at 293 K in D_2O, as assessed by ^{1}H NMR titration.

Table 1. Stability constant (K_s) of Pt(II) complex/β-cyclodextrin (β-CD), as determined from the ^{1}H-NMR chemical shift of Pt(II) complexes (0.2 mM) in different concentrations of β-CD (0.2–2 mM) in D_2O at 293 K.

Pt(II) Complex	K_s/M^{-1}
Oxaliplatin	n. d. [a]
5-H-Y	n. d. [a]
1	$(1.81 \pm 1.28) \times 10^3$
2	$(1.30 \pm 0.26) \times 10^4$
3	$(1.27 \pm 0.03) \times 10^5$

[a] Not determined since no definite directional downfield/upfield shift or no chemical shift change on protons originating from the guest compounds was observed.

2.2. In Vitro Cytotoxicity

We evaluated the in vitro cytotoxicity of four platinum(II) complexes—5-H-Y and complexes **1–3**—against the Colon-26 colorectal cancer cell line in the absence or presence of equimolar β-CD. Oxaliplatin was used as the positive control. The Colon-26 cells were exposed to each of the compounds for 24 or 48 h, after which an MTS (3-(4,5-dimethylthiazol-2-yl)-5-(3-carboxymethoxyphenyl)-2-(4-sulfophenyl)-2*H*-tetrazolium, inner salt) assay was used to determine the half-maximal inhibitory concentrations (IC_{50}) of the complexes (Table 2). As expected, the longer exposure time (48 h) provided lower IC_{50} values for all of the tested compounds in the absence of β-CD. When the shorter exposure time was used (24 h), in the absence of β-CD, 5-H-Y had the highest cytotoxicity and was approximately 20 times more cytotoxic than oxaliplatin; complex **2** had a level of cytotoxicity similar to that of oxaliplatin, complex **3** was moderately cytotoxic, and complex **1** had the least cytotoxicity. β-CD alone showed no cytotoxicity (data not shown). In the presence of β-CD, the cytotoxicity of each of the complexes decreased, particularly that of oxaliplatin (24-h exposure), complex **1** (48-h exposure) and **3** (both 24 h and 48 h exposure). Only complex **3** showed a marked decrease in cytotoxicity at both exposure times, probably because this complex possessed an adamantyl group and so likely formed a tight inclusion complex with β-CD, as indicated by the K_s value of the 1:1 inclusion complex. In contrast, for 5-H-Y, which did not have an additional substituent, and complex **2**, which had a phenyl group at tetrazole C5, there was little difference between the IC_{50} values in the absence or presence of β-CD (+β-CD/−β-CD ratio in Table 2).

Table 2. In vitro cytotoxicity (IC_{50}) of oxaliplatin, 5-H-Y, and derivatives of 5-H-Y with bulky substitutions at tetrazole C5 (complexes **1–3**) against Colon-26 colorectal cancer cells in the absence (−β-CD) or presence (+β-CD) of β-cyclodextrin (β-CD). +β-CD/−β-CD values are the ratios of the mean IC_{50} values in the presence or absence of β-CD.

Pt(II) Complex	Mean $IC_{50} \pm SD/\mu M$ ($n = 6$)		
	−β-CD	+β-CD	+β-CD/−β-CD
Oxaliplatin (24 h) [a]	11 ± 3	28 ± 3	2.5
Oxaliplatin (48 h) [b]	5.7 ± 1.2	7.9 ± 0.3	1.4
5-H-Y (24 h) [a]	0.59 ± 0.21	0.74 ± 0.03	1.3
5-H-Y (48 h) [b]	0.23 ± 0.09	0.25 ± 0.02	1.1
1 (24 h) [a]	>360	>270	-
1 (48 h) [b]	43 ± 1	>270	>6.3
2 (24 h) [a]	6.3 ± 0.7	5.6 ± 0.5	0.9
2 (48 h) [b]	4.9 ± 0.6	7.4 ± 0.2	1.5
3 (24 h) [a]	113 ± 8	>270	>2.4
3 (48 h) [b]	109 ± 10	>270	>2.5

[a] Exposed to the Pt(II) complex for 24 h; [b] exposed to the Pt(II) complex for 48 h.

2.3. In Vivo Antitumor Efficacy

We examined the in vivo antitumor efficacy of complexes **1–3** against homografted Colon-26 colorectal tumors in male BALB/c mice. This was a preliminary study, in which the same dosage was used for all of the compounds tested, to ensure that the animals survived for at least one week. Complexes **1–3**, oxaliplatin, or a 1:1 (molar ratio) mixture of complex **3** with β-CD were dissolved in 5% glucose and administered to the mice as a single dose (10 mg (Pt complex)/kg) via the tail vein on day 0, which was 7 days after their inoculation with the Colon-26 colorectal cancer cells. No mice in any of the groups died and none showed significant body weight loss (Figure 3a). No marked tumor growth inhibition was observed in the mice treated with complexes **1–3**: The mean terminal tumor volume in the mice treated with complexes **1–3** was 90%, 83%, and 89% of that in the control group, respectively, and the value for oxaliplatin was 54% (Figure 3b). However, when complex **3** was administered as a 1:1 (molar ratio) mixture with β-CD, marked tumor growth inhibition was observed five times greater than that when complex **3** was administered alone (mean terminal tumor volume, 44% of that in the control group) and was slightly more effective than oxaliplatin (Figure 3b).

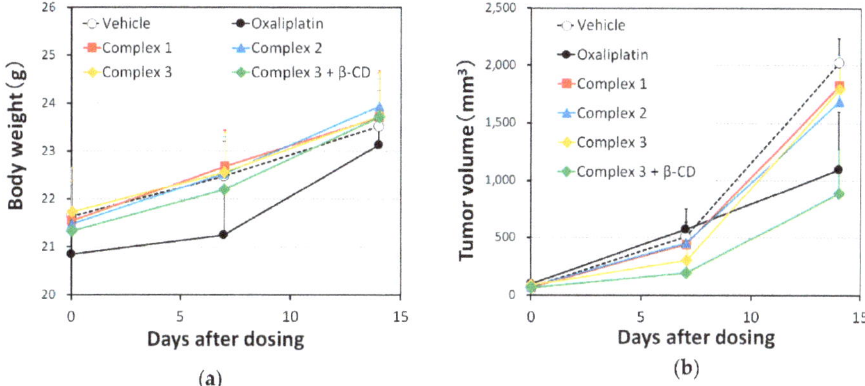

Figure 3. (**a**) Body weight and (**b**) tumor volume in male BALB/c mice laterally homografted with Colon-26 colorectal cancer cells and then treated with 10 mg kg^{-1} oxaliplatin, complexes **1–3**, a 1:1 (molar ratio) mixture of complex **3** and β-cyclodextrin (β-CD), or vehicle. Mice were treated with a single dose of the test compounds or vehicle on day 0, which was 7 days after their inoculation with the Colon-26 colorectal cancer cells. Body weights and tumor volumes were measured weekly, starting on day 0. Each data point represents the mean of six body weights or tumor volumes, and the error bars indicate standard deviations of the mean.

3. Discussion

Previously, we found that a variety of substitutions at tetrazole C5 increased the cytotoxicity of the tetrazolato-bridged complex, and that complexes with an ester group substituted at tetrazole C5, such as [{*cis*-Pt(NH₃)₂}₂(μ-OH)(μ-ethyl tetrazolato-5-carboxylate-*N2,N3*)](NO₃)₂ and [{*cis*-Pt(NH₃)₂}₂(μ-OH)(μ-propyl tetrazolato-5-acetate-*N2,N3*)](NO₃)₂, had much higher antitumor efficacies than oxaliplatin [35]. To build on this previous research, here we examined the efficacies of derivatives with bulky alkyl or aryl groups at tetrazole C5.

Compared with 5-H-Y, complexes **1–3** were much less cytotoxic against the Colon-26 cell line. Comparable results were obtained for cytotoxicity against L1210 murine leukemia cell lines [35]. Therefore, the introduction of bulky substituents tended to produce a lower in vitro cytotoxicity, possibly because the DNA interaction mode and cellular accumulation of complex **1** and **2** are distinct from those of 5-H-Y, and from other derivatives with a relatively small substituent at tetrazole C5 [34,45]. The cytotoxicity of oxaliplatin and complex **3** was markedly decreased in the presence of equimolar β-CD, whereas the cytotoxicity of 5-H-Y and complex **2** remained largely unchanged. Complex **3** possesses an adamantyl group that forms a tight inclusion complex with β-CD, as indicated by the stability constant. β-CD and other CDs, and their inclusion complexes, are unable to cross the cell membrane. Therefore, the reduction in cytotoxicity induced by β-CD could be due to a reduction in the speed of release of the platinum(II) complex from the β-CD inclusion complex.

Although the in vitro cytotoxicity results for complexes **1–3** suggest that the substitution of relatively bulky alkyl or aryl groups at tetrazole C5 is not a successful approach for improving antitumor efficacy, because β-CD decreased the cytotoxicity of the compounds, we did find in vivo that β-CD markedly enhanced the antitumor efficacy of complex **3** until it was higher than that of oxaliplatin. Some anticancer drugs complexed with a cyclodextrin have increased bioavailability and reduced toxicity, indicating that complexation with cyclodextrin changes the polarity and molecular dimensions of the compound. The present results suggest that the water solubility of complex **3** was increased by partial inclusion into β-CD, which covered the lipophilic part of the complex. Since opposite trends were observed in vitro and in vivo, the improved water solubility and increase in the molecular dimensions of complex **3** must have improved its delivery to the tumor site and reduced its cellular accumulation; the increase in the molecular dimensions likely means that the complex was unaffected by the enhanced permeability and retention effect. Together, the present results suggest that substitution with substituents bulkier than those examined here may be a promising means of identifying highly antitumor-active lead tetrazolato-bridged complexes.

With respect to methodology, the present in vivo study was a preliminary study in which the same dosage was used for all of the compounds tested to ensure that the animals survived for at least one week. Therefore, it is possible that the antitumor efficacy of the complexes can be increased by increasing the dosage. Furthermore, it may be possible to combine complex **3** with other β-CD derivatives, such as methyl-β-CD, (2-hydroxyalkyl)-β-CD, or other specially functionalized cyclodextrins. Further studies are warranted. We have already designed and synthesized tetrazolato-bridged complexes with an adamantyl group linked by an ester or alkyl group at tetrazole C5, and in future experiments, we intend to find out which derivative is most suitable for complexation with β-CD.

4. Materials and Methods

4.1. Materials

Tetrazolato-bridged complexes with the formula [{cis-Pt(NH$_3$)$_2$}$_2$(μ-OH)(μ-5-R-tetrazolato-$N2,N3$)](NO$_3$)$_2$ (complexes **1–3**, 5-H-Y) were prepared using previously reported methods [13,31,34,35]. Oxaliplatin and β-CD were purchased from Tokyo Chemical Industry (Tokyo, Japan) and Wako Pure Chemical (Osaka, Japan), respectively.

4.2. Determination of Stability Constant

The stability constant (K_s) of Pt(II) complex/β-CD was determined by analysis of the ^{1}H-NMR chemical shift of Pt(II) complexes (0.2 mM) in different concentrations of β-CD (0.2–2 mM) in D$_2$O at 293 K. To prevent changes of the resonance frequency due to the formation of an inclusion complex between β-CD and the reference compound, 3-(Trimethylsilyl)propionic-2,2,3,3-d4 acid sodium salt (TSP), a solution of TSP in D$_2$O, sealed in a capillary tube, was placed inside an NMR tube. The NMR tube was then placed in the sample solution and used as the reference ($\delta = 0$). The K_s value was obtained by using the Benesi–Hildebrand Equation (1), assuming a 1:1 guest/host interaction (Pt(II) complex/β-CD):

$$1/\Delta\delta = 1/(K_s\,\Delta\delta_{max}\,[H]_0) + 1/\Delta\delta_{max}, \tag{1}$$

where $\Delta\delta$ is the change in the ^{1}H-NMR chemical shift, $\Delta\delta_{max}$ is the maximum possible change in the ^{1}H-NMR chemical shift, $[H]_0$ is the total β-CD concentration, and K_s is the stability constant.

4.3. In Vitro Cytotoxicity Study

The murine colorectal cancer cell line Colon-26 was provided by RIKEN BioResource Center through the National Bio-Resource Project of the Ministry of Education, Culture, Sports, Science, and Technology of Japan. The Colon-26 mouse colorectal cells were maintained in RPMI-1640 containing 10% fetal bovine serum (Gibco; Life Technologies, Carlsbad, CA, USA), 100 U/mL penicillin, and 100 mg/mL streptomycin (Wako, Osaka, Japan) in a humidified atmosphere of 5% CO_2 at 310 K. The Colon-26 cells (3×10^4 cells/mL; 100 µL/well) were seeded onto 96-well microplates (Corning, Corning, NY, USA). All of the Pt(II) complexes tested, except for complex **2**, were dissolved in water to prepare 4 mM solutions. For assays in the presence of β-CD, each Pt(II) complex solution was mixed with 12 mM β-CD aqueous solution in a 1:1 molar ratio and incubated for at least 10 min. Due to the low solubility of complex **2**, the concentrations of the solution or suspension of complex **2** were 0.25 or 1 mM for assays in the absence or presence of β-CD, respectively. After sterilization by filtration, the solutions were diluted with water, and then 10 µL of each diluted solution was added to the wells of the microplate. After incubation of the microplate for 23 or 47 h at 310 K, 10 µL of Cell Counting Kit-8 solution (Nacalai Tesque, Kyoto, Japan) was added to each well, and the incubation was continued for an additional 1 h at 310 K. The absorbance of each well at a wavelength of 460 nm was measured with a Spectra Max M5 microplate reader (Molecular Device; Orleans Drive Sunnyvale, CA, USA). Each experiment was performed independently for six wells per drug concentration. Half-maximal inhibitory concentrations (IC_{50}) were calculated as the concentration that provided 50% formazan production, relative to the control (no complex added), using the KaleidaGraph analytical software (version 4; Synergy Software, Reading, PA, USA).

4.4. In Vivo Mouse Homografts

The homograft study was performed using BALB/c mice (male, 4 weeks old; Japan SLC, Inc., Hamamatsu, Japan). The Colon-26 cells were maintained in 55-cm^2 dishes containing RPMI-1640 medium (Sigma-Aldrich (Merck), Darmstadt, HE, Germany) supplemented with 10% fetal bovine serum (Gibco; Life Technologies, Carlsbad, CA, USA) containing 100 U/mL penicillin, and 100 mg/mL streptomycin. The cultures were grown in a humidified atmosphere of 5% CO_2 at 310 K. Cells were grown to 80%–90% confluence and then harvested with 0.25% trypsin/0.02% ethylenediaminetetraacetic acid (Sigma-Aldrich (Merck), Darmstadt, HE, Germany) before each subsequent passage.

The Colon-26 cells were subcutaneously injected into the lateral side of the mice (2 million cells/flank). About 7 days later (tumor diameter, ca. 8 mm), the animals were randomly assigned to the following eight study groups (n = 6 per group): Control, oxaliplatin, 5-H-Y, complexes **1**–**3**, and 1:1 (molar ratio) mixture of complex **3** and β-CD. Test substances were dissolved in 5% glucose and administered by single intravenous injection to the Colon-26 cell-bearing mice. The control group received 5% glucose (vehicle) only (total volume = 200 µL). The homograft tumor dimensions (d and D, shortest and longest dimensions of the tumor, respectively) were measured once a week with a digital caliper, and tumor volume (mm^3) was calculated by using the equation $d^2D/2$. The body weights were measured weekly and statistically analyzed by means of one-way analysis of variance. This animal study was carried out with approval from the Institutional Animal Care and Use Committee of Suzuka University of Medical Science (Permission number: 34, 9 August 2017) and in accordance with all applicable institutional animal experimentation regulations.

5. Conclusions

The approval of cisplatin for clinical use prompted a search for novel platinum coordination compounds with improved efficacy. The tetrazolato-bridged complex can be greatly modified from the basic structure that is in current clinical use, and these modified complexes are an important group of potential next-generation platinum-based drug candidates. In the present study, the introduction of a bulky adamantyl group at tetrazole C5 provided the interesting finding that while the in vitro cytotoxicity of the modified complex was reduced in the presence of β-CD, its in vivo antitumor efficacy increased and was greater than that of the currently used colorectal cancer treatment, oxaliplatin. This finding indicates a new direction for further drug discovery research to provide tetrazolato-bridged complexes with high in vivo antitumor activity and less toxicity.

Author Contributions: Conceptualization, S.K.; methodology, S.K., M.U., H.Y., S.H. and K.H.; validation, S.K., M.U., H.Y., S.H. and K.H.; formal analysis, S.K., M.U. and K.H.; investigation, S.K., M.U. and K.H.; resources, S.K., H.Y. and S.H.; data curation, S.K.; writing—original draft preparation, S.K.; writing—review and editing, S.K., M.U., H.Y., S.H. and K.H.; visualization, S.K., M.U., H.Y., S.H. and K.H.; supervision, S.K. and S.H.; project administration, S.K.; funding acquisition, S.K.

Funding: This research was funded by a Japan Society for the Promotion of Science grant-in-aid (KAKENHI; grant number 15K07905).

Conflicts of Interest: The authors declare no conflict of interest.

References

1. Rosenberg, B.; VanCamp, L.; Trosko, J.E.; Mansour, V.H. Platinum compounds: A new class of potent antitumour agents. *Nature* **1969**, *222*, 385–386. [CrossRef] [PubMed]
2. Rosenberg, B.; VanCamp, L. The successful regression of large solid sarcoma 180 tumors by platinum compounds. *Cancer Res.* **1970**, *30*, 1799–1802.
3. Gottlieb, J.A.; Drewinko, B. Review of the current clinical status of platinum coordination complexes in cancer chemotherapy. *Cancer Chemother. Rep.* **1975**, *59*, 621–628. [PubMed]
4. Calvert, A.H.; Harland, S.J.; Newell, D.R.; Siddik, Z.H.; Jones, A.C.; McElwain, T.J.; Raju, S.; Wiltshaw, E.; Smith, I.E.; Baker, J.M.; et al. Early clinical studies with *cis*-diammine-1,1-cyclobutane dicarboxylate platinum II. *Cancer Chemother. Pharmacol.* **1982**, *9*, 140–147. [CrossRef] [PubMed]
5. Sharma, H.; Thatcher, N.; Baer, J.; Zaki, A.; Smith, A.; McAucliffe, C.A.; Crowther, D.; Owens, S.; Fox, B.W. Blood clearance of radioactively labelled *cis*-diammine 1,1-cyclobutane dicarboxylate platinum(II) (CBDCA) in cancer patients. *Cancer Chemother. Pharmacol.* **1983**, *11*, 5–7. [CrossRef]
6. Tashiro, T.; Kawada, Y.; Sakurai, Y.; Kidani, Y. Antitumor activity of a new platinum complex, oxalato (trans-l-1,2-diaminocyclohexane)platinum(II): New experimental data. *Biomed. Pharmacother.* **1989**, *43*, 251–260. [CrossRef]
7. Cvitkovic, E. Ongoing and unsaid on oxaliplatin: The hope. *Br. J. Cancer* **1998**, *77* (Suppl. 4), 8–11. [CrossRef]
8. Seeber, S.; Osieka, R.; Schmidt, C.G.; Achterrath, W.; Crooke, S.T. In vivo resistance towards anthracyclines, etoposide, and *cis*-diamminedichloroplatinum(II). *Cancer Res.* **1982**, *42*, 4719–4725.
9. Eastman, A.; Illenye, S. Murine leukemia L1210 cell lines with different patterns of resistance to platinum coordination complexes. *Cancer Treat. Rep.* **1984**, *68*, 1189–1190.
10. Eastman, A.; Bresnick, E. Studies on the resistance of a murine leukemia L1210 cell line *cis*-diamminedichloroplatinum(II). *Biochem. Pharmacol.* **1981**, *30*, 2721–2723. [CrossRef]
11. Komeda, S.; Ohishi, H.; Yamane, H.; Harikawa, M.; Sakaguchi, K.-I.; Chikuma, M. An NMR study and crystal structure of [{cis-Pt(NH3)2(9EtG-kN7)2(m-pz)}[NO3]3 (9EtG = 9-ethylguanine) as a model compound for the 1,2-intrastrand GG crosslink. *J. Chem. Soc. Dalton Trans.* **1999**, *17*, 2959–2962. [CrossRef]
12. Komeda, S.; Lutz, M.; Spek, A.L.; Chikuma, M.; Reedijk, J. New antitumor-active azole-bridged dinuclear platinum(II) complexes: Synthesis, characterization, crystal structures, and cytotoxic studies. *Inorg. Chem.* **2000**, *39*, 4230–4236. [CrossRef] [PubMed]
13. Komeda, S.; Lin, Y.L.; Chikuma, M. A Tetrazolato-Bridged Dinuclear Platinum(II) Complex Exhibits Markedly High in vivo Antitumor Activity against Pancreatic Cancer. *ChemMedChem* **2011**, *6*, 987–990. [CrossRef] [PubMed]
14. Farrell, N.; Ha, T.T.; Souchard, J.P.; Wimmer, F.L.; Cros, S.; Johnson, N.P. Cytostatic trans-platinum(II) complexes. *J. Med. Chem.* **1989**, *32*, 2240–2241. [CrossRef] [PubMed]
15. Farrell, N.; Qu, Y. Chemistry of Bis(Platinum) Complexes–Formation of Trans Derivatives from Tetraamine Complexes. *Inorg. Chem.* **1989**, *28*, 3416–3420. [CrossRef]

16. Farrell, N.; Kiley, D.M.; Schmidt, W.; Hacker, M.P. Chemical-Properties and Antitumor-Activity of Complexes of Platinum Containing Substituted Sulfoxides [PtCl(R′R″SO)(Diamine)]NO₃. Chirality and Leaving-Group Ability of Sulfoxide Affecting Biological Activity. *Inorg. Chem.* **1990**, *29*, 397–403. [CrossRef]
17. Farrell, N.; Qu, Y.; Feng, L.; Van Houten, B. Comparison of chemical reactivity, cytotoxicity, interstrand cross-linking and DNA sequence specificity of bis(platinum) complexes containing monodentate or bidentate coordination spheres with their monomeric analogues. *Biochemistry* **1990**, *29*, 9522–9531. [CrossRef]
18. Farrell, N.; Qu, Y.; Hacker, M.P. Cytotoxicity and antitumor activity of bis(platinum) complexes. A novel class of platinum complexes active in cell lines resistant to both cisplatin and 1,2-diaminocyclohexane complexes. *J. Med. Chem.* **1990**, *33*, 2179–2184. [CrossRef]
19. Yang, D.; van Boom, S.S.G.E.; Reedijk, J.; van Boom, J.H.; Wang, A.H.J. Structure and isomerization of an intrastrand cisplatin-cross-linked octamer DNA duplex by NMR analysis. *Biochemistry* **1995**, *34*, 12912–12920. [CrossRef]
20. van Boom, S.S.; Yang, D.; Reedijk, J.; van der Marel, G.A.; Wang, A.H. Structural effect of intra-strand cisplatin-crosslink on palindromic DNA sequences. *J. Biomol. Struct. Dyn.* **1996**, *13*, 989–998. [CrossRef]
21. Shellard, S.A.; Fichtinger-Schepman, A.M.; Lazo, J.S.; Hill, B.T. Evidence of differential cisplatin-DNA adduct formation, removal and tolerance of DNA damage in three human lung carcinoma cell lines. *Anticancer Drugs* **1993**, *4*, 491–500. [CrossRef] [PubMed]
22. Takahara, P.M.; Rosenzweig, A.C.; Frederick, C.A.; Lippard, S.J. Crystal structure of double-stranded DNA containing the major adduct of the anticancer drug cisplatin. *Nature* **1995**, *377*, 649–652. [CrossRef] [PubMed]
23. Coste, F.; Malinge, J.M.; Serre, L.; Shepard, W.; Roth, M.; Leng, M.; Zelwer, C. Crystal structure of a double-stranded DNA containing a cisplatin interstrand cross-link at 1.63 A resolution: Hydration at the platinated site. *Nucleic Acids Res.* **1999**, *27*, 1837–1846. [CrossRef] [PubMed]
24. Jamieson, E.R.; Lippard, S.J. Structure, Recognition, and Processing of Cisplatin-DNA Adducts. *Chem. Rev.* **1999**, *99*, 2467–2498. [CrossRef] [PubMed]
25. Teletchea, S.; Komeda, S.; Teuben, J.M.; Elizondo-Riojas, M.A.; Reedijk, J.; Kozelka, J. A pyrazolato-bridged dinuclear platinum(II) complex induces only minor distortions upon DNA-binding. *Chemistry* **2006**, *12*, 3741–3753. [CrossRef]
26. Magistrato, A.; Ruggerone, P.; Spiegel, K.; Carloni, P.; Reedijk, J. Binding of novel azole-bridged dinuclear platinum(II) anticancer drugs to DNA: Insights from hybrid QM/MM molecular dynamics simulations. *J. Phys. Chem. B* **2006**, *110*, 3604–3613. [CrossRef]
27. Mlcouskova, J.; Kasparkova, J.; Suchankova, T.; Komeda, S.; Brabec, V. DNA conformation and repair of polymeric natural DNA damaged by antitumor azolato-bridged dinuclear Pt(II) complex. *J. Inorg. Biochem.* **2012**, *114*, 15–23. [CrossRef]
28. Mlcouskova, J.; Malina, J.; Novohradsky, V.; Kasparkova, J.; Komeda, S.; Brabec, V. Energetics, conformation, and recognition of DNA duplexes containing a major adduct of an anticancer azolato-bridged dinuclear Pt(II) complex. *Biochim. Biophys. Acta* **2012**, *1820*, 1502–1511. [CrossRef]
29. Imai, R.; Komeda, S.; Shimura, M.; Tamura, S.; Matsuyama, S.; Nishimura, K.; Rogge, R.; Matsunaga, A.; Hiratani, I.; Takata, H.; et al. Chromatin folding and DNA replication inhibition mediated by a highly antitumor-active tetrazolato-bridged dinuclear platinum(II) complex. *Sci. Rep.* **2016**, *6*, 24712. [CrossRef]
30. Uemura, M.; Yoshikawa, Y.; Yoshikawa, K.; Sato, T.; Mino, Y.; Chikuma, M.; Komeda, S. Second- and higher-order structural changes of DNA induced by antitumor-active tetrazolato-bridged dinuclear platinum(II) complexes with different types of 5-substituent. *J. Inorg. Biochem.* **2013**, *127*, 169–174. [CrossRef]
31. Komeda, S.; Takayama, H.; Suzuki, T.; Odani, A.; Yamori, T.; Chikuma, M. Synthesis of antitumor azolato-bridged dinuclear platinum(II) complexes with in vivo antitumor efficacy and unique in vitro cytotoxicity profiles. *Metallomics* **2013**, *5*, 461–468. [CrossRef] [PubMed]
32. Uemura, M.; Suzuki, T.; Nishio, K.; Chikuma, M.; Komeda, S. An in vivo highly antitumor-active tetrazolato-bridged dinuclear platinum(II) complex largely circumvents in vitro cisplatin resistance: Two linkage isomers yield the same product upon reaction with 9-ethylguanine but exhibit different cytotoxic profiles. *Metallomics* **2012**, *4*, 686–692. [CrossRef] [PubMed]
33. Uemura, M.; Hoshiyama, M.; Furukawa, A.; Sato, T.; Higuchi, Y.; Komeda, S. Highly efficient uptake into cisplatin-resistant cells and the isomerization upon coordinative DNA binding of anticancer tetrazolato-bridged dinuclear platinum(II) complexes. *Metallomics* **2015**, *7*, 1488–1496. [CrossRef]

34. Komeda, S.; Yoneyama, H.; Uemura, M.; Muramatsu, A.; Okamoto, N.; Konishi, H.; Takahashi, H.; Takagi, A.; Fukuda, W.; Imanaka, T.; et al. Specific Conformational Change in Giant DNA Caused by Anticancer Tetrazolato-Bridged Dinuclear Platinum(II) Complexes: Middle-Length Alkyl Substituents Exhibit Minimum Effect. *Inorg. Chem.* **2017**, *56*, 802–811. [CrossRef] [PubMed]

35. Komeda, S.; Yoneyama, H.; Uemura, M.; Tsuchiya, T.; Hoshiyama, M.; Sakazaki, T.; Hiramoto, K.; Harusawa, S. Synthesis and Structure–Activity Relationships of Tetrazolato-Bridged Dinuclear Platinum(II) Complexes: A Small Modification at Tetrazole C5 Markedly Influences the In Vivo Antitumor Efficacy. *J. Inorg. Biochem.* **2019**. [CrossRef] [PubMed]

36. Hanna, M.W.; Ashbaugh, A.L. Nuclear Magnetic Resonance Study of Molecular Complexes of 7,7,8,8-Tetracyanoquinodimethane and Aromatic Donors1, 2. *J. Phys. Chem.* **1964**, *68*, 811–816. [CrossRef]

37. Benesi, H.A.; Hildebrand, J. A spectrophotometric investigation of the interaction of iodine with aromatic hydrocarbons. *J. Am. Chem. Soc.* **1949**, *71*, 2703–2707. [CrossRef]

38. Mathur, R.; Becker, E.D.; Bradley, R.B.; Li, N.C. Proton magnetic resonance studies of hydrogen bonding of benzenethiol with several hydrogen acceptors. *J. Phys. Chem.* **1963**, *67*, 2190–2194. [CrossRef]

39. Eftink, M.R.; Andy, M.L.; Bystrom, K.; Perlmutter, H.D.; Kristol, D.S. Cyclodextrin inclusion complexes: Studies of the variation in the size of alicyclic guests. *J. Am. Chem. Soc.* **1989**, *111*, 6765–6772. [CrossRef]

40. Leong, N.J.; Prankerd, R.J.; Shackleford, D.M.; Mcintosh, M.P. The Effect of Intravenous Sulfobutylether7-β-Cyclodextrin on the Pharmacokinetics of a Series of Adamantane-Containing Compounds. *J. Pharm. Sci.* **2015**, *104*, 1492–1498. [CrossRef]

41. Carlstedt, J.; Bilalov, A.; Krivtsova, E.; Olsson, U.; Lindman, B.R. Cyclodextrin–surfactant coassembly depends on the cyclodextrin ability to crystallize. *Langmuir* **2012**, *28*, 2387–2394. [CrossRef] [PubMed]

42. Trofymchuk, I.; Belyakova, L.; Grebenyuk, A. Study of complex formation between β-cyclodextrin and benzene. *J. Incl. Phenom. Macrocycl. Chem.* **2011**, *69*, 371–375. [CrossRef]

43. Kahle, C.; Holzgrabe, U. Determination of binding constants of cyclodextrin inclusion complexes with amino acids and dipeptides by potentiometric titration. *Chirality* **2004**, *16*, 509–515. [CrossRef] [PubMed]

44. Jansook, P.; Ogawa, N.; Loftsson, T. Cyclodextrins: Structure, physicochemical properties and pharmaceutical applications. *Int. J. Pharm.* **2018**, *535*, 272–284. [CrossRef] [PubMed]

45. Uemura, M.; Yoshikawa, Y.; Chikuma, M.; Komeda, S. A circular dichroism study uncovers a two-step interaction of antitumor azolato-bridged dinuclear platinum(II) complexes with calf thymus DNA. *Metallomics* **2012**, *4*, 641–644. [CrossRef]

 *inorganics*

Article

Oxidative Assets Toward Biomolecules and Cytotoxicity of New Oxindolimine-Copper(II) and Zinc(II) Complexes

Maurício Cavicchioli [1], Aline Monteiro Lino Zaballa [1], Queite Antonia de Paula [1],
Marcela Bach Prieto [2], Carla Columbano Oliveira [2], Patrizia Civitareale [3], Maria Rosa Ciriolo [3]
and Ana Maria Da Costa Ferreira [1,*]

[1] Departamento de Química, Instituto de Química, Universidade de São Paulo, São Paulo 05508-000, SP, Brazil;
mauriciocavicchioli@gmail.com (M.C.); alinezaballa@usp.br (A.M.L.Z.); qdepaula@gmail.com (Q.A.d.P.)

[2] Departamento de Bioquímica, Instituto de Química, Universidade de São Paulo,
São Paulo 05508-000, SP, Brazil; marbp@usp.br (M.B.P.); ccoliv@iq.usp.br (C.C.O.)

[3] Dipartamento di Biologia, Università Tor Vergata di Roma, 00133 Roma, Italy; civitare@uniroma2.it (P.C.);
ciriolo@uniroma2.it (M.R.C.)

* Correspondence: amdcferr@iq.usp.br; Tel.: +55-11-3091-9147 or +55-11-2648-1681

Received: 30 October 2018; Accepted: 17 January 2019; Published: 26 January 2019

Abstract: A new oxindolimine ligand derived from isatin (1*H*-indole-2,3-dione) and 2-aminomethylbenzimidazole was synthesized, leading to two novel complexes after metalation with copper(II) perchlorate or zinc(II) chloride, [Cu(*isambz*)$_2$](ClO$_4$)$_2$ (complex **1**) and [Zn(*isambz*)Cl$_2$] (complex **2**). This new ligand was designed as a more lipophilic compound, in a series of oxindolimine–metal complexes with antitumor properties, having DNA, mitochondria, and some proteins, such as CDK1 kinase and topoisomerase IB, as key targets. The new complexes had their reactivity to human serum albumin (HSA) and DNA, and their cytotoxicity toward tumor cells investigated. The binding to CT-DNA was monitored by circular dichroism (CD) spectroscopy and fluorescence measurements using ethidium bromide in a competitive assay. Consequent DNA cleavage was verified by gel electrophoresis with complex **1**, in nmolar concentrations, with formation of linear DNA (form III) after 60 min incubation at 37 °C, in the presence of hydrogen peroxide, which acts as a reducing agent. Formation of reactive oxygen species (ROS) was observed, monitored by spin trapping EPR. Interaction with HSA lead to α-helix structure disturbance, and formation of a stable radical species (HSA–Tyr·) and carbonyl groups in the protein. Despite showing oxidative ability to damage vital biomolecules such as HSA and DNA, these new complexes showed moderate cytotoxicity against hepatocellular carcinoma (HepG$_2$) and neuroblastoma (SHSY5Y) cells, similarly to previous compounds in this series. These results confirm DNA as an important target for these compounds, and additionally indicate that oxidative damage is not the leading mechanism responsible for their cytotoxicity. Additionally, this work emphasizes the importance of ligand characteristics and of speciation in activity of metal complexes.

Keywords: isatin-derived ligands; oxindolimine–metal complexes; DNA cleavage; HSA oxidation; cytotoxicity; antiproliferative activity

---

1. Introduction

Cancer is a very important cause of death globally, according to the World Health Organization (WHO) [1]. It corresponds to a large group of diseases, affecting different organs, with slow through to very quick progression. Although platinum complexes are the only metallodrugs approved by Food and Drug Administration (FDA) for clinical use against cancer [2], different metal complexes have been developed as alternatives, trying to bypass induced resistance and severe toxic effects [3]. Besides platinum, ruthenium, copper, gold, and vanadium compounds are being extensively investigated. Furthermore, different targets have been identified in wide-ranging studies, depending on the metal and on the ligand.

A primary target in such studies is DNA, and different metal complexes exhibit quite different binding constants, depending on the nature of the metal ion and the structural features of the ligands. Usually, three types of interactions can occur: intercalation, covalent bonds, and electrostatic interactions [4]. For complexes of redox active metal ions, such as iron or copper, oxidative stress plays an important rule, usually causing single and double strand scission by reactive oxygen species (ROS), and can lead to a wide array of DNA lesions implicated in the etiology of many human diseases [5]. Many metal complexes reported as anticancer agents act by an oxidative mechanism involving ROS, in a process modulated by the ligand that is responsible for changes in charge, redox potential, and geometry around the metal, affecting different organelles [6].

However, although most of the reported studies focus on DNA, considerable efforts have been made in unravelling mechanistic details of new metal-based drugs [7], including speciation of the most active compounds, identification of critical regulatory genes, and search for alternative targets. An array of different strategies are being developed to circumvent cancer hallmarks [8]. Most of the potential metallodrugs investigated are multifunctional, interacting with diverse biomolecules. As alternative targets, kinases [9], topoisomerases [10], and histone deacetylases (HDACs) [11] have received high interest in the literature, since these proteins play crucial roles in cell-cycle progression and differentiation, and several types of cancer are associated with deregulation in these activities.

Among the new developed anticancer metal complexes, copper has a deserved special interest, coordinated to a wide range of ligands [12], some of them acting as intercalators to DNA [13] or groove binders [14]. In this scenario, Schiff base complexes have allowed important progress in the field. Classical diimines or Schiff bases, such as phenanthroline and correlated planar ligands, are common examples of DNA intercalators, acting efficiently as artificial metallonucleases [15]. Since many copper compounds show preference for interaction with guanine moieties [16], investigations into new complexes able to act as G-quadruplex DNA binders have increased significantly [17]. Further, there is evidence that G-quadruplex structures are improved in cancer cells compared to healthy ones [18]. Therefore, different metal complexes have been reported interacting with both duplex and quadruplex DNA, especially with Schiff base ligands [19]. Recently, two dinuclear copper(II) complexes based on phenanthroline ligands were shown to promote oxidative damage to DNA and mitochondria through the formation of singlet oxygen and superoxide radicals, causing double strand breaks and mitochondrial membrane depolarization [20]. Furthermore, these compounds are able to discriminate oligomer sequences, mediating Z-like DNA formation.

Isatin (1*H*-indole-2,3-dione, or 2,3-dioxoindoline), an endogenous oxindole widely distributed in different tissues in mammals, is formed in the metabolism of amino acids such as tryptophan, and some of its derivatives have demonstrated a wide range of effects in biological media, including inhibition of monoamine oxidase, anti-viral, fungicidal, bactericidal, and anti-proliferative activities [21]. Schiff base ligands containing an oxindole group also exhibited pharmacological properties, especially anti-convulsing, anti-depressive, analgesic, and anti-inflamatory activities [22], and these properties were improved when coordinated to metal ions [23]. Further, synthetic isatin derivatives were developed as potent kinase inhibitors, exhibiting good antineoplastic and anti-angiogenesis activities [24,25], and some of them entered into clinical tests [26], being approved by FDA, such as, for instance, sunitinib malate (SU11248; Sutent®; Pfizer Inc., New York, NY, USA).

With the aim of preparing new metal complexes that exhibit promising pharmacological or medicinal activity in neoplastic processes, we have designed and isolated some oxindolimines and their corresponding metal complexes, inspired by such oxindole derivatives [27,28]. Therefore, we have investigated a series of oxindolimine–metal complexes as pro-apoptotic compounds toward different tumor cells, trying to elucidate their possible modes of action and verifying their potential as alternative antitumor agents.

As delocalized lipophilic cations [29,30], they are able to enter the cell and generate reactive oxygen species (ROS), causing oxidative damage and triggering AMPK-dependent apoptosis in different tumor cells. DNA and mitochondria were reported as their main targets. In more recent studies, a significant inhibition of crucial proteins, kinase (CDK1/cyclin B) [31] and topoisomerase IB [32], was additionally verified, emphasizing the contribution of the ligand to the reactivity of such complexes, and their behavior as multifunctional compounds. Interactions of the ligands at the active cavity of the proteins, mostly through hydrogen bonds and stacking, modulate the activity of such complexes [31,32]. These compounds exhibited high thermodynamic stability when tested using human serum albumin (has) as a competitive biological ligand. The relative stability constants determined for this series of compounds are very similar to those of copper ions inserted at the N-terminal of the protein (pK = 12) or of zinc ions bound at Cys34 residue (pK = 7) [31].

Herein, a copper(II) and an analogous zinc(II) complex of a new oxindolimine ligand, derived from isatin and 2-aminomethylbenzimidazole and designed to facilitate intercalative interactions, were investigated for their oxidative ability and cytotoxicity toward human hepatocellular carcinoma (HepG$_2$) and neuroblastomas (SH-SY5Y), trying to improve their antiproliferative properties in relation to previously studied related compounds.

2. Results and Discussion

The new complexes, named here [Cu(isambz)$_2$]$^{2+}$ **1** and [Zn(isambz)]$^{2+}$ **2**, (see Figure 1) were prepared according to a similar procedure used in a series of other oxindolimine–metal compounds previously studied [27,28]. After characterization by UV/Vis and IR spectroscopies (see Experimental Section), their reactivity versus human serum albumin (HSA) and CT-DNA was investigated.

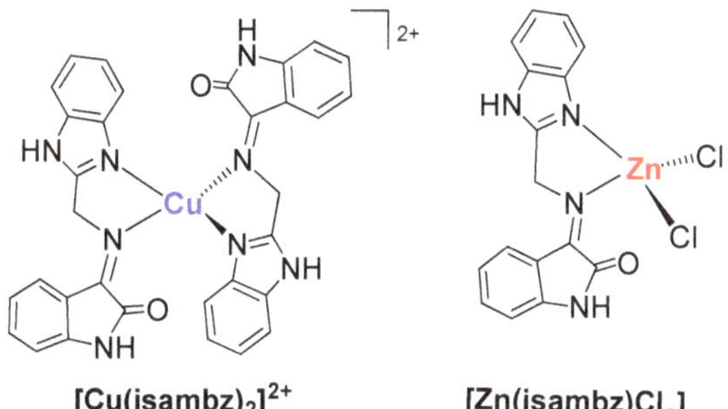

Figure 1. Scheme of the studied complexes, [Cu(isambz)$_2$]$^{2+}$ **1**, and [Zn(isambz)]$^{2+}$ **2**.

The copper(II) complex was isolated as 1:2 species, with ligands in the keto form, while the analogous zinc(II) was obtained as the 1:1 species, with the ligands in the enol form. Both species can be obtained by controlling the pH of the solution during the syntheses. However, in solution there is an equilibrium (Scheme 1) between those two forms depending on the pH, involving tautomeric forms of the ligands, similarly to what was observed in correlated complexes [28].

In further experiments, the analogous copper(II) compound [Cu(isambz)]ClO$_4$ was isolated and the EPR spectra of both copper(II) species were compared in order to clarify their structural features. Based on the spectroscopic parameters determined, the compound 1:2 showed a more tetrahedral distorted structure (g$_{//}$/A$_{//}$ = 182 cm), while the complex 1:1 exhibited a tetragonal configuration (g$_{//}$/A$_{//}$ = 116 cm), according to a semi-empirical approach [33].

Initially, the ability of the copper complex 1 in generating reactive oxygen species (ROS) was tested after 10 min incubation of this complex with hydrogen peroxide, by EPR measurements, using the spin trapping method. In Figure 2, it can be seen that up to 150 µM this complex is not very oxidative, forming a limited amount of hydroxyl radicals, as detected by DMPO–OH adducts. However, in complementary experiments at much lower concentrations, it was able to oxidize HSA, as monitored by carbonyl groups formation through reaction with DNPH.

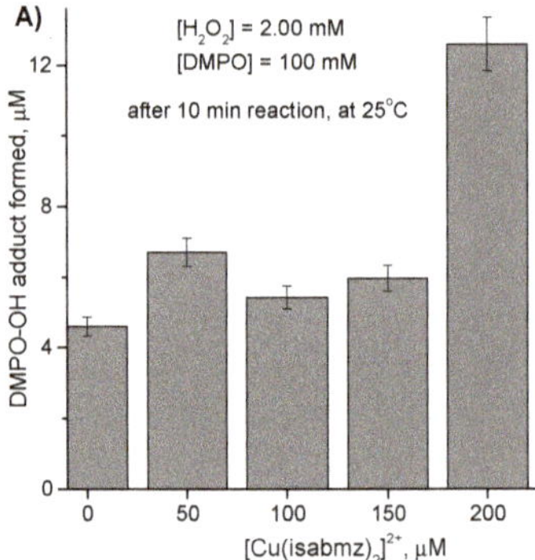

Scheme 1. Equilibrium involving tautomeric forms of oxindolimine ligands coordinated to copper(II) ions.

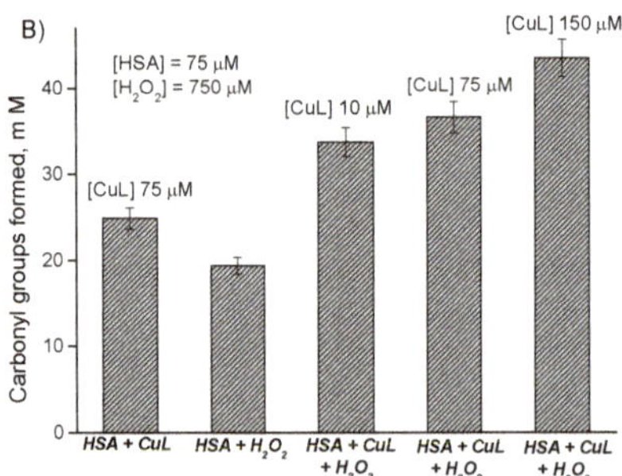

Figure 2. (**A**) Double integrated signal of 5,5-Dimethyl-1-pyrroline N-oxide (DMPO)–OH adduct detected by EPR spectroscopy after 10 min reaction in phosphate buffer at pH 7.4, at room temperature, using DMPO (100 mM) as spin trap, in the presence of complex $[Cu(isambz)_2]^{2+}$ **1** (50 to 200 μM) and H_2O_2 (2.0 mM); (**B**) Oxidative damage to human serum albumin (HSA, 75 μM) induced by complex $[Cu(isambz)_2]^{2+}$ **1** after incubation for 30 min at 37 °C in the presence of hydrogen peroxide (750 μM), monitored by the formation of carbonyl groups in the protein.

By incubation of the protein with complex **1** and hydrogen peroxide, and using EPR spectroscopy, a stable protein–radical species was detected under air, at g = 1.989 (line width of 52 G), as shown in Figure 3. A very similar species (g = 1.996, line width of 45 G) was obtaining by using an analogous complex, $[Cu(isaepy)_2]^{2+}$, one of the most reactive in a series of such copper(II) complexes [29,30].

Studies on the oxidation of bovine serum albumin (BSA) by Cu,Zn–superoxide dismutase (SOD1) and hydrogen peroxide in the presence of nitrite, have identified a similar radical (g = 2.007; total line width of 54 G) [34]. This protein-bound radical is very stable, detectable at room temperature and under air, and was identified as a protein–Tyr· species. It was also reported in the heme detoxification process by HSA, inhibiting further destructive or irreversible oxidation of the protein [35].

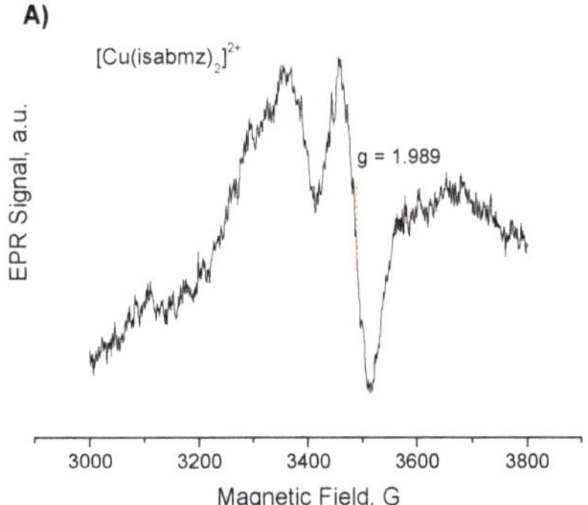

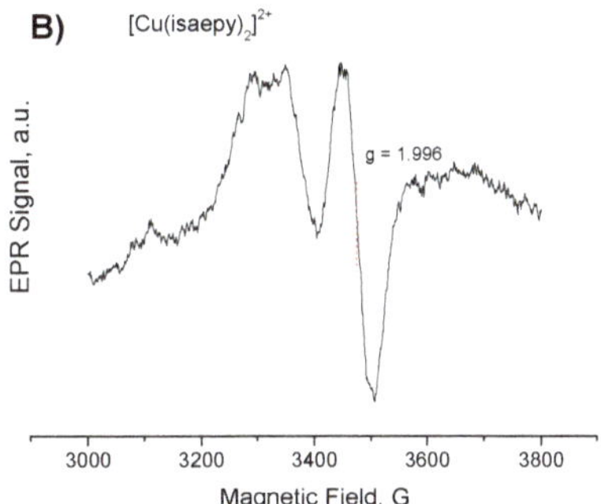

Figure 3. EPR spectra of radical species formed by oxidation of HSA (0.600 mM), in the presence of $[H_2O_2] = 2.5$ mM and each of the complexes (100 µM), $[Cu(isambz)_2]^{2+}$ **1** (**A**) or $[Cu(isaepy)_2]^{2+}$ (**B**).

Further studies by CD spectroscopy showed that by titrating HSA with this copper complex **1**, the protein α-helix structure is substantially altered, indicating an efficient binding of the complex to the protein (see Figure 4) that probably favors its oxidation.

However, complementary experiments by SDS PAGE (Figure S1, supplementary material) of HSA (75 µM) incubated for 30 min at 37 °C with each of these copper(II) complexes (75 µM), in the presence or in the absence of hydrogen peroxide, did not show degradative damage (cleavage) of the protein.

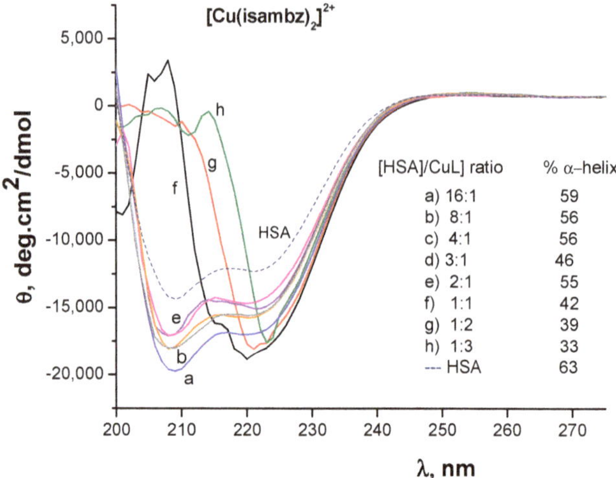

Figure 4. CD spectra registered (double scanning) after increasing concentrations of complex **1** (0.15 to 7.5 μM), to HSA solution (2.5 μM), in phosphate buffer 50 mM containing 0.10 M NaCl, at pH 7.4, at room temperature.

2.1. Quenching of DNA–EtBr Fluorescence

Since one of the most studied targets for metal complexes is DNA, we verified by fluorescence measurements the ability of both complexes to bind to CT-DNA saturated with ethidium bromide (EtBr), a known nucleic acid intercalator.

Results in this competitive assay revealed that changes in fluorescence were more noticeable for complex **1** than for complex **2** at high concentrations of quencher [Q], as shown in Figure 5. By using two methods, Stern–Volmer plots of F_0/F versus [Q], and logarithmic Scatchard graphs, $\log(F_0 - F)/F$ versus $\log[Q]$, the corresponding binding constants K_{SV} as well as K_a and binding site number n were determined. The values of K_{SV} was obtained by linear fitting up to 400 μM [Q] and was higher for the copper complex **1** (4.52×10^2 M^{-1}) than for the zinc complex **2** (2.75×10^2 M^{-1}), suggesting a stacking or intercalative interaction. The difference observed can be probably attributed to a more planar structure of the copper compound relative to that of zinc, enabling an intercalation among bases.

However, for both complexes a non-collisional quenching seemed to work, as indicated by the curves obtained in Scatchard graphs (see Figure 5). A linear plot indicates a single quenching mechanism, while a positive deviation points to different binding sites in the biomolecule available for the quenchers, usually with different affinities.

The corresponding parameters, K_a and n, were determined and were very low when compared to analogous values previously obtained for a series of similar Schiff base–copper(II) complexes, [Cu(isaepy)$_2$]$^{2+}$, [Cu(enim)H$_2$O]$^{2+}$, [Cu(isaenim)]$^{2+}$, and [Cu(o-phen)$_2$]$^{2+}$ in competitive experiments with EtBr intercalated in plasmidial DNA (see Table 1) [36]. Furthermore, in the case of plasmidial DNA, the observed K_a values are usually lower (100-fold) than for CT-DNA. For all the complexes in previous series, an efficient interaction with DNA structure was verified, with K_a in the range of 10^2 and n between 0.5 and 1. However, for the new complexes **1** and **2**, these parameters are too low, with K_a in the range of 10^{-1} and n between 0.2 and 0.3, indicating little binding to DNA, that is, little competitive substitution of EtBr as DNA intercalator.

The small decrease of fluorescence intensity of EB bound to CT-DNA observed upon increasing addition of the complexes **1** and **2** (up to 400 μM), shown in Figure S2 (supplementary material), corroborates this fact. Additionally, this was reinforced by CD spectra, monitoring the decrease in the typical DNA band at 270–280 nm by increasing addition of the metal complexes, up to 250 μM (see Figure 6). Most likely, in the case of complexes focused on herein, the main interactions occur at minor or major grooves, with slight intercalation.

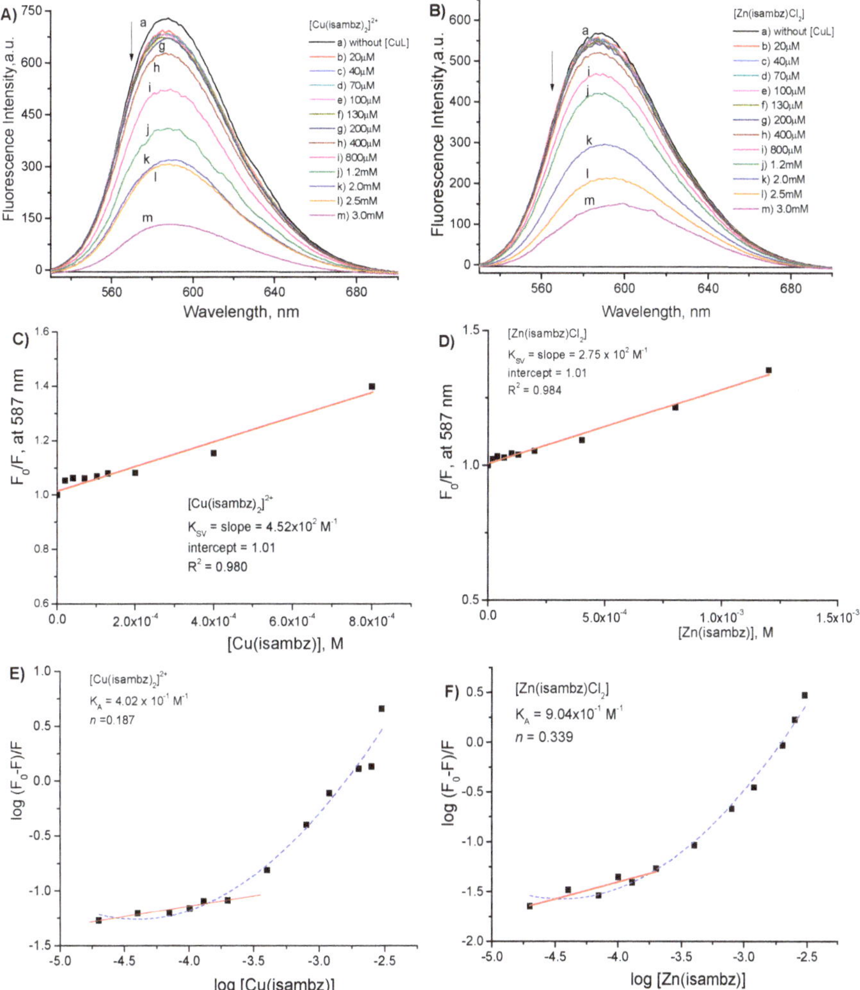

Figure 5. (**A**,**B**) Fluorescence emission spectra of EtBr intercalated to CT-DNA, in the absence and in the presence of the complexes $[Cu(isambz)_2]^{2+}$ **1** or $[Zn(isambz)]^{2+}$ **2**, respectively. [EtBr] = 50 μM; [CT-DNA] = 30 μM, and quencher concentration varying from 20–3000 μM; (**C**,**D**) Stern–Volmer equation dependence on the concentration of quenchers, complexes $[Cu(isambz)]^{2+}$ **1** or $[Zn(isambz)]^{2+}$ **2**, respectively; (**E**,**F**) Scathchard plots and linear regression fittings to estimate association binding constants, and corresponding binding site number for quenchers, $[Cu(isambz)_2]^{2+}$ **1** or $[Zn(isambz)]^{2+}$ **2**, respectively.

Table 1. The quenching constants and binding site number of CT-DNA/EtBr or plasmidial DNA/EtBr*
by some Schiff base–copper(II) or zinc(II) complexes.

Complex	K_{SV} (M^{-1})	K_a, M^{-1}	N
$[Cu(isambz)_2]^{2+}$	5.83×10^2	4.02×10^{-1}	0.19
$[Zn(isambz)]^{2+}$	2.75×10^2	9.04×10^{-1}	0.34
$[Cu(isaepy)_2]^{2+}$	-	3.64×10^2	0.68 *
$[Cu(enim)H_2O]^{2+}$	-	4.13×10^2	0.45 *
$[Cu(isaenim)]^{2+}$	-	8.85×10^2	0.73 *
$[Cu(o\text{-}phen)_2]^{2+}$	-	7.55×10^2	0.60 *

* Previous work, ref. [29].

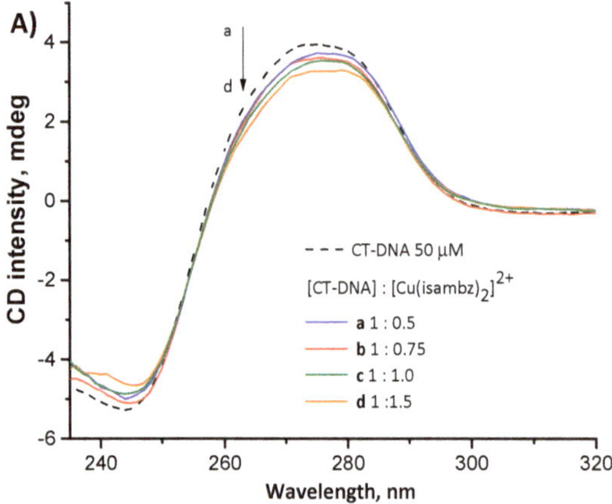

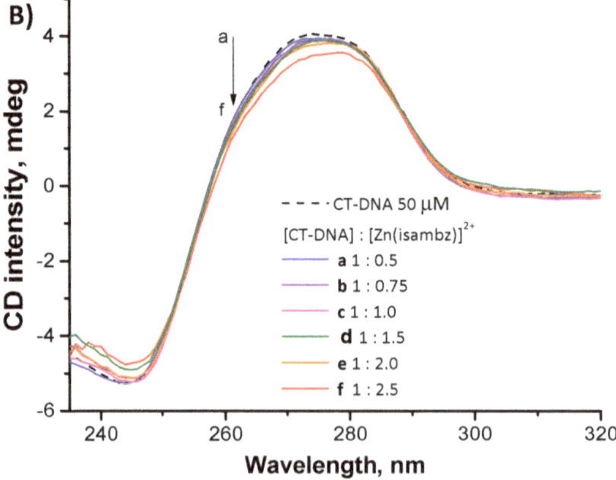

Figure 6. CD spectra of CT-DNA (50 µM) in phosphate buffer 50 mM containing NaCl 100 mM, in the
absence and presence of (**A**) varied concentrations of the copper(II) complex, $[Cu(isambz)]^{2+}$ **1** and
(**B**) varied concentrations of the zinc complex $[Zn(isambz)]^{2+}$ **2**.

2.2. Oxidative Damage to DNA

Despite not showing very significant interaction with DNA (at low concentrations), significant
single and double cleavage of CT-DNA was verified in the presence of hydrogen peroxide (50 µM) and
complex **1**, after incubation for 30 or 60 min at 37 °C (see Figure 7). This was the most reactive copper

complex in an already investigated series of complexes regarding its nuclease activity [37], leading to the linear DNA form III after 60 min, at 50 nM. In different experimental conditions, by using 125 mM H_2O_2, formation of linear DNA form was observed at 5 µM after 12 h incubation, as also shown in Figure 7.

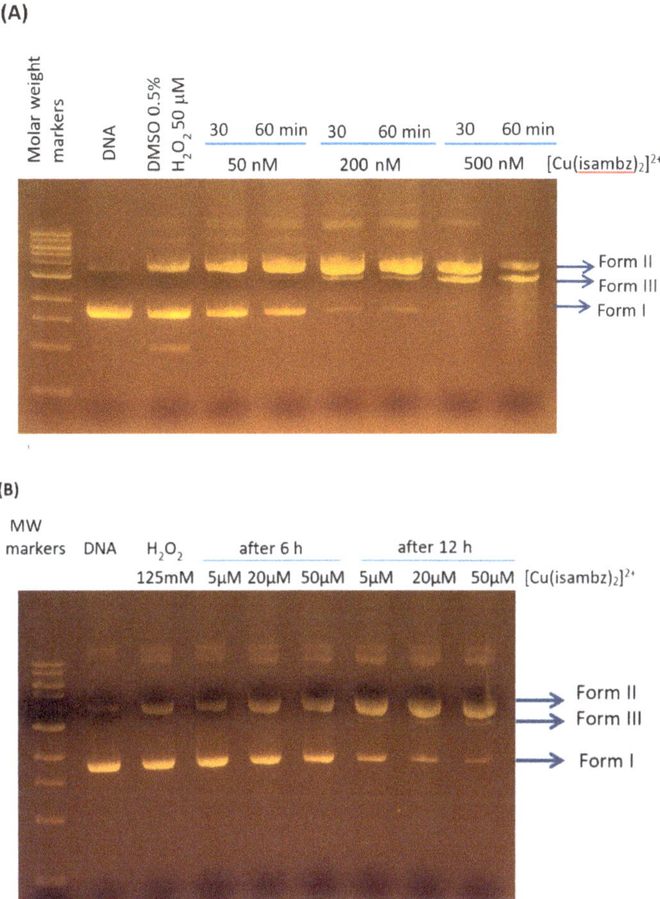

Figure 7. DNA strand double cleavage observed in the presence of (**A**) [Cu(isambz)₂]²⁺ (50, 200, or 500 nM), and H_2O_2 (50 µM), after 30 or 60 min incubation, at 37 °C; (**B**) [Cu(isambz)₂]²⁺ (5, 20, or 50 µM), and H_2O_2 (125 mM), after 6 or 12 h incubation, at 37 °C. Initial [DNA] = 240 ng of supercoiled DNA (Form I), in 20 µL phosphate buffer (50 mM, pH 7.4).

2.3. Cytotoxicity Toward Tumor Cells

The antiproliferative properties of the new complexes **1** and **2** were tested against hepatocellular carcinoma cells (HepG₂) and neuroblastoma cells (SHSY5Y), up to 50 µM concentration. The compounds showed different cytotoxic effects toward SHSY5Y cells which were more susceptible with respect to the HepG2 cells (Figure 8). In particular, the zinc complex **2** did not significantly affects cell viability of HepG2 cells, but is significantly toxic for SHSY5Y cells only at 50 µM ($p < 0.01$). On the contrary, copper complex **1** showed slightly toxic effect at both 25 µM and 50 µM for HepG2 cells ($p < 0.05$), and a more pronounced toxicity for the SHSY5Y cells at both 25 µM and 50 µM ($p < 0.01$). Therefore, the cytotoxicity of both complexes was cell type dependent. This behavior is very similar to that observed with copper(II) complexes with related ligands derived from isatin [28,29]. Consequently, the introduction of a more lipophilic moiety in the ligand (benzimidazole group) apparently did not significantly improve the cytotoxicity of these complexes, as expected.

In the literature [38], $[Cu(o\text{-phen})_2]^{2+}$ complex has been reported to induce apoptosis in HepG2 cells by generation of ROS, with IC_{50} around 5 µM after 24 h incubation at 37 °C. Its higher cytotoxicity can be explained by the almost planar arrangement of the phenanthroline rings that helps its intercalation with DNA [39]. For both of our complexes, the main interactions with DNA probably occurred at minor or major grooves due their structure (distorted tetrahedral geometry).

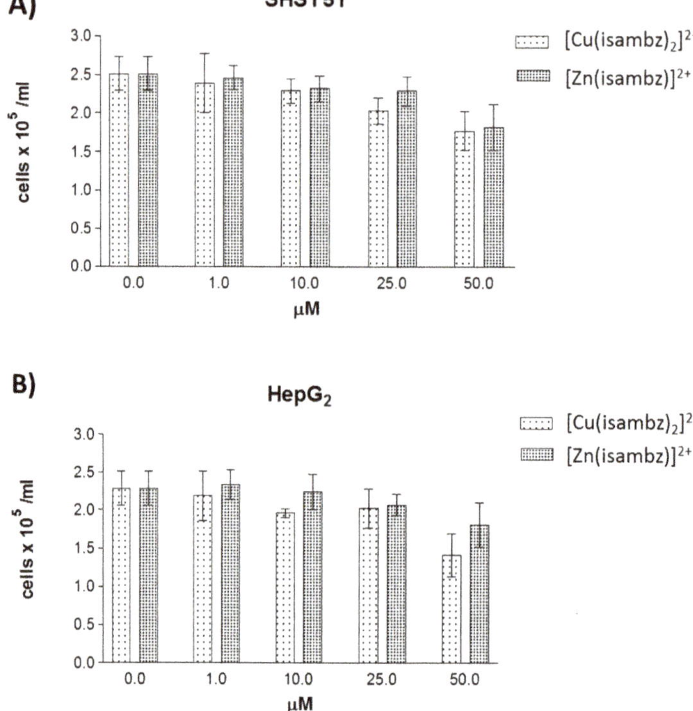

Figure 8. Viability of (**A**) neuroblastoma SHSY5Y cells, and (**B**) hepatocellular carcinoma HepG2 cells, after 24 h incubation at 37 °C with complexes $[Cu(isambz)_2]^{2+}$ **1** or $[Zn(isambz)]^{2+}$ **2**.

3. Experimental Section

3.1. Materials

The reagents isatin (1*H*-indole-2,3-dione, 98%), 2-(2-aminoethyl)pyridine (95%), 1,3-diaminopropane (98%), histamine dihydrochloride (99%), 2-aminomethylbenzimidazole dihydrochloride (99%), copper(II) perchlorate hexahydrated (98%), were purchased from Aldrich Chemical Co. Calf thymus DNA as sodium salt (CT-DNA) and human serum albumin (HSA) were from Sigma-Aldrich (St. Louis, MO, USA), while most of the solvents, ethanol, methanol, and dichloromethane were from Merck Chemical Co. (Darmstadt, Germany). All the solutions were prepared with water purified in a deionizer Barnstead apparatus, model D470.

Syntheses of the Complexes

The new metal complexes with oxindolimine ligands were prepared by methodology already developed in previous studies, with suitable modifications [27,28]. Depending on the pH during the metalation step, tautomeric forms (keto or enol) of the ligand are preferentially formed and the corresponding metal complexes can be isolated. For complex **1**, the species [CuL$_2$] with the ligand in the keto form was isolated at apparent pH in the range 5 to 7, while the analogous [ZnL] complex **2**, with the ligand in the enol form, was isolated at pH 9. In previous studies with similar oxindolimine ligands, both complex species 1:2 (ligand in keto form) and 1:1 (ligand in enol form) could also be isolated and characterized [36].

Complex [Cu(isambz)$_2$](ClO$_4$)$_2$—complex **1**: To a solution of isatin 1.47 g (10.0 mmols) dissolved in 25 mL ethanol and 25 mL CH$_2$Cl$_2$, 2.20g (10.0 mmols) of 2-aminomethylbenzimidazol were added, adjusting the pH to 5.5 with drops of HCl. The solution was kept under constant stirring for 24 h. Afterwards, the metallation of the formed ligand was carried out by addition of 1.85g (5.00 mmols) copper(II) perchlorate hexahydrate dissolved in 5 mL water. In this step, the pH was adjusted around 7 by adding some drops of NaOH solution, and a brown precipitate was formed. This precipitate was collected after cooling the final solution, washed with cold ethanol and ethyl ether, and finally dried in a desiccator under reduced pressure. The keto form of the ligand coordinated to the copper ion, in a 1:2 metal:ligand complex, was isolated. The yield was 74% (3.02 g). Elemental analyses: *Calc. for* C$_{32}$H$_{24}$N$_8$O$_2$Cu(ClO$_4$)$_2$ (MW= 815.05): *C, 47.16; H, 2.97; N, 13.75; Experim.: C, 48.51; H, 3.05; N, 14.30.* UV/Vis: λ_{max} = 256 nm (ε = 1.61 $\times$ 10^{-4} M^{-1}·cm^{-1}), λ_{max} = 284 nm (ε = 1.87 $\times$ 10^{-4} M^{-1}·cm^{-1}). IR (cm^{-1}): 3343, ν(N–H); 2976, ν(C$_{sp2}$–H and C=C); 1722, ν(C=O), 1614 and 1597, ν(C=N Schiff base, isatin ring and N–H); 741, ν(C–H phenyl). EPR spectra, in DMSO/water at 77K: g$_\perp$ = 2.096; g$_{//}$ = 2.427; A$_{//}$ = 118 G = 133 $\times$ 10^{-4} cm^{-1} (distorted tetrahedral symmetry). For analogous complex, isolated as 1:1 species [Cu(isambz)]ClO$_4$, EPR spectra, in DMSO/water at 77K: g$_\perp$ = 2.107; g$_{//}$ = 2.450; A$_{//}$ = 186 G = 212 $\times$ 10^{-4} cm^{-1} (tetragonal symmetry).

Complex [Zn(isambz)Cl$_2$]·H$_2$O—complex **2**: By a similar procedure, the analogous dark orange [Zn(isambz)Cl$_2$] was prepared, using zinc(II) chloride instead of copper salt. In this case, a 1:1 metal:ligand complex was isolated, maintaining the pH around 9 during metalation. The yield was 58% (1.25 g). Elemental analyses: *Calc. for* C$_{16}$H$_{12}$N$_4$OZnCl$_2$ (MW = 430.50): *C, 44.64; H, 3.28; N, 13.01; Experim.: C, 44.24; H, 3.11; N, 12.77.* UV/Vis: λ_{max} = 259 nm (ε = 8.85 $\times$ 10^{-3} M^{-1}·cm^{-1}), λ_{max} = 284 nm (ε = 8.67 $\times$ 10^{-3} M^{-1}·cm^{-1}). IR (cm^{-1}): 3344, ν(N–H); 2978, νC$_{sp2}$–H and C=C); 1721, ν(C=O), 1613 and 1471, ν(C=N Schiff base, isatin ring and N–H); 744, ν(C–H phenyl).

3.2. Methods

Elemental analyses (C, H, and N) were performed using a Perkin Elmer model 2400 analyzer (Perkin-Elmer, Billerica, MA, USA), and metal components were determined in duplicate by ICP emission spectroscopy in a Spectro Analytical Instruments equipment (Spectro/AMETEK, Kleve, Germany), at *Central Analítica* of our institution. UV/Vis spectra were registered in a UV-1650PC spectrophotometer from Shimadzu (Shimadzu Corp., Kyoto, Japan), using 1.00 cm optical length quartz cells. Fourier transform infrared (FTIR) spectra of samples were recorded in the 4000–400 cm^{-1} range on a Bruker FTIR ALPHA (Bruker, Billerica, USA) equipped with a single reflection diamond ATR module with a spectral resolution of 4 cm^{-1}. EPR experiments were conducted in a Bruker EMX spectrometer operating at X-band (standard conditions: frequency 9.51 GHz, microwave power 20.12 mW, modulation frequency 100 kHz) (Bruker, Karlsruhe, Germany), using Wilmad flat cells. HSA-radical was registered directly, at room temperature, after incubation of the protein with complex **1** or [Cu(isaepy)] (100 μM), and immediately after addition of hydrogen peroxide (2.5 mM). 5,5-Dimethyl-1-pyrroline N-oxide (DMPO, 100 mM) and 4-hydroxy-1-oxyl-2,2,6,6-tetramethylpiperidine (Tempol, 36 μM) were used as spin trap and calibrant respectively, in spin trapping EPR experiments, to detect hydroxyl radical generation in the presence of hydrogen peroxide and complex **1**.

3.2.1. Interactions with Biomolecules Monitored by CD Spectroscopy

Interactions of complex **1** with HSA were monitored by circular dichroism (CD) spectroscopy, in the range 190 to 300 nm, in a JASCO J-720 spectropolarimeter. The spectra (double scan) were registered in a spherical quartz cell (0.10 cm optical path), after increasing addition of complex **1** (stock solution 1.0 mM) to HSA solution (2.5 µM), in phosphate buffer 50 mM containing 0.10 M NaCl, at pH 7.4 at room temperature. Results were expressed as residual elipsicity (θ), defined as θ (deg cm^2/dmol) = CD (mdeg)/10 $\times$ *n* $\times$ *l* $\times$ **C_p**, where CD is the measure in mdeg, *n* is the number of amino acid residues (for HSA, n = 585), *l* is the optical path, and **C_p** is the protein concentration [40].

Binding to DNA was monitored through a typical DNA band at 270–280 nm, using the same instrument, a JASCO J-720 spectropolarimeter (JASCO Inc., Easton, MD, USA). Decreasing intensity of this band (DNA 50 µM) was verified by addition of increasing concentrations of complex **1** or **2**, up to 250 µM.

3.2.2. Monitoring HSA Damage

Oxidation of the protein was detected by following the formation of carbonyl groups through reaction with dinitrophenylhydrazine (DNPH) [41], after incubation of HSA with the copper complex **1** for 30 min at 37 °C, in the presence of hydrogen peroxide.

Samples of HSA (75 µM) were incubated with hydrogen peroxide (750 µM) and different concentrations of copper complex **1** at 37 °C for 30 min. To each sample was added 1 mL of 0.10 M DNPH solution in HCl 1 M, and the mixture was incubated for more 30 min at the same temperature. After cooling, 1.5 mL of 1 M NaOH solution was added, and the formation of corresponding dinitrophenylhydrazones was monitored spectrophotometrically, at 370 nm (ε = 22,000 M^{-1}·cm^{-1}) [42].

3.2.3. Fluorescence Measurements with DNA and Ethidium Bromide

The fluorescence assays were performed by using fixed concentrations of CT-DNA (30 µM) and ethidium bromide (EtBr, 50 µM), and adding varying concentration of each metal complex as a quencher (20–3000 µM). The CT-DNA concentration per nucleotide was determined by absorption spectrometry with a molar extinction coefficient value of 6600 M^{-1}·cm^{-1} at 260 nm. A SPEX-Fluorolog 2 spectrometer (Horiba Scientific, Kyoto, Japan) was used in fluorescence intensity measurements, with excitation wavelength at 510 nm and emission wavelength in the range of 530–800 nm. All these measurements were made by taking fresh solution each time in a quartz cell with 1.00 cm optical path, and the acquisitions were performed at 37 °C under lower and constant stirring. The maximum fluorescence resembles at 587 nm, and the first acquisition for each set of data was made in the absence of the quencher (complex). The initial fluorescence intensity was called F_0, and the subsequent fluorescence intensities at a fixed solution concentration were entitled as F.

To describe the fluorescence quenching of the EtBr in a competitive reaction between CT-DNA and the complexes [Cu(isambz)$_2$]$^{2+}$ **1** and [Zn(isambz)]$^{2+}$ **2**, two different mathematical methods were used. The first method describes a dynamic process in which the quenching mechanism is mainly due to collision and can be described by the linear Stern–Volmer equation [43], $F_0/F = 1 + K_q\tau_0[Q] = 1 + K_{SV}[Q]$, where F_0 and F represent the fluorescence intensities in the absence and in the presence of different quencher concentration [Q] in mole per liter, respectively. The product $K_q\tau_0$ is known as the Stern–Volmer constant, K_{SV}, and it is a parameter that responds to the availability of the quencher to the fluorophore. The fluorophore quenching rate constant is given by K_q, K_{SV} is the quenching constant, and τ_0 is the lifetime of the fluorophore in the absence of a quencher. When non linearity occurs in the experimental data, the quenching mechanism is not purely collisional. This variation may be due either to the ground state complex formation or to the static quenching model [44]. The second method, described using the expression $\log[(F_0 - F)/F] = \log K_A + n \log[Q]$, expresses the binding constant and the number of binding sites per nucleotide (n) for the new species formed between the studied compound and the CT-DNA. The linear correlation $\log[(F_0 - F)/F]$ versus $\log[Q]$ gives an equation where the slope corresponds to the binding stoichiometry (n), and the intercept gives the K_a ($10^{intecept} = K_a$). In our experiments, fitting was restricted up to 400 μM quencher concentration.

3.2.4. DNA Cleavage Studies

In these experiments, the plasmid pBluescript II (Stratagene, San Diego, USA) was used after purification, using Qiagen plasmid purification kit (Qiagen, Hilden, Germany). Mixtures of 240 ng of supercoiled DNA (Form I), in 20 μL total volume containing 50 mM phosphate buffer (pH 7.4), in the presence or absence of hydrogen peroxide and varying concentrations of complex **1** were incubated at 37 °C for different periods of time. Stock solutions of the complexes (1.0 mM) were prepared by dissolving them in small amounts of dimethylsulfoxide (DMSO, 0.5 to 1 mL), and immediately diluting appropriate samples to the desired concentration with buffer solution. The final amount of DMSO was ≤1%. After incubation, a quench buffer solution (4 μL) was added, and the final solution was subjected to electrophoresis on an 1% agarose gel in 1× TAE buffer (40 mM Tris–acetate, 1 mM EDTA) at 100 V, for 2 h.

3.2.5. Cytotoxicity Assays In Vitro

Human hepatoma cells, HepG2, and neuroblastoma cells, SHSY5Y, were purchased from the European Cell Culture Collection and grown in RPMI 1640 supplemented with 10% fetal calf serum, and in Dulbecco's modified Eagle's/F12 medium supplemented with 15% fetal calf serum, respectively, at 37°C in an atmosphere of 5% CO_2 in air.

Cells were routinely trypsinized and plated at $30 \times 10^5/75$ cm^2 flasks until sub-confluence. For the experiments, cells were plated on multiwell at 2×10^5 cells/ml and treated after 24 h. The complexes were dissolved in DMSO (stock solution 1.0 mM), and used at different concentrations (1 μM, 10 μM, 25 μM, and 50 μM). Cytotoxicity assay was performed at 24 h. Cells were collected in the medium by a scraper and harvested by centrifugation at 1000 rpm × 5 min. Cell pellets were washed once in phosphate buffer solution, pH 7.4. Cell viability was estimated by direct count with a Toma's chamber upon Trypan blue exclusion.

4. Conclusions

Two novel complexes were obtained with an oxindolimine ligand, and had their oxidative properties toward HSA and DNA verified. Their cytotoxicities against tumor cells (HepG2 and SHSY5Y) were verified to be only moderate, in comparison to similar copper(II) complexes in the literature, and were then correlated with these oxidative properties, to clarify the main targets or modes of action of these studied complexes. This type of ligand exhibits tautomers in solution, depending on the pH, leading to corresponding keto and enol forms of copper(II) or zinc(II) complexes (Figure 1, Scheme 1). In the case of zinc, the 1:1 metal:ligand complex was isolated, while for copper the 1:2 species was obtained by careful control of the pH during the synthesis. According to the overall results, in both cases the 1:1 species is probably the most active toward biomolecules at pH 7.4, as already observed for other complexes in this series. However, in our studies we should consider the possibility of having both complexes 1:1 and 1:2 in equilibria, based on our previous studies with similar copper complexes, and on reported data in the literature for the speciation of different metallodrugs [45]. In this case, the predominant species detected in solution depends primarily on the pH. The enol form of these oxindolimine ligands has a more delocalized electronic dispersion, in addition to a negative charge, since it is more easily deprotonated, and this usually leads to more stabilized and less sterically hindered metal center.

Both complexes **1** and **2** interact with HSA as well as with DNA, as detected by CD spectroscopy and fluorescence measurements, and are able to damage these biomolecules. Complex **1** is able to reasonably generate reactive oxygen species (ROS) up to 150 μM concentration, but to a remarkable extent at concentration $\geq$200 μM, as detected by spin trapping EPR measurements. It causes oxidative damage to has, as monitored through carbonyl group formation, and significantly affects its α-helix structure to a high extent. Additionally, a very stable protein–radical species was detected at g = 1.989 (width 52G) by EPR spectroscopy after treatment with complex **1** in the presence of H_2O_2 at room temperature. This radical was attributed to a protein-bound radical, HAS–Tyr· species, by comparison to a similar one, already identified in the literature, by treatment of bovine serum albumin (BSA) with Cu,Zn–superoxide dismutase, and hydrogen peroxide in the presence of nitrite [40]. Furthermore, this radical has been implicated in heme detoxification process by HSA, and seems to protect the protein against other potentially more toxic effects of this oxidant [41]. Therefore, the formation of such intermediary species attested to the remarkable oxidative properties of both copper(II) complexes investigated, complex **1** and $[Cu(isaepy)_2]^{2+}$. Interactions with HSA and other extracellular proteins can collaborate to decrease the amount of complexes entering the cells, and therefore contribute to their lower observed cytotoxicity.

The copper complex **1** is also capable of oxidizing DNA at very low IC_{50}, in nM concentration range, in the presence of hydrogen peroxide (50 μM). It leads to single and double strand cleavage after 60 min incubation at 37 °C, despite not binding remarkably to DNA by intercalation, as attested by fluorescence quenching in competitive measurements with EtBr. Compared to other copper complexes with analogous ligands, this new copper compound showed much better oxidative properties, although it exhibited only moderate anti-proliferative effect against hepatocellular carcinoma (HepG$_2$) and neuroblastoma (SHSY5Y) tumor cells [29]. Additionally, the analogous zinc complex **2**, with no redox properties, exhibited similar cytotoxic effects against the same tumor cells. The ligand structure can be crucial for metal activity toward DNA, as exemplified in comparative studies of copper(II) complexes with correlated ligands, 2,2′-bipyridine (bpy), 2,2′-dipyridylamine (dpa) or dipicolylamine (dpca), where $[Cu(bpy)_2]^{2+}$ complex was the most active in double cleavage reaction [46].

These data can also be indicative of alternative targets inside the cell in addition to DNA, in a process modulated by the metal ion, by the ligand, as well as by the predominant metal–ligand species formed. This class of metal complexes has already shown good inhibition activity toward kinases [31] and topoisomerases [32], and the remarkable ability to damage mitochondria [30].

These results point to other factors than oxidative assets in determining the reactivity of these compounds. A more voluminous and sterically hindered ligand can be responsible for the verified decreased toxicity of complexes **1** and **2**, due to difficulties in entering the cell or intercalating at DNA structure. Their low solubility in aqueous solution is also a factor to consider. More recent studies in the development of metallodrugs use the strategy of inserting the active compounds into nanostructured materials, trying to increase their stability and efficiency in achieving their biological targets, as well enhancing anticancer activity. $[Cu(phen)]Cl_2$ complex, an inhibitor of aquaporin, was successfully incorporated into different liposomes, preserving its cytotoxicity against different tumor cells [47]. These copper loaded liposomes seem to have therapeutic potential for the treatment of solid tumors, based on their preferential accumulation. In other investigations, two correlated oxindolimine complexes based on copper and zinc, for which the anticancer properties had been already verified [29] were immobilized in functionalized nanoporous silica, and the obtained materials exhibited increased antiproliferative activity against melanomas [48]. They were monitored inside the cells, and released the active compounds in a kinetic process that depended on the metal. For the copper based materials, a 98% release was observed after 24 h at 37 °C, while for the analogous zinc based materials an 80% or 55% release could be verified, depending on the matrix used.

Our results confirmed DNA as an important target for these oxindolimine–metal complexes, as already observed [29], and additionally pointed out that oxidative damage is not the leading mechanism responsible for the cytotoxicity of this class of compounds.

Supplementary Materials: The following are available online at http://www.mdpi.com/2304-6740/7/2/12/s1. Figure S1: Monitoring HSA damage in the presence of hydrogen peroxide and copper complexes. Figure S2: Quenching of CT-DNA/EB fluorescence by complexes **1** or **2**.

Author Contributions: All the co-authors contributed to this work. In conceptualization, A.M.D.C.F. and M.R.C.; methodology, M.C., Q.A.d.P., and P.C.; formal analysis, M.C., A.M.L.Z., Q.A.d.P., M.B.P., and P.C.; investigation, M.C., A.M.L.Z., Q.A.d.P., M.B.P., and P.C.; resources, A.M.D.C.F., C.C.O. and M.R.C.; data curation, A.M.D.C.F., C.C.O., and M.R.C.; writing—original draft preparation, M.C., A.M.L.Z., Q.A.d.P., A.M.D.C.F., and M.R.C.; writing—review and editing, A.M.D.C.F., C.C.O. and M.R.C.; project administration, A.M.D.C.F.; funding acquisition, A.M.D.C.F., C.C.O., and M.R.C.

Funding: This research was funded by Fundação de Amparo à Pesquisa do Estado de São Paulo (FAPESP, grant 11/50318-1); Coordenação de Aperfeiçoamento de Pessoal de Nível Superior (CAPES, Finance code 001); Programa Executivo de Cooperação Científica e Tecnológica Brasil-Itália, Conselho Nacional para o Desenvolvimento Científico e Tecnológico (CNPq)/Ministeri degli Affari Esteri (grant 490021/2008-5), and network CEPID Redoxoma (Redox Processes in Biomedicine, grant FAPESP 2013/07937-8).

Acknowledgments: Financial support from the Brazilian agencies (FAPESP, CAPES and CNPq), International Program for Scientific Collaboration Brazil-Italy (CNPq/Ministeri degli Affari Esteri), and network CEPID Redoxoma is gratefully acknowledged. M.C. and Q.A.P. also thank FAPESP for fellowships (2008/54470-0 and 2011/50204-6) during Post-doctoral studies.

Conflicts of Interest: The authors declare no conflict of interest.

References

1. World Health Organization. Cancer. Available online: http://www.who.int/news-room/fact-sheets/detail/cancer (accessed on 18 October 2018).

2. Barry, N.P.E.; Sadler, P.J. Exploration of the medical periodic table: Towards new targets. *Chem. Commun.* **2013**, *49*, 5106–5131. [CrossRef] [PubMed]

3. Muhammad, N.; Guo, Z. Metal-based anticancer chemotherapeutic agents. *Curr. Opin. Chem. Biol.* **2014**, *19*, 144–153. [CrossRef] [PubMed]

4. Barone, G.; Terenzi, A.; Lauria, A.; Almerico, A.M.; Leal, J.M.; Busto, N.; García, B. DNA-binding of nickel(II), cooper(II) and zinc(II) complexes: Structure-affinity relationships. *Coord. Chem. Rev.* **2013**, *257*, 2848–2862. [CrossRef]

5. Yu, Y.; Cui, Y.; Niedernhofer, L.J.; Wang, Y. Occurrence, biological consequences, and human health relevance of oxidative stress-induced DNA damage. *Chem. Res. Toxicol.* **2016**, *29*, 2008–2039. [CrossRef] [PubMed]

6. Qiu, K.; Chen, Y.; Rees, T.W.; Ji, L.; Chao, H. Organelle-targeting metal complexes: From molecular design to bio-applications. *Coord. Chem. Rev.* **2019**, *378*, 66–86. [CrossRef]

7. Bergamo, A.; Dyson, P.J.; Sava, G. The mechanism of tumour cell death by metal-based anticancer drugs is not only a matter of DNA interactions. *Coord. Chem. Rev.* **2018**, *360*, 17–33. [CrossRef]

8. Hanahan, D.; Weinberg, R.A. Hallmarks of cancer: The next generation. *Cell* **2011**, *144*, 646–674. [CrossRef] [PubMed]

9. Feng, L.; Geisselbrecht, Y.; Blanck, S.; Wilbuer, A.; Atilla-Gokcumen, G.E.; Filippakopoulos, P.; Kräling, K.; Celik, M.A.; Harms, K.; Maksimoska, J.; et al. Structurally sophisticated octahedral metal complexes as highly selective protein kinase inhibitors. *J. Am. Chem. Soc.* **2011**, *133*, 5976–5986. [CrossRef]

10. Neves, A.P.; Pereira, M.X.; Peterson, E.J.; Kipping, R.; Vargas, M.D.; Silva, F.P., Jr.; Carneiro, J.W.; Farrell, N.P. Exploring the DNA binding/cleavage, cellular accumulation and topoisomerase inhibition of 2-hydroxy-3-(aminomethyl)-1,4-naphthoquinone Mannich bases and their platinum(II) complexes. *J. Inorg. Biochem.* **2013**, *119*, 54–64. [CrossRef]

11. Kelly, W.K.; Connor, O.A.; Marks, P.A. Histone Deacetylase Inhibitors: From Target to Clinical Trials. *Expert Opin. Investig. Drugs* **2002**, *11*, 1695–1713. [CrossRef]

12. Santini, C.; Pellei, M.; Gandin, V.; Porchia, M.; Tisato, F.; Marzano, C. Advances in copper complexes as anticancer agents. *Chem. Rev.* **2014**, *114*, 815–862. [CrossRef] [PubMed]

13. Deo, K.M.; Pages, B.J.; Ang, D.L.; Gordon, C.P.; Aldrich-Wright, J.R. Transition metal intercalators as anticancer agents—Recent advances. *Int. J. Mol. Sci.* **2016**, *17*, 1818. [CrossRef] [PubMed]

14. Fu, X.B.; Liu, D.D.; Lin, Y.; Hu, W.; Mao, Z.W.; Le, X.Y. Water-soluble DNA minor groove binders as potential chemotherapeutic agents: Synthesis, characterization, DNA binding and cleavage, antioxidation, cytotoxicity and HSA interactions. *Dalton Trans.* **2014**, *43*, 8721–8737. [CrossRef] [PubMed]

15. McGivern, T.J.P.; Afsharpour, S.; Marmion, C.J. Copper complexes as artificial DNA metallonucleases: From Sigman's reagent to next generation anti-cancer agent? *Inorg. Chim. Acta* **2018**, *472*, 12–39. [CrossRef]

16. Erxleben, A. Interactions of copper complexes with nucleic acids. *Coord. Chem. Rev.* **2018**, *360*, 92–121. [CrossRef]

17. Cao, Q.; Li, Y.; Freisinger, E.; Qin, P.Z.; Sigel, R.K.O.; Mao, Z.W. G-quadruplex DNA targeted metal complexes acting as potential anticancer drugs. *Inorg. Chem. Front.* **2017**, *4*, 10–32. [CrossRef]

18. Hansel-Hertsch, R.; Di Antonio, M.; Balasubramanian, S. DNA G-quadruplexes in the human genome: Detection, functions and therapeutic potential. *Nat. Rev. Mol. Cell Biol.* **2017**, *18*, 279–284. [CrossRef]

19. Bonsignore, R.; Russo, F.; Terenzi, A.; Spinello, A.; Lauria, A.; Gennaro, G.; Almerico, A.M.; Keppler, B.K.; Barone, G. The interaction of Schiff Base complexes of nickel(II) and zinc(II) with duplex and G-quadruplex DNA. *J. Inorg. Biochem.* **2018**, *178*, 106–114. [CrossRef]

20. Slator, C.; Molphy, Z.; McKee, V.; Long, C.; Brown, T.; Kellett, A. Di-copper metallodrugs promote NCI-60 chemotherapy via singlet oxygen and superoxide production with tandem TA/TA and AT/AT oligonucleotide discrimination. *Nucleic Acids Res.* **2018**, *1*. [CrossRef]

21. Vine, K.L.; Matesic, L.; Locke, J.M.; Skropeta, D. Recent Highlights in the Development of Isatin-Based Anticancer Agents. In *Advances in Anticancer Agents in Medicinal Chemistry*; Bentham Science: Oak Park, IL, USA, 2013; Volume 2, pp. 254–312. ISBN 978-1-60805-715-3.

22. Sridhar, S.K.; Ramesh, A. Synthesis and pharmacological activities of hydrazones, Schiff and mannich bases of isatin derivatives. *Biol. Pharm. Bull.* **2001**, *24*, 1149–1152. [CrossRef]

23. Cerchiaro, G.; Ferreira, A.M.D. Oxindoles and copper complexes with oxindole-derivatives as potential pharmacological agents. *J. Braz. Chem. Soc.* **2006**, *17*, 1473–1485. [CrossRef]

24. Karali, N. Synthesis and primary cytotoxicity evaluation of new 5-nitroindole-2,3-dione derivatives. *Eur. J Med. Chem.* **2002**, *37*, 909–918. [CrossRef]

25. Evdokimov, N.M.; Magedov, I.V.; McBrayer, D.; Kornienko, A. Isatin derivatives with activity against apoptosis-resistant cancer cells. *Bioorg. Med. Chem. Lett.* **2016**, *26*, 1558–1560. [CrossRef] [PubMed]

26. Prakash, C.R.; Theivendren, P.; Raja, S. Indolin-2-Ones in Clinical Trials as Potential Kinase Inhibitors: A Review. *Pharmacol. Pharm.* **2012**, *3*, 62–71. [CrossRef]

27. Cerchiaro, G.; Micke, G.A.; Tavares, M.F.M.; da Costa Ferreira, A.M. Kinetic studies of carbohydrate oxidation catalyzed by novel isatin–Schiff base copper(II) complexes. *J. Mol. Catal. A Chem.* **2004**, *221*, 29–39. [CrossRef]

28. Cerchiaro, G.; Aquilano, K.; Filomeni, G.; Rotilio, G.; Ciriolo, M.R.; Ferreira, A.M.D. Isatin-Schiff base copper(II) complexes and their influence on cellular viability. *J. Inorg. Biochem.* **2005**, *99*, 1433–1440. [CrossRef]

29. Filomeni, G.; Cerchiaro, G.; Ferreira, A.M.D.; De Martino, A.; Pedersen, J.Z.; Rotilio, G.; Ciriolo, M.R. Pro-apoptotic activity of novel isatin-Schiff base copper(II) complexes depends on oxidative stress induction and organelle-selective damage. *J. Biol. Chem.* **2007**, *282*, 12010–12021. [CrossRef]

30. Filomeni, G.; Piccirillo, S.; Graziani, I.; Cardaci, S.; Da Costa Ferreira, A.M.; Rotilio, G.; Ciriolo, M.R. The isatin-Schiff base copper(II) complex Cu(isaepy)$_2$ acts as delocalized lipophilic cation, yields widespread mitochondrial oxidative damage and induces AMP-activated protein kinase-dependent apoptosis. *Carcinogenesis* **2009**, *30*, 1115–1124. [CrossRef]

31. Miguel, R.B.; Petersen, P.A.D.; Gonzales-Zubiate, F.A.; Oliveira, C.C.; Kumar, N.; Nascimento, R.R.; Petrilli, H.M.; Da Costa Ferreira, A.M. Inhibition of cyclin-dependent kinase CDK1 by oxindolimine ligands and corresponding copper and zinc complexes. *J. Biol. Inorg. Chem.* **2015**, *20*, 1205–1217. [CrossRef]

32. Castelli, S.; Gonçalves, M.B.; Katkar, P.; Stuchi, G.C.; Couto, R.A.A.; Petrilli, H.M.; Da Costa Ferreira, A.M. Comparative studies of oxindolimine-metal complexes as inhibitors of human DNA topoisomerase IB. *J. Inorg. Biochem.* **2018**, *186*, 85–94. [CrossRef]

33. Sakaguchi, U.; Addison, A.W. Spectroscopic and redox studies of some copper(II) complexes with biomimetic donor atoms: Implications for protein copper centres. *J. Chem. Soc. Dalton Trans.* **1979**, 600–608. [CrossRef]

34. Bonini, M.G.; Fernandes, D.C.; Augusto, O. Albumin oxidation to diverse radicals by the peroxidase activity of Cu,Zn-superoxide dismutase in the presence of bicarbonate or nitrite: Diffusible radicals produce cysteinyl and solvent-exposed and -unexposed tyrosyl radicals. *Biochemistry* **2004**, *43*, 344–351. [CrossRef]

35. Huang, Y.; Shuai, Y.; Li, H.; Gao, Z. Tyrosine residues play an important role in heme detoxification by serum albumin. *Biochim. Biophys. Acta* **2014**, *1840*, 970–976. [CrossRef] [PubMed]

36. Da Silveira, V.C.; Luz, J.S.; Oliveira, C.C.; Graziani, I.; Ciriolo, M.R.; Ferreira, A.M.D. Double-strand DNA cleavage induced by oxindole-Schiff base copper(II) complexes with potential antitumor activity. *J. Inorg. Biochem.* **2008**, *102*, 1090–1103. [CrossRef]

37. Da Silveira, V.C.; Benezra, H.; Luz, J.S.; Georg, R.C.; Oliveira, C.C.; Ferreira, A.M.D. Binding of oxindole-Schiff base copper(II) complexes to DNA and its modulation by the ligand. *J. Inorg. Biochem.* **2011**, *105*, 1692–1703. [CrossRef] [PubMed]

38. Wu, J.; Chen, W.; Yin, Y.; Zheng, Z.; Zou, G. Probing the cell death signaling pathway of HepG2 cell line induced by copper-1,10-phenanthroline complex. *Biometals* **2014**, *27*, 445–458. [CrossRef] [PubMed]

39. Robertazzi, A.; Magistrato, A.; de Hoog, P.; Carloni, P.; Reedjik, J. Density functional theory studies on copper phenanthroline complexes. *Inorg. Chem.* **2007**, *46*, 5873–5881. [CrossRef]

40. Kumar, Y.; Tayyab, S.; Muzammil, S. Molten-globule like partially folded states of human serum albumin induced by fluoro and alkyl alcohols at low pH. *Arch. Biochem. Biophys.* **2004**, *426*, 3–10. [CrossRef]

41. Requena, J.R.; Levine, R.L.; Stadtman, E.R. Recent advances in the analysis of oxidized proteins. *Amino Acids* **2003**, *25*, 221–226. [CrossRef]

42. Davies, M.J. Protein oxidation and peroxidation. *Biochem. J.* **2016**, *473*, 805–825. [CrossRef]

43. Lakowicz, J.R.; Weber, G. Quenching of fluorescence by oxygen. Probe for structural fluctuations in macromolecules. *Biochemistry* **1973**, *12*, 4161–4170. [CrossRef] [PubMed]

44. Giraddi, T.P.; Kadadevarmath, J.S.; Malimath, G.H.; Chikkur, G.C. Effect of solvent on the fluorescence quenching of organic liquid scintillators by aniline and carbon tetrachloride. *Appl. Radiat. Isot.* **1996**, *47*, 461–466. [CrossRef]

45. Levina, A.; Crans, D.C.; Lay, P.A. Speciation of metal drugs, supplements and toxins in media and bodily fluids controls in vitro activities. *Coord. Chem. Rev.* **2017**, *352*, 473–498. [CrossRef]

46. Kwon, J.H.; Park, H.-J.; Chitrapriya, N.; Cho, T.-S.; Kim, S.; Kim, J.; Hwang, I.H.; Kim, C.; Kim, S.K. DNA cleavage induced by [Cu(L)$_x$(NO$_3$)$_2$] (L = 2,2′-dipyridylamine, 2,2′-bipyridine, dipicolylamine, *x* = 1 or 2): Effect of the ligand structure. *J. Inorg. Biochem.* **2014**, *131*, 79–86. [CrossRef] [PubMed]

47. Nave, M.; Castro, R.E.; Rodrigues, C.M.P.; Casini, A.; Soveral, G.; Gaspar, M.M. Nanoformulations of a potent copper-based aquaporin inhibitor with cytotoxic effect against cancer cells. *Nanomedicine* **2016**, *11*, 1817–1830. [CrossRef] [PubMed]

48. Vieira, E.G.; Miguel, R.B.; da Silva, D.R.; Fazzi, R.B.; de Couto, R.A.A.; Marin, J.H.; Temperini, M.L.A.; Shinohara, J.S.; Toma, H.E.; Russo, L.C.; et al. Functionalized nanoparticles as adjuvant to increase the cytotoxicity of metallodrugs toward tumor cells. *New J. Chem.* **2019**, *43*, 386–398. [CrossRef]

Article

Cytotoxic Gold(I) Complexes with Amidophosphine Ligands Containing Thiophene Moieties

Helen Goitia, M. Dolores Villacampa, Antonio Laguna and M. Concepción Gimeno *

Departamento de Química Inorgánica, Instituto de Síntesis Química y Catálisis Homogénea (ISQCH), CSIC-Universidad de Zaragoza, 50009 Zaragoza, Spain; hgoitia@gmail.com (H.G.); dvilla@unizar.es (M.D.V.); alaguna@unizar.es (A.L.)
* Correspondence: gimeno@unizar.es; Tel.: +34-976-762-291

Received: 14 December 2018; Accepted: 24 January 2019; Published: 29 January 2019

Abstract: A new phosphine ligand bearing a thiophene moiety, $C_4H_3SNHCOCH_2CH_2PPh_2$ **(L)**, has been prepared by reaction of the aminophosphine $Ph_2PCH_2CH_2NH_2$ with thiophenecarbonylchloride in the presence of triethylamine. The coordination behavior towards gold(I), gold(III) and silver(I) species has been studied and several metal compounds of different stoichiometry have been achieved, such as $[AuL_2]OTf$, $[AuXL]$ (X = Cl, C_6F_5), $[Au(C_6F_5)_3L]$, $[AgL_2]OTf$ or $[Ag(OTf)L]$. Additionally, the reactivity of the chloride gold(I) species with biologically relevant thiolates was explored, thus obtaining the neutral thiolate compounds $[AuL(SR)]$ (SR = 2-thiocitosine, 2-thiolpyridine, 2-thiouracil, 2-thionicotinic acid, 2,3,4,6-tetra-6-acetyl-1-thiol-β-D-glucopyranosato or thiopurine). The antitumor activity of the compounds was measured by the MTT method in several cancer cells and the complexes exhibit excellent cytotoxic activity.

Keywords: gold; silver; amidophosphine; thiophene; cytotoxicity; cancer

---

1. Introduction

Metal complexes have received increasing interest in the development of new chemotherapeutical drugs after the great success of cisplatin and the second or third generation drugs Carboplatin, Paraplatin and Oxaliplatin [1–3]. Platinum drugs are extensively used in cancer treatment, but the side effects and the development of platinum drug resistance on several cancer cell lines have led to research with other metallic complexes, as for example ruthenium compounds, which have been entered into clinical trials with great success [4]. Gold derivatives have been known as antiarthritic drugs for a long time, such as the well-known Auranofin. The discovery of the strong antiproliferative activity of this compound prompted to deepen in further studies about its wide biological activity, and nowadays is on clinical trials for ovarian cancer [5]. Many gold compounds, including gold(I) or gold(III) derivatives [6–9], have revealed as promising cytotoxic agents, in many cases overcoming cisplatin resistance to specific types of cancer. Additionally, they have exhibited a mechanism of action very different from that of platinum drugs. Several targets have been identified for gold complexes being the inhibition of Thioredoxin reductase one of the most important [10–12].

Five-membered aromatic rings are very important building blocks as may confer an improvement in the pharmacokinetic and pharmacodynamic drug properties. In particular, thiophene, a five-membered aromatic sulfur-containing heterocycle, is encountered in many therapeutically active agent [13], which show a number of pharmacological properties, such as antipsychotic, antidepressive, antithrombolytic, antifungal, antiviral, antiallergic, prostaglandin, dopamine receptor antagonist, and 5-lipoxygenase inhibitor [14–16]. Examples of some drugs containing a thiophene ring are tiagabine (Gabitril), raloxifene (Evista), olanzapine (Zyprexa) or clopidogrel (Plavix) (Figure 1). These compounds are widely used in the treatment of several diseases, the first is an anticonvulsant, the second is an oral selective estrogen receptor modulator with estrogenic actions on bone (prevention of osteoporosis) [17], the third is an antipsychotic [18] and the latter is one of the most successful platelet aggregation inhibitors [19].

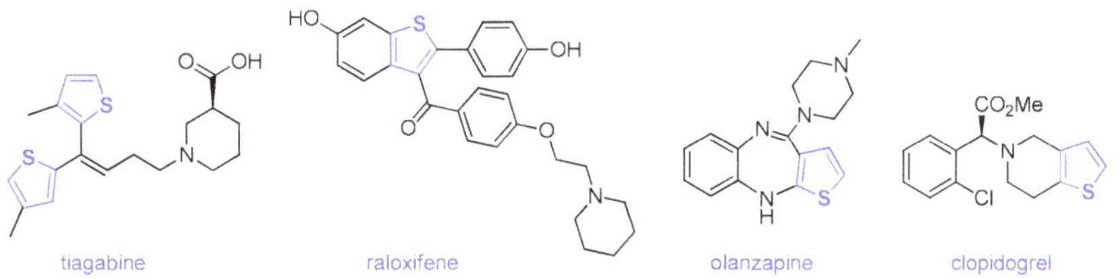

Figure 1. Drugs containing the thiophene ring.

Within our research line in gold and silver compounds as antitumoral agents [20–24], here we report on the synthesis of a thiophene-substituted phosphine ligand and the studies of coordination to gold or silver complexes. The cytotoxicity of these derivatives in several cancer cell lines has been carried out, showing excellent activities in all of them.

2. Results and Discussion

2.1. Synthesis and Characterization

The synthesis of the new ligand containing the thiophene moiety was achieved by reaction of thiophenecarbonylchloride with the aminophosphine $Ph_2PCH_2CH_2NH_2$ in a 1:1 ratio in the presence of triethylamine, which results in the formation of the ligand $C_4H_3SNHCOCH_2CH_2PPh_2$ (**L**) in good yield (see Scheme 1). In the ^1H NMR spectrum of **L** the signals of the protons H3 and H5 of the thiophene group are overlapped with the resonances due to the phenyl protons. The resonance for the H4 of the thiophene appears at 7.02 ppm as a triplet, and the NH proton of the amide group is observed at 6.25 as a broad doublet. The protons of the methylene groups appear at 3.61 and 2.42 ppm. The $^{13}C\{^1H\}$ APT presents the expected resonances for all functional groups, the carbon of the CO group appears at 161.76 ppm, the carbon atoms of the thiophene at 138.99, 129.98, 128.01 and 127.64 ppm and the methylene carbons at 37.56 and 28.63 ppm. The $^{31}P\{^1H\}$ spectrum presents only one resonance for the free phosphorus atom at −20.6 ppm.

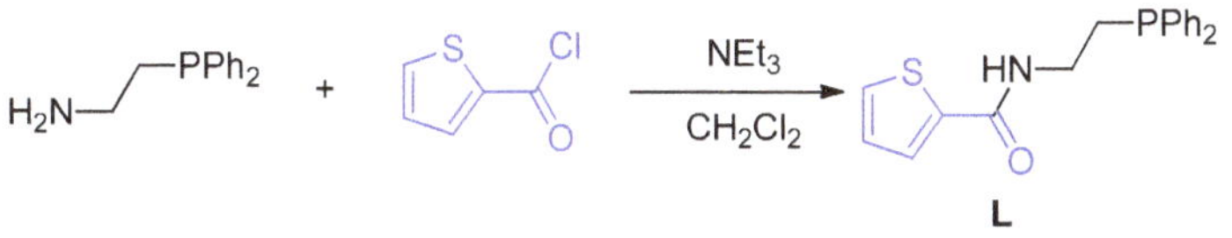

Scheme 1. Synthesis of the ligand **L**.

The coordination behavior of **L** with several gold and silver complexes has been studied. In first place the reaction of the complex $[Au(tht)_2]OTf$ (tht = tetrahydrothiophene; OTf = trifluoromethanesulphonate) with **L** in a 1:2 ratio in dichloromethane results in the formation of complex $[AuL_2]OTf$ (**1**) with good yields (see Scheme 2). The $^{31}P\{^1H\}$ NMR spectrum showed a signal at 37.0 ppm, which confirms a low field shift of 57 ppm relative to the free ligand, due to the coordination of the metal fragment to the phosphorus atom of the ligand. All resonances of the ligand were observed in the proton NMR spectrum and, as expected, the methylene groups near the phosphine move slightly to a low field due to the coordination of the metal. The proton of the amide group –NH– undergoes a low field shift of 1.54 ppm, probably due to the formation of intermolecular hydrogen bonds with the carbonyl group of another molecule.

The reaction of **L** with the gold(I) species [AuCl(tht)] or [Au(C$_6$F$_5$)(tht)] or the gold(III) derivative [Au(C$_6$F$_5$)$_3$(tht)] in dichloromethane in a 1:1 molar ratio afforded the mononuclear linear gold(I) complexes [AuX(L)] (X = Cl (**2**), C$_6$F$_5$ (**3**)) or the square planar gold(III) compound [Au(C$_6$F$_5$)$_3$(L)] in good yields (see Scheme 2). The complexes have been characterized by IR, NMR spectroscopy and mass spectrometry. In the IR spectra the absorptions for the ν(Au–Cl) in complex **2** and the vibrations due to the pentafluorophenyl rings bonded to gold(I) or gold(III) for compounds **3** or **4** were observed. In the ^{31}P{^{1}H} NMR spectra a downfield displacement for the phosphorus atoms is observed as a consequence of the coordination to the gold center, with resonances at 24 ppm for the gold(I) species and at 11 ppm for the gold(III) derivative. The ^{19}F spectra for compounds **3** and **4** are also characteristic of the coordination of the ligand to the AuC$_6$F$_5$ or the Au(C$_6$F$_5$)$_3$ fragments, with three or six resonances for the *ortho*, *meta* and *para* fluorine of the pentafluorophenyl groups, respectively.

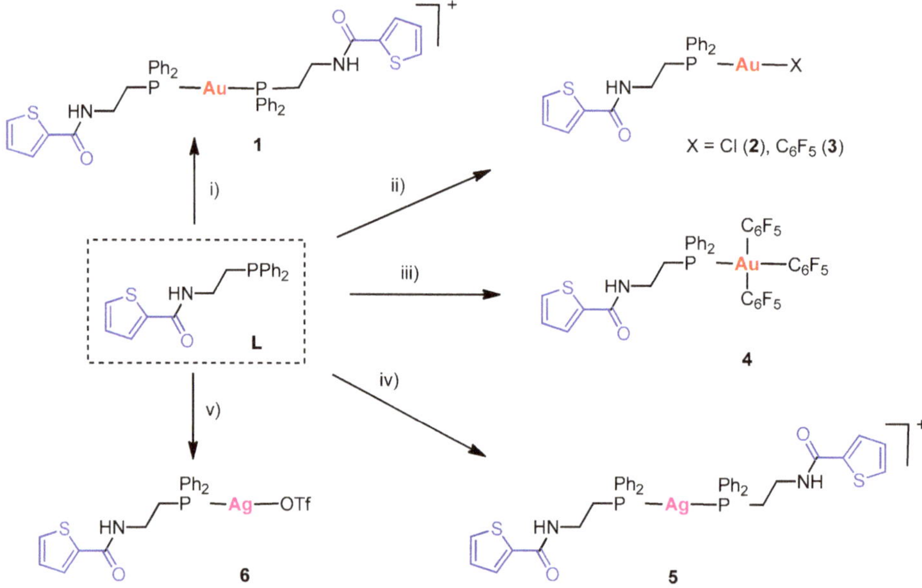

Scheme 2. Coordination studies of ligand **L**. Reaction conditions: CH$_2$Cl$_2$, r.t., (i) [Au(tht)$_2$]OTf: (ii) [AuCl(tht)] or [Au(C$_6$F$_5$)(tht)]; (iii) [Au(C$_6$F$_5$)$_3$(tht)]; (iv) 1/2 Ag(OTf); (v) Ag(OTf).

The reaction of the chloro-gold derivative **2** with different thiols derived from DNA-bases or related species, such as 2-thiocitosine, 2-thiolpyridine, 2-thiouracil, 2-thionicotinic acid, 2,3,4,6-tetra-6-acetyl-1-thiol-μ-D-glucopyranosato or thiopurine were carried out in dichloromethane in the presence of potassium carbonate to afford the phosphine-thiolate gold derivatives **7–12** (see Scheme 3).

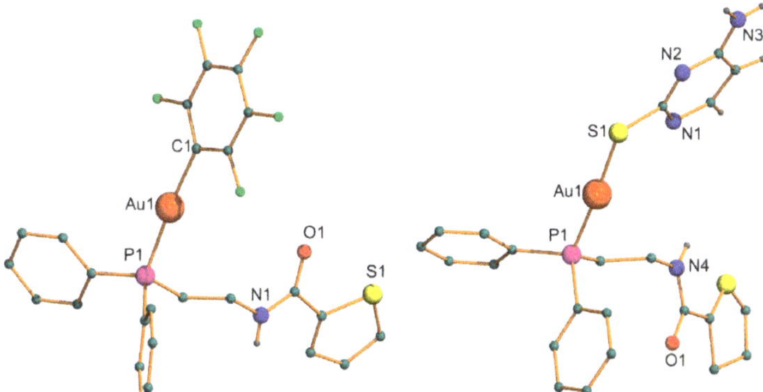

Scheme 3. Synthesis of the thiolate–gold(I) derivatives.

The NMR data for these complexes are in accordance with the substitution of the chlorine ligand for the corresponding thiolate. The resonances in the ^{1}H and ^{13}C NMR spectra corroborate the presence of both the amide-phosphine and the thiolate ligand. In the ^{31}P{^{1}H} NMR the resonances for the phosphorus atom are displaced downfield.

2.2. X-ray Diffraction Studies

Molecular structures of complexes **3**, **6** and **7** were confirmed by an X-ray crystallographic study. Complexes **3** and **6** crystallized in the monoclinic space group P2$_1$/n with two or one molecules by an asymmetric unit, respectively. Compound **7** crystallized in the triclinic space group $P-1$ with two molecules by asymmetric unit. Figure 2 collects the molecular structures for the pentafluorophenyl and thiolate gold(I) derivatives.

Figure 2. Molecular structure of the gold compounds **3** and **7**, with the atom labelling scheme (some of the hydrogen atoms are omitted for clarity).

The gold(I) atoms in complexes **3** and **7** lie in an almost linear geometry, as expected for gold(I) derivatives, defined by the P atom and the C of the pentafluorophenyl (**3**) or the S atom of the thiolate ligand (**7**). The P–Au–C bond angles in **3** are 175.9(3)° and 170.6(3)°, and the P–Au–S bond angles are 176.02(7)° and 173.55(8)° in **7**. The Au–P, Au–C, and Au–S bond distances are unexceptional for both complexes and compared well with most of the distances found in related complexes with the same C$_6$F$_5$–Au–P or S–Au–P core [25,26]. The longer Au–P bond distances in the pentafluorophenyl derivative **3** (2.276(3) Å, 2.273(3) Å) compared with those in complex **7** (2.256(2) Å, 2.256(2) Å) and are in agreement with the higher *trans* influence of the pentafluorophenyl ligand compared with the thiolate ligand.

In the crystal of **3** the molecules are associated into chains parallel to the crystallographic b-axis via short hydrogen bonds between the amide NH proton and the oxygen atom of the CO group of neighbouring molecules as a proton acceptor (Figure 3). The distances and angles of the D⋯A moieties are N2–H2⋯O1_$1 ($1 $x - 1/2$, $y + 1/2$, $z - 1/2$) 2.897(11) Å and 167(9)° and N1–H1⋯O2_$2 ($2 $-x + 1$, $-y + 1$, $-z + 1$) 2.918(11) Å and 148.1°.

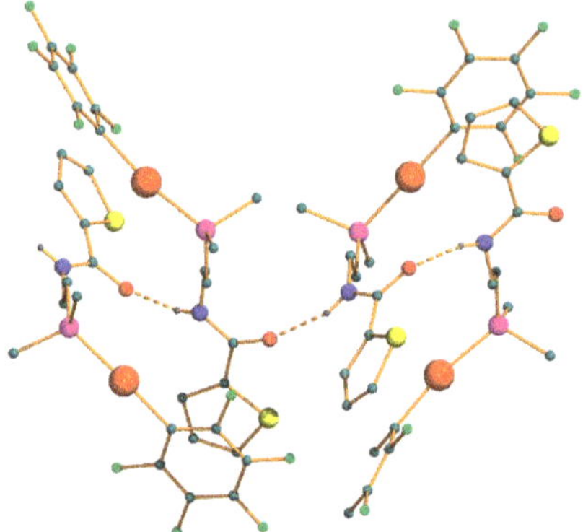

Figure 3. Association of molecules in **3** via hydrogen bonding.

In the crystal of **7** the molecules are organized into dimers though classical H-bonds (N–H⋯N) between the thiocytosine groups of neighbouring molecules, with D⋯A distances of 3.103(9) Å for N3–H3b⋯N5 with an angle of 147.5° and a distance of 3.059(9) Å for N3–H3a⋯N2_$1 ($1 $-x + 1$, $-y + 2$, $-z + 1$) with an angle of 169.1 C. Moreover, dimers are organized in a 3D array through additional H-bond (N–H⋯N), N7–H7a⋯N6_$2 ($2 $-x + 2$, $-y + 1$, $-z - 1$) distance of 3.124(10) Å with an angle of 170.3° (Figure 4).

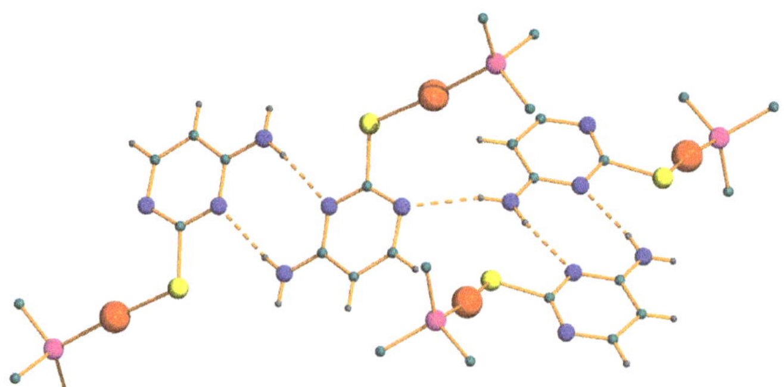

Figure 4. Association of molecules in complex **7** via hydrogen bonding. Phenyl rings are omitted for clarity, only *ipso* carbon is maintained.

Complex **6** can be seen as chains of silver atoms in which the thiophene-phosphine group is acting as a bidentate bridging ligand through the P atom and the O of the carboxamide group atom (Figure 5). In addition, each triflate anion is bonded to two silver atoms of different chains that lead to a 3D array. The silver atom is in a highly distorted tetrahedral geometry. The angles around the silver center range from 77.4(2)° to 155.0(1)°. The bond lenghts are normal in tetra-coordinated silver compounds, with Ag–P distance of 2.380(2) Å and Ag–O carboxamide distance of 2.265(4) Å, whereas the Ag–O triflate bonds (2.444(4) Å and 2.574(4) Å) are longer than the carboxamide oxygen. However, these Ag–O distances are shorter than the usually found in silver-triflate complexes with weak bonds [27,28].

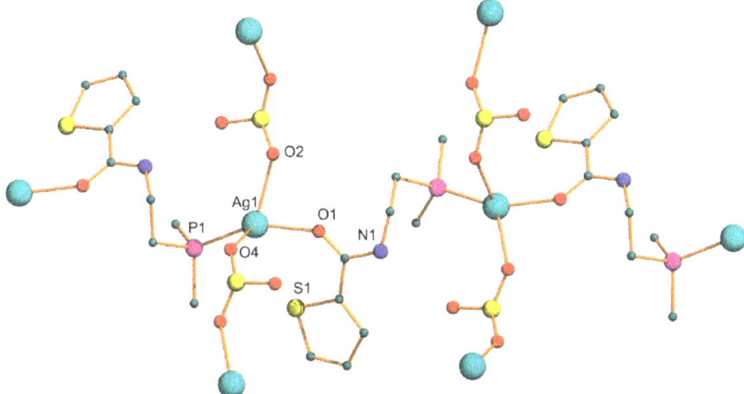

Figure 5. Structure of compound **6**, with the atom labelling scheme, showing the tridimensional network (hydrogen atoms, phenyl rings and the CF_3 units from the triflate are omitted for clarity).

2.3. Biological Studies

The cytotoxic activity of ligand **L** and the gold and silver complexes **1–12** were tested against four different tumor cell lines: A-549 (human lung carcinoma), Hep-G2 (human liver cancer), NIH-3T3 (mouse embryonic fibroblast) or PC12 (pheochromocytoma of the rat adrenal medulla) and comparison with that of cisplatin has been performed [29–31].

Compounds **L** and **1–10** are not water-soluble, but they are soluble in DMSO and in the DMSO/water mixtures used in to perform the tests, containing a small amount of DMSO, concentrations up to 100 µM. No precipitation of the compounds, or metallic gold or silver was observed while performing the tests. Their DMSO-d_6 solutions are stable at room temperature, according to the ^{1}H NMR spectra in which the same resonances were present for more than two weeks. Cells were exposed to various concentrations of each compound for a total of 48 h. The IC_{50} values were calculated using the colorimetric MTT viability assay, (MTT = 3-(4,5-dimethyl-thiazol-2-yl)-2,5-diphenyltetrazolium bromide). The final concentration of DMSO was <0.1%. IC_{50} values are necessary concentrations of a drug to inhibit tumor cell proliferation by 50%, compared to the control cells treated with DMSO alone. The IC_{50} values for **L** and complexes **1–12** are collected in Table 1.

Table 1. IC_{50} of **L** and complexes **1–12** at 48 h (µM).

	IC$_{50}$ (µM)			
Complex	**A-549**	**Hep-G2**	**NIH-3T3**	**PC-12**
L	>100	>100	>100	>100
1	48.9 ± 2.1	17.1 ± 1.1	1.4 ± 0.09	0.5 ± 0.1
2	>100	16.9 ± 1.2	2.8 ± 0.9	1.6 ± 0.8
3	10.6 ± 1.4	56 ± 3.5	39.37 ± 2.2	1.6 ± 0.7
4	>100	>100	>100	>100
5	20.8 ± 2.1	10.0 ± 1.1	4.0 ± 0.8	1.1 ± 0.6
6	12.5 ± 1.2	78.9 ± 2.4	20.8 ± 1.8	6.9 ± 1.0
7	45.1 ± 2.7	5.9 ± 1.7	3.0 ± 1.2	1.0 ± 0.5
8	24.5 ± 1.8	6.7 ± 2.3	2.1 ± 0.4	0.5 ± 0.2
9	>100	14.1 ± 2.5	6.1 ± 1.9	1.3 ± 0.4
10	>100	15.8 ± 2.4	6.3 ± 1.7	2.7 ± 1.0
11	9.9 ± 1.4	5.9 ± 1.1	2.9 ± 0.9	-
12	65.7 ± 3.4	>100	13.7 ± 2.4	-
Cisplatin	64	25 ± 3.1	145 ± 13	-

The IC_{50} values listed in Table 1 showed that the tested thiophene amidophosphine ligand (**L**) did not show significant antiproliferative activity and neither the gold(III) complex bearing pentafluorophenyl groups (**4**). However, the rest of the compounds were quite effective as cytotoxic agents against in vitro growth of the tested cancer cell lines. They showed certain selectivity in their cytotoxic action, being the A549 and the Hep-G2 cell lines the more resistant and the PC-12 the most sensitive in all the tested compounds.

The bis(amidophosphine) gold (**1**) or silver (**5**) complexes exhibited excellent activities with low IC_{50} values in all the cell lines, especially the silver species **5** showed IC_{50} values ranging from 1.1 to 20.8 cells. The chloro-gold derivative **2** did not show significant activity in A549 cancer cells, whereas presented excellent results in the other cell lines. Surprisingly, the gold(I) species with a pentafluorophenyl moiety **3** showed very good activity in A549 cancer cells, in contrast to the lower activity observed in our group for other gold complexes with pentafluorophenyl ligands. The best results in the thiolate gold derivatives were achieved with the 2,3,4,6-tetra-6-acetyl-1-thiol-β-D-glucopyranosato ligand, complex **11**, for which excellent activities were obtained in all the cancer cell lines.

With these results is possible to get some interesting structure activity relationships. The tested thiophene amidophosphine derivatives showed excellent activities in PC-12 and NIH-3T3 cell lines. Interestingly, the gold complexes exhibited slightly better activity than the silver compounds with the exception of the A549 cell line for which the silver species are more active. The best results were obtained for the thiolate-gold derivatives, specifically with the 2,3,4,6-tetra-6-acetyl-1-thiol-β-D-glucopyranosato ligand, although in general the other thiolate gold complexes presented dissimilar IC_{50} values depending on the cell line. Excellent values were also achieved with the bis(phosphine) gold and silver complexes, probably because of their cationic character. Practically in most of the cases the antiproliferative activity was superior to that presented by cisplatin.

3. Experimental Section

3.1. Instrumentation

Mass spectra were recorded on a BRUKER ESQUIRE 3000 PLUS (Bruker, Bremen, Germany), with the electrospray (ESI) technique and on a BRUKER MICROFLEX (MALDI-TOF) (Bruker, Bremen, Germany), using Dithranol or a T-2-(3-(4-tbutyl-phenyl)-2-methyl-2-propenylidene)malononitrile as a matrix. NMR spectra were recorded with a Bruker ARX 400 spectrometer (Bruker, Bremen, Germany) (^{1}H, 400 MHz and ^{19}F, 376.5 MHz). Chemical shifts are reported in ppm relative to the residual solvent peak [^{1}H (CD$_3$)$_2$CO: 2.05] and CFCl$_3$, respectively.

3.2. Starting Materials

[AuCl(tht)], [Au(C$_6$F$_5$)(tht)] [32], [Au(tht)$_2$]OTf [33,34] and [Au(C$_6$F$_5$)$_3$(tht)] [32], were prepared according to literature procedures. Other starting materials and solvents are commercially available.

3.3. General Procedure for the Synthesis of Ligand **L** and Complexes **1–6**

Synthesis of 2-(diphenylphosphino)ethylamine-2-carbonylthiophene (L): To a solution of 2-(diphenylphosphino)ethylamine (46 μL, 0.2 mmol) in CH$_2$Cl$_2$ (20 mL) was added triethylamine (60 μL, 0.43 mmol). After stirring the solution for 10 min at room temperature, it was poured onto an ice bath and a solution of 2-chlorocarbonylthiophene in CH$_2$Cl$_2$ (20 mL) was added dropwise and stirred overnight. An aqueous solution of NaHCO$_3$ (30 mL) was added and the two phases were separated. The organic phase was dried with anhydrous Na$_2$SO$_4$. The solvent was evaporated to ca. 5 mL and hexane (20 mL) was added to obtain a white solid. 80% yield. ^{1}H NMR (CDCl$_3$) δ (ppm): 7.45 (m, 6H, Ph, H3, H5), 7.34 (m, 6H, Ph), 7.02 (dd, J = 4.8 Hz, 3.6 Hz, 1H, H4), 6.25 (m, br, 1H, NH), 3.61 (m, 2H, –CH$_2$N–), 2.42 (t, J = 7.2 Hz, 2H, –CH$_2$P–). ^{13}C{^{1}H} NMR (CDCl$_3$) δ (ppm): 161.76 (C=O),

138.99 (C2), 137.70 (d, $^1J_{PC}$ = 11.7 Hz, *ipso*-C_6H_5), 132.87 (d, $^2J_{PC}$ = 18.8 Hz, *o*-C_6H_5), 129.98 (C3), 129.04 (*p*-C_6H_5), 128.79 (d, $^3J_{PC}$ = 6.9 Hz, *m*-C_6H_5), 128.01 (C5), 127.64 (C4), 37.56 (d, $^2J_{C-P}$ = 19.9 Hz, –CH_2N–), 28.63 (d, $^1J_{C-P}$ = 13.3 Hz, –CH_2P–). $^{31}P\{^1H\}$ NMR ($CDCl_3$) δ (ppm): −20.64.

Synthesis [AuL₂]OTf (1): To a solution of **L** (0.1018 g, 0.30 mmol) in CH_2Cl_2 (20 mL), [Au(tht)₂]OTf (0.0783 g, 0.15 mmol) was added and the solution was stirred for 1 h. Evaporation of the solvent to ca. 5 mL and addition of hexane (15 mL) gave a white solid of **1**. 73% yield. ^1H NMR ($CDCl_3$) δ (ppm): 7.79 (m, 2H, NH), 7.53 (m, 10H, Ph, H3), 7.39 (m, 14H, Ph, H5), 6.92 (dd, *J* = 4.9, 3.9 Hz, 2H, H4), 3.70 (m, 4H, –CH_2N–), 2.82 (t, *J* = 6.4 Hz, 4H, –CH_2P–). $^{13}C\{^1H\}$ NMR ($CDCl_3$) δ (ppm): 162.96 (C=O), 138.84 (C2), 133.08 (*o*-C_5H_4), 131.28 (*p*-C_5H_4), 130.37 (C3), 129.47 (*m*-C_5H_4), 128.66 (C5), 127.96 (C4), 36.83 (–CH_2N–), 27.70 (–CH_2P–). $^{31}P\{^1H\}$ NMR ($CDCl_3$) δ (ppm): 36.99. ^{19}F NMR ($CDCl_3$) δ (ppm): −78.4. MS (ESI⁺) *m/z*: [M]⁺ 875.3 (100%). $C_{39}H_{36}AuF_3N_2O_5P_2S_3$ (1024.8) calcd. C 45.70, H 3.54, N 2.73, S 9.37; found C 45.48, H 3.35, N 2.73, S 8.97.

Synthesis of [AuClL] (2): To a solution of **L** (0.15 mmol) in CH_2Cl_2 (20 mL), [AuCl(tht)] (0.15 mmol) was added and the mixture was stirred for 1 h. Then, the solution was evaporated to ca. 5 mL and addition of hexane (15 mL) afforded a white solid of **2**. 94% yield. ^1H NMR ($CDCl_3$) δ (ppm): 7.69 (m, 4H, Ph), 7.45 (m, 8H, Ph, H3, H5), 7.03 (dd, *J* = 4.7, 4.0 Hz, 1H, H4), 6.69 (m, 1H, NH), 3.74 (m, 2H, –CH_2N–), 2.91 (dt, *J* = 10.9 Hz, 7.1 Hz, 2H, –CH_2P–). $^{13}C\{^1H\}$ NMR ($CDCl_3$) δ (ppm): 162.44 (C=O), 138.22 (C2), 133.30 (d, $^2J_{PC}$ = 13.3 Hz, *o*-C_6H_5), 132.28 (*p*-C_6H_5), 130.62 (C3), 129.55 (d, $^3J_{PC}$ = 11.8 Hz, *m*-C_6H_5), 128.84 (d, $^1J_{PC}$ = 42.2 Hz, *ipso*-C_6H_5), 128.44 (C5), 127.90 (C4), 36.98 (d, $^2J_{C-P}$ = 5.6 Hz, –CH_2N–), 28.35 (d, $^1J_{C-P}$ = 39.1 Hz, –CH_2P–). $^{31}P\{^1H\}$ NMR ($CDCl_3$) δ (ppm): 24.55. $C_{19}H_{18}AuClNOPS$ (571.8) calcd. C 39.91, H 3.17, N 2.45, S 5.61; found C 39.61, H 3.08, N 2.60, S 5.39.

Synthesis of [Au(C_6F_5)L] (3): To a solution of **L** (0.15 mmol) in CH_2Cl_2 (20 mL), [Au(C_6F_5)(tht)] (0.15 mmol) was added and the solution was stirred for 1 h. Evaporation of the solvent to ca. 5 mL and addition of hexane (15 mL) gave a white solid of **3**. 75% yield. ^1H NMR ($CDCl_3$) δ (ppm): 7.66 (m, 4H, Ph), 7.51 (d, *J* = 3.3 Hz, 1H, H3), 7.40 (m, 7H, Ph, H5), 7.07 (t, *J* = 5.5 Hz, 1H, NH), 7.0 (m, 1H, H4), 3.70 (m, 1H, –CH_2N–), 2.89 (m, 1H, –CH_2P–). $^{13}C\{^1H\}$ NMR ($CDCl_3$) δ (ppm): 162.36 (C=O), 137.15 (C2), 133.48 (d, $^2J_{PC}$ = 13.4 Hz, *o*-C_6H_5), 132.01 (d, $^4J_{PC}$ = 2.1 Hz, *p*-C_6H_5), 130.55 (C3), 130.06 (d, $^1J_{PC}$ = 53.8 Hz, *ipso*-C_6H_5), 129.61 (d, $^3J_{PC}$ = 11.2 Hz, *m*-C_6H_5), 128.36 (C5), 127.67 (C4), 36.88 (d, $^2J_{C-P}$ = 6.2 Hz, –CH_2N–), 28.57 (d, $^1J_{C-P}$ = 33.3 Hz, –CH_2P–). $^{31}P\{^1H\}$-NMR ($CDCl_3$) δ (ppm): 24.24. ^{19}F NMR ($CDCl_3$) δ (ppm): −116.13 (m, 2F, *o*-C_6F_5), −157.81 (t, J_{F-F} = 19.96 Hz, 1F, *p*-C_6F_5), −161.97 (m, 2F, *m*-C_6F_5). $C_{25}H_{18}AuF_5NOPS$ (703.4) calcd. C 42.68, H 2.58, N 1.99, S 4.56; found C 42.65, H 2.84, N 2.01, S 4.73.

Synthesis of [Au(C_6F_5)₃L] (4): To a solution of **L** (0.15 mmol) in CH_2Cl_2 (20 mL), [Au(C_6F_5)₃(tht)] (0.15 mmol) was added and the mixture stirred for 1 h. Then the solution was evaporated to ca. 5 mL and added hexane (15 mL) to obtain a white solid of **4**. 72% yield. ^1H NMR ($CDCl_3$) δ (ppm): 7.53 (m, 11H, Ph, H3), 7.41 (dd, *J* = 3.7, 1.0 Hz, 1H, H5), 7.08 (dd, *J* = 4.9, 3.8 Hz, 1H, H4), 6.14 (t, *J* = 5.9 Hz, 1H, NH), 3.50 (m, 2H, –CH_2N–), 2.90 (m, 2H, –CH_2P–). $^{13}C\{^1H\}$ NMR ($CDCl_3$) δ (ppm): 162.35 (C=O), 137.62 (C2), 133.11 (d, $^4J_{PC}$ = 2.7 Hz, *p*-C_6H_5), 132.76 (d, $^2J_{PC}$ = 10.4 Hz, *o*-C_6H_5), 130.94 (C3), 129.53 (d, $^3J_{PC}$ = 11.6 Hz, *m*-C_6H_5), 128.77 (C5), 128.01 (C4), 123.18 (d, $^1J_{PC}$ = 57.7 Hz, *ipso*-C_6H_5), 35.93 (–CH_2N–), 25.59 (d, $^1J_{C-P}$ = 29.6 Hz, –CH_2P–). $^{31}P\{^1H\}$ NMR ($CDCl_3$) δ (ppm): 11.26 (q, J_{P-F} = 11.10 Hz). ^{19}F NMR ($CDCl_3$) δ (ppm): −120.34 (m, 4F, *o*-C_6F_5), −121.51 (m, 2F, *o*-C_6F_5), −155.95 (t, J_{F-F} = 19.92 Hz, 2F, *p*-C_6F_5), −156.80 (t, J_{F-F} = 19.85 Hz, 1F, *p*-C_6F_5), −159.94 (m, 4F, *m*-C_6F_5), −160.79 (m, 2F, *m*-C_6F_5). MS (ESI⁺) *m/z*: [M−C_6F_5]⁺ 870.1 (25.8%). $C_{37}H_{18}AuF_{15}NOPS$ (1037.5) calcd. C 42.83, H 1.75, N 1.35, S 3.09; found C 43.01, H 1.79, N 1.38, S 3.16.

Synthesis of [Ag(L)₂]OTf (5): To a solution of **L** (0.30 mmol) in CH_2Cl_2 (20 mL), Ag(OTf) (0.15 mmol) was added and the mixture stirred for 1 h. The solution was evaporated to ca. 5 mL and addition of hexane (15 mL) afforded a white solid of **5**. 77% yield. ^1H NMR ($CDCl_3$) δ (ppm): 8.35 (t, *J* = 5.5 Hz, 2H, NH), 7.67 (m, 8H, Ph), 7.48 (m, 14H, Ph, H3), 7.36 (d, *J* = 4.9 Hz, 2H, H5), 6.96 (dd, *J* = 4.8, 3.9 Hz, 2H,

H4), 3.82 (m, 4H, –CH$_2$N–), 3.12 (m, br, 4H, –CH$_2$P–). ^{13}C{^{1}H} NMR (CDCl$_3$) δ (ppm): 163.05 (C=O), 139.04 (C2), 133.37 (d, $^2J_{PC}$ = 6.9 Hz, *o*-C$_6$H$_5$), 132.39 (*p*-C$_6$H$_5$), 130.37 (C3), 129.87 (d, $^3J_{PC}$ = 5.4 Hz, *m*-C$_6$H$_5$), 128.89 (C5), 128.19 (d, $^1J_{PC}$ = 28.8 Hz, *ipso*-C$_6$H$_5$), 128.14 (C4), 36.20 (–CH$_2$N–), 27.82 (2d, J_{C-P}= 16.9 Hz, –CH$_2$P–). ^{31}P{^{1}H} NMR (CD$_2$Cl$_2$, −85 °C) δ (ppm): 6.44 (d, J_{P-Ag} = 527.1 Hz). ^{19}F NMR (CDCl$_3$) δ (ppm): −77.75. MS (MALDI$^+$) *m/z*: [M]$^+$ 787.2 (100%). C$_{39}$H$_{36}$AgF$_5$N$_2$O$_5$P$_2$S$_3$ (935.7) calcd. C 50.06, H 3.88, N 2.99, S 10.28; found C 49.98, H 4.01, N 2.92, S 9.90.

Synthesis of [Ag(OTf)L] (6): To a solution of **L** (0.15 mmol) in CH$_2$Cl$_2$ (20 mL), Ag(OTf) (0.15 mmol) was added and the solution was stirred for 1 h. Then the solution was evaporated to ca. 5 mL and addition of hexane (15 mL) led to a white solid. 62% yield. ^{1}H NMR (CDCl$_3$) δ (ppm): 7.82 (m, br, 1H, NH), 7.55 (m, br, 4H, Ph), 7.46 (m, br, 1H, H3), 7.35 (m, br, 7H, H5), 6.90 (m, 1H, H4), 3.76 (m, br, 2H, –CH$_2$N–), 2.80 (m, br, 2H, –CH$_2$P–). ^{31}P{^{1}H} NMR (Acetone-d_6, −85 °C) δ (ppm): 3.27 (d, J_{P-Ag} = 708.8 Hz). ^{19}F NMR (CDCl$_3$) δ (ppm): −77.49. C$_{20}$H$_{18}$AgF$_5$NO$_4$PS$_2$ (596.31) calcd. C 40.28, H 3.04, N 2.35, S 10.75; found C 40.15, H 3.16, N 2.42, S 11.04.

3.4. General Procedure for the Synthesis of Phosphinegold(I) Thionucleobase Analogues 7–12

A suspension of thionucleobase (0.2 mmol) and K$_2$CO$_3$ (1 mmol) in CH$_2$Cl$_2$ (15 mL) was stirred for 30 minutes. Then, the chlorophosphinegold(I) complex (0.2 mmol) was added and the suspension was stirred overnight. The solution was filtrated and the solvent was evaporated until ca. 5 mL and added hexane (15 mL) to obtain a solid of the corresponding complex.

[Au(2-thiocytosine)L] (7): White solid, 60% yield. ^{1}H NMR (Acetone-d_6) δ (ppm): 8.70 (m, br, 1H, NH), 7.92 (m, 4H, Ph), 7.76 (d, *J* = 5.77 Hz, 1H, H6), 7.60 (m, 2H, H3',H5'), 7.53 (m, 6H, Ph), 7.02 (dd, *J* = 5.0, 3.8 Hz, 1H, H4'), 6.13 (d, *J* = 5.79 Hz, 1H, H5), 5.93 (m, br, 2H, NH$_2$), 3.79 (m, 2H, –CH$_2$N–), 3.08 (m, 2H, –CH$_2$P–). ^{13}C{^{1}H} NMR (Acetone-d_6) δ (ppm): 163.89 (C4), 162.58 (C=O), 155.36 (C6), 140.64 (C2'), 134.29 (d, $^2J_{PC}$ = 13.6 Hz, *o*-C$_6$H$_5$), 132.47 (d, $^4J_{PC}$ = 2.3 Hz, *p*-C$_6$H$_5$), 131.85 (d, $^1J_{PC}$ = 55.1 Hz, *ipso*-C$_6$H$_5$), 131.05 (C3'), 130.12 (d, $^3J_{PC}$ = 11.3 Hz, *m*-C$_6$H$_5$), 128.86 (C5'), 128.35 (C4'), 100.70 (C5), 37.37 (d, *J* = 7.8 Hz, –CH$_2$N–), 28.85 (d, *J* = 35.2 Hz, –CH$_2$P–). ^{31}P{^{1}H} NMR (Acetone-d_6) δ (ppm): 31.03. MS (ESI$^+$) *m/z*: [M]$^+$ 663 (100%). C$_{23}$H$_{22}$AuN$_4$OPS$_2$ (662.5) calcd. C 41.69, H 3.35, N 8.46, S 9.68; found C 41.55, H 3.13, N 8.42, S 10.15.

[Au(2-thiopyridine)L] (8): Yellow solid, 55% yield. ^{1}H NMR (Acetone-d_6) δ (ppm): 8.59 (m, br, 1H, NH), 8.18 (d, br, *J* = 5.0 Hz, 1H, H6), 7.93 (m, 4H, Ph), 7.60 (dd, *J* = 5.0, 1.1 Hz, 1H, H4), 7.57 (dd, *J* = 3.8, 1.1 Hz, 1H, H3), 7.53 (m, 6H, Ph), 7.38 (m, 2H, H5', H3'), 7.00 (dd, *J* = 5.0, 3.8 Hz, 1H, H5), 6.88 (td, *J* = 4.9, 3.6 Hz, 1H, H4'), 3.79 (m, 2H, –CH$_2$N–), 3.11 (m, 2H, –CH$_2$P–). ^{13}C{^{1}H} NMR (Acetone-d_6) δ (ppm): 169.33 (C2), 162.50 (C=O), 148.97 (C6), 140.65 (C2'), 136.46 (C4), 134.28 (d, $^2J_{PC}$ = 13.6 Hz, *o*-C$_6$H$_5$), 132.52 (d, $^4J_{PC}$ = 2.3 Hz, *p*-C$_6$H$_5$), 131.69 (d, $^1J_{PC}$ = 55.4 Hz, *ipso*-C$_6$H$_5$), 131.03 (C3'), 130.13 (d, $^3J_{PC}$ = 11.3 Hz, *m*-C$_6$H$_5$), 128.79 (C5'), 128.30 (C4'), 127.06 (C5), 118.79 (C3), 37.32 (d, *J* = 7.4 Hz, –CH$_2$N–), 28.72 (d, *J* = 35.3 Hz, –CH$_2$P–). ^{31}P{^{1}H} NMR (Acetone-d_6) δ (ppm): 35.31. C$_{24}$H$_{22}$AuN$_2$OPS$_2$ (646.5) calcd. C 44.59, H 3.43, N 4.33, S 9.90; found C 44.65, H 3.28, N 4.34, S 9.93.

[Au(2-thiouracil)L] (9): White solid, 62% yield. ^{1}H NMR (CDCl$_3$) δ (ppm): 8.70 (m, br, 1H, NH), 7.63 (m, 4H, Ph), 7.50 (m, 2H, H6, H3'), 7.37 (m, 7H, Ph, H5'), 6.92 (dd, *J* = 5.0, 3.8 Hz, 1H, H4'), 6.05 (d, *J* = 5.6 Hz, 1H, H5), 3.69 (m, 2H, –CH$_2$N–), 2.88 (m, 2H, –CH$_2$P–). ^{13}C{^{1}H} NMR (CDCl$_3$) δ (ppm): 162.84 (C=O), 138.84 (C2'), 133.24 (d, $^2J_{PC}$ = 13.4 Hz, *o*-C$_6$H$_5$), 131.85 (*p*-C$_6$H$_5$), 130.26 (C3'), 129.64 (d, $^1J_{PC}$ = 57.4 Hz, *ipso*-C$_6$H$_5$), 129.35 (d, $^3J_{PC}$ = 11.5 Hz, *m*-C$_6$H$_5$), 128.70 (C5'), 127.72 (C4'), 109.89 (C5), 36.95 (d, *J* = 3.5 Hz, –CH$_2$N–), 28.72 (d, *J* = 36.8 Hz, –CH$_2$P–). ^{31}P{^{1}H} NMR (CDCl$_3$) δ (ppm): 30.59. MS (ESI$^+$) *m/z*: [M]$^+$ 663.8 (38.34%). C$_{23}$H$_{21}$AuN$_3$O$_2$PS$_2$ (663.5) calcd. C 41.63, H 3.19, N 6.33, S 9.67; found C 41.55, H 3.18, N 6.50, S 10.11.

[Au(2-thionicotinic acid)L] (10): Yellow solid, 57% yield. ^{1}H NMR (CDCl$_3$) δ (ppm): 9.36 (m, br, 1H, OH), 7.70–7.00 (m, 14H, Ph, H6, H4, H5, H3', H5'), 6.72 (m, br, 1H, H4'), 6.31 (m, br, 1H, NH), 3.50–3.45 (m, 2H, –CH$_2$N–), 2.56 (m, br, 2H, –CH$_2$P–). ^{31}P{^{1}H} NMR (CDCl$_3$) δ (ppm): 28.91. MS (ESI$^+$) *m/z*: [M]$^+$

691 (100%). $C_{25}H_{22}AuN_2O_3PS_2$ (690.1) calcd. C 43.48, H 3.21, N 4.06, S 9.27; found C 43.52, H 3.03, N 3.93, S 8.55.

[Au(thioglucose)L] (11): White solid, 65% yield. ^{1}H NMR (Acetone-d_6) δ (ppm): 7.94 (m, 5H, Ph, NH), 7.64 (dd, J = 5.0, 1.0 Hz, 1H, H3′), 7.55 (m, 7H, Ph, H5′), 7.07 (dd, J = 5.0, 3.8 Hz, 1H, H4′), 5.18 (t, J = 9.4 Hz, 1H, H5), 5.07 (t, J = 9.7 Hz, 1H, H4), 5.00 (m, 2H, H6, H1), 4.24 (dd, J = 12.3, 4.8 Hz, 1H, –CH$_2$–), 4.09 (dd, J = 12.3, 2.2 Hz, 1H, –CH$_2$–), 3.92 (m, br, 1H, H3), 3.77 (m, 2H, –CH$_2$N–), 3.17 (m, 1H, –CH$_2$P–), 3.00 (m, 1H, –CH$_2$P–), 1.98 (s, 6H, –COCH$_3$), 1.91 (s, 6H, –COCH$_3$). ^{13}C{^{1}H} NMR (CDCl$_3$) δ (ppm): 170.29, 169.69 (O–C=O), 162.50 (N–C=O), 138.80 (C3′), 133.49 (d, $^2J_{PC}$ = 13.6 Hz, o-C$_6$H$_5$), 133.11 (d, $^2J_{PC}$ = 13.4 Hz, o-C$_6$H$_5$), 131.73 (d, $^4J_{PC}$ = 2.5 Hz, p-C$_6$H$_5$), 131.65 (d, $^4J_{PC}$ = 2.5 Hz, p-C$_6$H$_5$), 130.37 (d, $^1J_{PC}$ = 55.4 Hz, $ipso$-C$_6$H$_5$), 130.21 (C3′), 130.04 (d, $^1J_{PC}$ = 55.6 Hz, $ipso$-C$_6$H$_5$), 129.43 (d, $^3J_{PC}$ = 5.9 Hz, o-C$_6$H$_5$), 129.28 (d, $^3J_{PC}$ = 5.9 Hz, m-C$_6$H$_5$), 128.24 (C5′), 127.43 (C4′), 76.16 (C3), 74.29 (C5, C6, C1), 68.84 (C4), 62.60 (–CH$_2$–), 37.66 (d, J = 4.6 Hz, –CH$_2$N–), 28.10 (d, J = 34.9 Hz, –CH$_2$P–), 21.42, 20.77 (CH_3–COOCH$_2$). ^{31}P{^{1}H} NMR (Acetone-d_6) δ (ppm): 31.49. MS (ESI$^+$) m/z: [M]$^+$ 900.3 (45.70%). $C_{33}H_{37}AuNO_{10}PS_2$ (899.7) calcd. C 44.04, H 4.15, N 1.56, S 7.13; found C 43.96, H 4.14, N 1.47, S 6.99.

[Au(6-thiopurine)L] (12): White solid, 57% yield. ^{1}H NMR (Acetone-d_6) δ (ppm): 9.77 (m, br, 1H, NH), 8.34 (s, 1H, H9), 8.03 (s, 1H, H5), 7.96 (m, 4H, Ph), 7.50 (m, 8H, Ph, H5′, NH, H3′), 6.79 (dd, J = 4.9, 3.8 Hz, 1H, H4′), 3.84 (m, 2H, –CH$_2$N–), 3.08 (m, 2H, –CH$_2$P–). ^{13}C{^{1}H} NMR (Acetone-d_6) δ (ppm): 162.55 (C=O), 134.31 (d, $^2J_{PC}$ = 11.5 Hz, o-C$_6$H$_5$), 132.36 (s, p-C$_6$H$_5$), 130.23 (C3′), 130.06 (d, $^3J_{PC}$ = 9.4 Hz, m-C$_6$H$_5$), 129.91 (C5′), 128.51 (C4′). ^{31}P{^{1}H} NMR (Acetone-d_6) δ (ppm): 26.82. MS (ESI$^+$) m/z: [M]$^+$ 691 (100%). $C_{24}H_{21}AuN_5OPS_2$ (687.1) calcd. C 41.93, H 3.08, N 10.17, S 9.33; found C 42.24, H 3.12, N 9.71, S 8.55.

3.5. Cristallography

Crystals were mounted in inert oil on glass fibers and transferred to the cold gas stream of Xcalibur Oxford Diffraction (**3**, **6**) or a SMART Apex CCD (**7**) diffractometer equipped with a low-temperature attachment. Data were collected using monochromated Mo Kα radiation (λ = 0.71073 Å). Scan type ω. Absorption correction based on multiple scans was applied using spherical harmonics implemented in SCALE3 ABSPACK [35] scaling algorithm or SADABS. The structures were solved by direct methods and refined on F^2 using the program *SHELX* [36]. All non-hydrogen atoms were refined anisotropically, with the exception of some solvent atoms. Refinements were carried out by full-matrix least-squares on F^2 for all data. CCDC 1881668 (**6**), 1881669 (**7**) and 1881670 (**3**) contains the Supplementary crystallographic data for this paper. These data can be obtained free of charge from The Cambridge Crystallographic Data Centre via www.ccdc.cam.ac.uk/data_request/cif.

3.6. Cell Culture

HEP-G2 cells (human hepatocellular carcinoma cells), NIH-3T3 cells (mouse fibroblasts), A-549 cells (human lung carcinoma cells), and PC12 (pheochromocytoma of the rat adrenal medulla). All cell lines were cultivated in Dulbecco's Modified Eagle's medium (DMEM, D7777 Sigma-Aldrich, Madrid, Spain) supplemented with with 10% fetal bovine serum (FBS) and 1% antibiotics (penicillin/ 100U/mL and streptomycin/ 100µg/mL) in a humidified environment at 37 °C/5% CO$_2$. After three to four days the cells were then detached with trypsin and cultured in another cell culture flask. Wells of a 96 well plate were seeded with the cells and incubated for three days. At day 4 the medium of each well was removed and 200 µL of the gold and silver complexes in different concentrations in the cell culture medium were added to the wells (1–100 µM). The cells were cultivated for two additional days and then, the viability of the cells was determined by the MTT-assay.

3.7. Cytotoxicity MTT Assay

The MTT assay was used to calculate cell viability. Exponentially growing cells were seeded at a density of ~4 $\times$ 10^5 cells/mL, in a 96-well microplate, and 48 h later they were incubated with the complexes. The compounds were dissolved in DMSO and tested in concentrations ranging from 1 to 100 μM.

Cytotoxicity of tested compounds was evaluated by the MTT method [37]. The optical density was measured at 570 nm using a 96-well multiscanner autoreader. The IC_{50} values were calculated by non-linear regression analysis using Origin software (Origin Software, Electronic Arts, Redwood City, CA, USA)

4. Conclusions

A new amidophosphine ligand bearing a thiophene moiety has been synthesized and the coordination behaviour towards gold and silver complexes has been studied. The coordination of the gold(I/III) or silver(I) fragments takes place through the phosphorus atom. In contrast for the silver species coordination to the phosphorus atom together with the oxygen atoms of the carbonyl and triflate groups is observed. The crystal structures of some of these derivatives present supramolecular structures through hydrogen bonding. Cytotoxicity studies in several cancer lines, such as A-549, Hep-G2 or NIH-3T3 and PC-12 have been performed for the ligand and all the synthesized complexes, showing that whereas the thiophene amidophosphine lacks activity, the gold and silver complexes exhibit moderate to excellent activities in all the cell lines. From the structure activity relationship perspective is possible to conclude that the best activities are achieved with the cationic bis(amidophosphine) species and with the thiolate derived from the 2,3,4,6-tetra-6-acetyl-1-thiol-β-D-glucopyranosato ligand. In general, gold complexes are more cytotoxic than the silver ones, with the exception of the A-549 cell line. The cytotoxicity values are much better than those exhibited by cisplatin.

Supplementary Materials: The following are available online at http://www.mdpi.com/2304-6740/7/2/13/s1, CIF and checkCIF files.

Author Contributions: Conceptualization, M.C.G. and A.L.; methodology, H.G.; software, H.G. and M.D.V.; formal analysis, H.G.; investigation, H.G.; writing—original draft preparation, M.C.G. and M.D.V.; writing—review and editing, M.C.G., M.D.V. and A.L.; project administration, M.C.G.; funding acquisition, M.C.G.

Funding: This research was funded by Ministerio de Ciencia, Innovación y Universidades, grant number CTQ2016-75816-C2-1-P and and Gobierno de Aragón-Fondo Social Europeo, grant number E07_17R.

Acknowledgments: The authors thank the Ministerio de Ciencia, Innovación y Universidades (CTQ2016-75816-C2-1-P) and Gobierno de Aragón-Fondo Social Europeo (E07_17R) for financial support.

Conflicts of Interest: The authors declare no conflict of interest.

References

1. Rosenberg, B.; VanCamp, L.; Trosko, J.E.; Monsour, V.H. Platinum Compounds: A New Class of Potent Antitumour Agents. *Nature* **1969**, *222*, 385–386. [CrossRef]
2. Lippert, B. *Cisplatin: Chemistry and Biochemistry of a Leading Anticancer Drug*; Wiley-VCH: Weinheim, Germany, 1999.
3. Johnstone, T.C.; Suntharalingam, K.; Lippard, S.J. The Next Generation of Platinum Drugs: Targeted Pt(II) Agents, Nanoparticle Delivery, and Pt(IV) Prodrugs. *Chem. Rev.* **2016**, *116*, 3436–3486. [CrossRef]
4. Bergamo, A.; Sava, G. Ruthenium anticancer compounds: Myths and realities of the emerging metal-based drugs. *Dalton Trans.* **2011**, *40*, 7817–7823. [CrossRef]
5. Phase I and II Study of Auranofin in Chronic Lymphocytic Leukemia (CLL). Available online: https://clinicaltrials.gov/ct2/show/NCT01419691 (accessed on 2 September 2018).
6. Ott, I. On the medicinal chemistry of gold complexes as anticancer drugs. *Coord. Chem. Rev.* **2009**, *253*, 1670–1681. [CrossRef]

7. Nobili, S.; Mini, E.; Landini, I.; Gabbiani, C.; Casini, A.; Messori, L. Gold compounds as anticancer agents: Chemistry, cellular pharmacology, and preclinical studies. *Med. Res. Rev.* **2010**, *30*, 550–580. [CrossRef]

8. Bertrand, B.B.; Casini, A. A golden future in medicinal inorganic chemistry: The promise of anticancer gold organometallic compounds. *Dalton Trans.* **2014**, *43*, 4209–4219. [CrossRef]

9. Zou, T.; Lum, C.T.; Lok, C.-N.; Zhang, J.-J.; Che, C.M. Chemical biology of anticancer gold(III) and gold(I) complexes. *Chem. Soc. Rev.* **2015**, *44*, 8786–8801. [CrossRef]

10. Barnard, P.T.; Berners-Price, S.J. Targeting the mitochondrial cell death pathway with gold compounds. *Coord. Chem. Rev.* **2007**, *251*, 1889–1902. [CrossRef]

11. Bindoli, A.; Rigobello, M.P.; Scutari, G.; Gabbiani, C.; Casini, A.; Messori, L. Thioredoxin reductase: A target for gold compounds acting as potential anticancer drugs. *Coord. Chem. Rev.* **2009**, *253*, 1692–1707. [CrossRef]

12. Dalla Via, L.; Nardon, C.; Fregona, D. Targeting the ubiquitin-proteasome pathway with inorganic compounds to fight cancer: A challenge for the future. *Future Med. Chem.* **2012**, *4*, 525–543. [CrossRef]

13. Baumann, M.; Baxendale, I.R.; Ley, S.V.; Nikbin, N. An overview of the key routes to the best selling 5-membered ring heterocyclic pharmaceuticals. *Beilstein J. Org. Chem.* **2011**, *7*, 442–495. [CrossRef]

14. Hartman, H.B.; Roehr, J.E.; Rogers, K.L.; Rush, D.K.; Szczepanik, A.M.; Szewczak, M.R.; Wilmot, C.A.; Conway, P.G. 3[4-[1-(6-Fluorobenzo[b]thiophen-3-yl)-4-piperazinyl]butyl]-2,5,5-trimethyl-4-thiazolidinone: A new atypical antipsychotic agent for the treatment of schizophrenia. *J. Med. Chem.* **1992**, *35*, 2712–2715.

15. Bohlman, F.; Zdero, C. *Thiophene and Its Derivatives*; Gronowitz, S., Ed.; John Wiley & Sons, Ltd.: Chichester, UK, 1986; Part 3, pp. 261–323.

16. Hafez, H.N.; El-Gazzar, A.B.A. Design and synthesis of 3-pyrazolyl-thiophene, thieno[2,3-d]pyrimidines as new bioactive and pharmacological activities. *Bioorg. Med. Chem. Lett.* **2008**, *18*, 5222–5227. [CrossRef]

17. Vogel, V.G.; Costantino, J.P.; Wickerham, D.L.; Cronin, W.M.; Cecchini, R.S.; Atkins, J.N.; Bevers, T.B.; Fehrenbacher, L.; Pajon, E.R., Jr.; Wade, J.L., III; et al. Effects of tamoxifen vs raloxifene on the risk of developing invasive breast cancer and other disease outcomes: The NSABP Study of Tamoxifen and Raloxifene (STAR) P-2 trial. *J. Am. Med. Assoc.* **2006**, *295*, 2727–2741. [CrossRef]

18. Chakrabarti, J.K.; Hotten, T.M.; Tupper, D.E. 2-Methylthienobenzodiazepine. U.S. Patent 5,627,178, 6 May 1997.

19. Wang, L.; Shen, J.; Tang, Y.; Chen, Y.; Wang, W.; Cai, Z.; Du, Z. Synthetic Improvements in the Preparation of Clopidogrel. *Org. Process Res. Dev.* **2007**, *11*, 487–489. [CrossRef]

20. Gimeno, M.C.; Goitia, H.; Laguna, A.; Luque, M.E.; Villacampa, M.D.; Sepúlveda, C.; Meireles, M. Conjugates of ferrocene with biological compounds. Coordination to gold complexes and antitumoral properties. *J. Inorg. Biochem.* **2011**, *105*, 1373–1382. [CrossRef]

21. Goitia, H.; Nieto, Y.; Villacampa, M.D.; Kasper, C.; Laguna, A.; Gimeno, M.C. Antitumoral Gold and Silver Complexes with Ferrocenyl-Amide Phosphines. *Organometallics* **2013**, *32*, 6069–6078. [CrossRef]

22. Gutiérrez, A.; Gracia-Fleta, L.; Marzo, I.; Cativiela, C.; Laguna, A.; Gimeno, M.C. Gold(I) thiolates containing amino acid moieties. Cytotoxicity and structure–activity relationship studies. *Dalton Trans.* **2014**, *43*, 17054–17066. [CrossRef]

23. Gutiérrez, A.; Marzo, I.; Cativiela, C.; Laguna, A.; Gimeno, M.C. Highly Cytotoxic Bioconjugated Gold(I) Complexes with Cysteine-Containing Dipeptides. *Chem. Eur. J.* **2015**, *21*, 11088–11095. [CrossRef]

24. Ortego, L.; Meireles, M.; Kasper, C.; Laguna, A.; Villacampa, M.D.; Gimeno, M.C. Group 11 complexes with amino acid derivatives: Synthesis and antitumoral studies. *J. Inorg. Biochem.* **2016**, *156*, 133–144. [CrossRef]

25. Bardají, M.; Jones, P.G.; Laguna, A.; Villacampa, M.D.; Villaverde, N. Synthesis and Structural Characterization of Luminescent Gold(I) Derivatives with an Unsymmetric Diphosphine. *Dalton Trans.* **2003**, *23*, 4529–4536. [CrossRef]

26. Fillat, M.F.; Gimeno, M.C.; Laguna, A.; Latorre, E.; Ortego, L.; Villacampa, M.D. Synthesis, structure and bactericide activity of (aminophosphane)gold(I) thiolate complexes. *Eur. J. Inorg. Chem.* **2011**, *9*, 1487–1495. [CrossRef]

27. Artigas, M.M.; Crespo, O.; Gimeno, M.C.; Jones, P.G.; Laguna, A.; Villacampa, M.D. Synthesis and characterization of [Ag$_4$(μ_3-SC$_2$B$_{10}$H$_{11}$)$_2$(μ-O$_3$SCF$_3$)$_2$(PPh$_3$)$_4$]: A silver complex with a μ_3-thiolate ligand. *Inorg. Chem.* **1997**, *36*, 6454–6456. [CrossRef]

28. Bardají, M.; Crespo, O.; Laguna, A.; Fischer, A.K. Structural characterization of silver(I) complexes [Ag(O$_3$SCF$_3$)(L)] (L = PPh$_3$, PPh$_2$Me, SC$_4$H$_8$) and [AgL$_n$](CF$_3$SO$_3$) (n = 2–4), (L = PPh$_3$, PPh$_2$Me). *Inorg. Chim. Acta* **2000**, *304*, 7–16. [CrossRef]

29. Zhang, P.; Gao, W.Y.; Turner, S.; Ducatman, B.S. Gleevec (STI-571) inhibits lung cancer cell growth (A549) and potentiates the cisplatin effect in vitro. *Mol. Cancer* **2003**, *2*, 1. [CrossRef]

30. Shao, J.; Ma, Z.-Y.; Li, A.; Liu, Y.-H.; Xie, C.-Z.; Qiang, Z.-Y.; Xu, J.-Y. Thiosemicarbazone Cu(II) and Zn(II) complexes as potential anticancer agents: Syntheses, crystal structure, DNA cleavage, cytotoxicity and apoptosis induction activity. *J. Inorg. Biochem.* **2014**, *136*, 13–23. [CrossRef]

31. Suntharalingam, K.; Mendoza, O.; Duarte, A.A.; Mann, D.J.; Vilar, R. A platinum complex that binds non-covalently to DNA and induces cell death via a different mechanism than cisplatin. *Metallomics* **2013**, *5*, 514–523. [CrossRef]

32. Usón, R.; Laguna, A. Polyaryl Derivatives of Gold(I), Silver(I) and Gold(III). In *Organometallic Syntheses*; King, R.B., Eisch, J.J., Eds.; Elsevier: Amsterdam, The Netherlands, 1986; Volume 3, pp. 322–342.

33. Usón, R.; Laguna, A.; Navarro, A.; Parish, R.V.; Moore, L.S. Synthesis and reactivity of perchlorate bis(tetrahydrothiophen)gold(I). ^{197}Au Mössbauer spectra of three-coordinate gold(I) complexes. *Inorg. Chim. Acta* **1986**, *112*, 205–208. [CrossRef]

34. Usón, R.; Laguna, A.; Laguna, M.; Jiménez, J.; Gómez, M.P.; Sainz, A.; Jones, P.G. Gold complexes with heterocyclic thiones as ligands. X-Ray structure determination of [Au(C$_5$H$_5$NS)$_2$]ClO$_4$. *J. Chem. Soc. Dalton Trans.* **1990**, 3457–3463. [CrossRef]

35. *CrysAlisPro, Version 1.171.35.11*; Agilent Technologies: Yarnton, UK, 2011.

36. Sheldrick, G.M. Crystal structure refinement with *SHELXL*. *Acta Cryst.* **2015**, *C71*, 3–8.

37. Mosmann, T.J. Rapid colorimetric assay for cellular growth and survival: Application to proliferation and cytotoxicity assays. *Immunol. Methods* **1983**, *65*, 55–63. [CrossRef]

Review

Designing Ruthenium Anticancer Drugs: What Have We Learnt from the Key Drug Candidates?

James P. C. Coverdale [1], Thaisa Laroiya-McCarron [2] and Isolda Romero-Canelón [2,*]

[1] Department of Chemistry, University of Warwick, Coventry CV4 7AL, UK; J.Coverdale@warwick.ac.uk
[2] School of Pharmacy, Institute of Clinical Sciences, University of Birmingham, Birmingham B15 2TT, UK; thaisalm@hotmail.co.uk
* Correspondence: i.romerocanelon@bham.ac.uk

Received: 29 November 2018; Accepted: 13 February 2019; Published: 1 March 2019

Abstract: After nearly 20 years of research on the use of ruthenium in the fight against cancer, only two Ru(III) coordination complexes have advanced to clinical trials. During this time, the field has produced excellent candidate drugs with outstanding in vivo and in vitro activity; however, we have yet to find a ruthenium complex that would be a viable alternative to platinum drugs currently used in the clinic. We aimed to explore what we have learned from the most prominent complexes in the area, and to challenge new concepts in chemical design. Particularly relevant are studies involving **NKP1339**, **NAMI-A**, **RM175**, and **RAPTA-C**, which have paved the way for current research. We explored the development of the ruthenium anticancer field considering that the mechanism of action of complexes no longer focuses solely on DNA interactions, but explores a diverse range of cellular targets involving multiple chemical strategies.

Keywords: ruthenium; anticancer; metal complex

---

1. Introduction

Cancer is a major health burden in the developed world. Although many breakthroughs in biological targeted approaches have occurred (specifically, immunological therapy), an unmet clinical need remains for a large section of the patient population, so wide-spectrum chemotherapy agents are still required. This space is currently addressed with the use of platinum(II) coordination complexes, namely Cisplatin (CDDP), Oxaliplatin and Carboplatin (worldwide approval), alongside Nedaplatin and Lobaplatin, which are restricted to selected markets. Such platinum agents lack cellular selectivity, so rapidly proliferating cells (hair follicles, bone marrow, and gastrointestinal tract) are particularly vulnerable to off-target effects. This lack of cellular selectivity can lead to the development of severe side effects not limited to nausea, vomiting, hair loss, nephrotoxicity, neurotoxicity, and hearing loss [1,2].

The prevalence of platinum resistance in the clinic, both intrinsic and acquired [3–5], is an escalating clinical concern, especially considering that platinum drugs are currently used in over half of all chemotherapy regimens [3]. This has sparked the development of a new generation of cytotoxics based on different metals, and corresponding coordination complexes, which excel in cellular selectivity and retain their use against a wide range of malignancies, while exhibiting unique mechanism(s) of action. Given the chemical similarity of the platinum group of metals (Pt, Pd, Rh, Ir, Ru, and Os), given the successes of platinum therapies, interest in ruthenium(II/III) complexes has increased considerably.

2. Ruthenium Complexes as Anticancer Agents

Ruthenium compounds have been proven to be a starting point in the search for alternatives to platinum drugs in the clinic, as these compounds have both comparable ligand exchange kinetics and a greater number of accessible coordination geometries. Such comparison considers that Pt(II) compounds are limited to square planar geometries, akin to cisplatin, and that octahedral geometries comparable to ruthenium complexes can be only be achieved by a higher oxidation state, e.g., Pt(IV). The investigation around such complexes has been intense, with highly-renowned and international-leading research groups pouring resources into this fast growing field. The chemistry of Ru(II) and Ru(III) complexes is well established in both the materials and medicinal chemistry fields, and research into their use as prospective anticancer agents is widely spread. The latter oxidation state complexes are typically considered to be more inert, which is, to some extent, attributable to the higher effective nuclear charge. In combination with the highly reducing environment of a cancer cell, an "activation by reduction" mechanism has been speculated for many Ru(III) complexes [6], which are reduced to their more active Ru(II) form by cellular reductants such as ascorbate [7].

The in-cell mechanism of action (MoA) for many ruthenium complexes differs from the DNA-binding mechanism typically associated with platinum drugs [8]. With a wider range of intracellular targets, ruthenium anticancer complexes are well established, and many examples have shown promise in chemical model systems, in vitro, and in vivo [9–11]. However, despite widespread efforts, only three ruthenium complexes (**NKP1339**, **NAMI-A**, and **TLD1433**) have reached clinical trials. **NKP1339**, and **NAMI-A** are both being developed as chemotherapeutic agents, while octahedral Ru(II) complex **TLD1433** has potential as a photosensitizer for photo-dynamic therapy [12]. Research into future anticancer agents should learn from both successes and difficulties encountered during the development of previous complexes in order to identify desirable physical, chemical, and biological properties associated with a successful future Ru drug candidate.

In this review, we critically evaluate four of the most significant Ru coordination complexes, **NAMI-A**, **NKP1339**, **RM175**, and **RAPTA-C**, and assess their contribution to the advancement of future ruthenium-based anticancer agents.

3. Case Study: RM175

RM175 [Ru(biphenyl)Cl(en)]$^+$, where en = 1,2-ethylenediamine, was amongst the first ruthenium(II) complexes to be explored for anticancer activity (Figure 1). This pseudo-octahedral organometallic complex was developed by the Sadler group in 2001 [13]. **RM175** is comprised of monodentate (chloride), bidentate (diamine), and arene (biphenyl) ligands, typical of a 3-legged "piano-stool" geometry. Originally synthesized with the aim of targeting DNA, it is intended to take advantage of the lower 2+ oxidation state that would not require activation by cellular reduction [14].

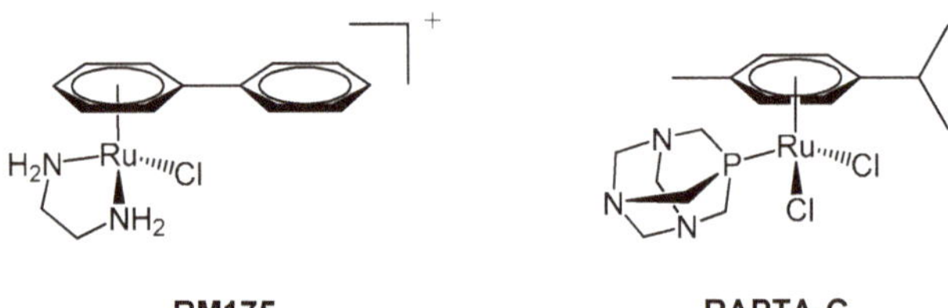

RM175   RAPTA-C

Figure 1. **RM175** (**left**) and **RAPTA-C** (**right**) are typical examples of 18-electron ruthenium arene "piano-stool" complexes, in which an η^6-arene ring stabilises the 2+ oxidation state of the ruthenium metal centre.

The arene substituent provides a hydrophobic surface, allowing for cellular diffusion through the lipophilic plasma membrane [14]. After entry into the cell, but prior to binding with DNA, complex activation is thought to occur by ligand exchange at the monodentate site [15,16]. The complex's design envisaged the halogen atom acting like a leaving group, drawing similarities with the activation of cisplatin. The aquation reaction would then generate a coordinative vacancy, which, in turn, would allow for covalent binding to the N7 of guanine in the DNA double helix [14]. Although

Ru(II) complexes are known to bind to the guanine residues in DNA [17], it is thought that the extended arene in this complex enables hydrophobic interactions to occur between **RM175** and DNA via arene-intercalation between base pairs [18]. As a consequence of the free rotation of biphenyl about the Ru(II) centre, the structure is relatively flexible, which is anticipated to limit steric hindrance and increase the complex's DNA-binding affinity [19]. Conformational flexibility may allow the complex to intercalate into the base pairs and bind to guanine simultaneously, as has been shown possible with other ruthenium complexes [19]. This may explain why the resulting **RM175**-DNA adduct is more resistant to DNA repair than platinated DNA; therefore, such observation begins to explain the lack of cross-resistance with platinum [20].

Further biological studies into the mechanism of action of **RM175** revealed new cellular targets in addition to DNA binding, most notably the inhibition of matrix metalloproteinase-2 (MMP-2) [14]. MMPs are the most important members of the metalloproteinase family that contribute to tumor progression. They govern the conditions of the tumor microenvironment and use signalling pathways to assist in the modulation of cell growth and angiogenesis [21]. MMPs facilitate the migration and invasion of cancerous cells, through degradation of the extracellular matrix (ECM), suppression of the adaptive immune system, and other processes that promote cell survival and protect malignant cells from apoptosis [21,22]. Apoptosis, as a form of programmed cell death, does not procure an inflammatory response in principle; thus, MMPs are a noteworthy target for anticancer agents [23].

RM175 also exerts activity by disrupting the processes of cell invasion and migration. In order to assess the effect of metallodrugs on cellular detachment and re-adhesion, several proteins have been investigated including collagen IV, fibronectin, and poly-L-lysine. All these proteins may come into contact with malignant cells in vivo, since poly-L-lysine attracts cells through electrostatic interactions, and both fibronectin and collagen IV are constituents of the extracellular matrix. Upon exposure to **RM175**, invasive MDA-MB-231 breast adenocarcinoma cells became more resistant to detachment from either fibronectin or poly-L-lysine, depending on the substance on which the cells were originally grown. This effect was absent in non-tumorigenic and non-invasive cell lines (HBL-100 and MCF-7, respectively), suggesting that the reduction in cellular detachment after exposure to **RM175** was selective for invasive cells that would otherwise metastasize. Adherence to cellular components is a vital stage in the formation of metastases, so the Ru(II) complex has the potential to expand this selective action across other cell lines. **RM175** significantly inhibits cellular contact-induced movement (haptotaxis) in MDA-MB231 and HLB-100 cells, preventing metastatic formation induced by this stimulus [14].

The described effects on cell invasion and migration explain why the use of **RM175** was directed toward preventing and treating metastases as opposed to the reduction in primary tumor volume. Though in vivo studies demonstrated a 50% reduction in primary mammary carcinoma (MCa) tumor mass, upon withdrawal of **RM175**, tumor growth resumed, demonstrating the limited effect of **RM175** on primary tumors. MCa cells are also liable to spontaneously spread to lung tissue; therefore, the anti-metastatic activity was also assessed. The ruthenium complex was found to have greater potency against metastases over primary tumors, and the size of the administered dose was found to significantly impact the extent of metastatic growth. Although a high dose (10 mg/kg/day) resulted in an 85–95% reduction in metastatic mass, only 70% reduction was achieved at a lower dosage (7.5 mg/kg/day). Supplementation of the dose with human serum albumin (HSA) in ratios from 1:1 to 1:10 were found to elevate the drug's cytotoxic profile and reduce cell viability to a greater extent, indicating that efficacy of **RM175** may be greater in vivo than in vitro [14].

RM175 has been reported to up-regulate the tumor suppressor *p53* and the pro-apoptotic protein *Bax* in HCT-116-wt cells, contributing to apoptosis. Significant apoptosis only occurred in cells that were not *p53/Bax*-null, demonstrating that both proteins are vital for early **RM175**-induced apoptosis (<48 h). Treatment with **RM175** also induced long-term loss of cellular replication, independent of *p53*, *p21 waf1/Cip1*, and *Bax* [17].

RM175 also induces G_1/G_2-phase cell cycle arrest in vitro [17]. G_1 inhibition of cell cycle progression was found to be dependent on tumor suppressors and cell cycle regulators, *p53* and *p21*, which binds to CDK1, as well as to, CDK2 [24], yet independent from the *Bcl-2* family member, *Bax* [17]. This highlights the vital role of both tumor suppressors in initiating cell cycle arrest in contrast to **RM175**-mediated apoptosis, which only requires *p53* and *Bax*. Exposure resulted in the accumulation of the aforementioned genetic components [17], particularly *p21*, as expression of this cell cycle regulator typically leads to cell cycle arrest [25]. Activation of *p53* initiates a cascade of anti-mitogenic signals, promoting the *p21*-induced inhibition of CDK2 activity. This consequently prevents expression of various genes (some of which are involved in DNA replication) [25] and inhibits progression to S phase. Alternatively, the cell cycle arrest in the G_2 phase could be attributed to *p53*-mediated activation of either the GADD45 protein or *p21* [24], both of which inhibit CDK1, thereby inducing G_2/M cell cycle arrest [26]. **RM175** exerts its anticancer activity by both preventing progression through the cell cycle leading to apoptosis, and "traditional" disruption of DNA replication by strand intercalation and nucleobase binding.

4. Case Study: RAPTA-C

A second notable piano-stool complex is **RAPTA-C**, a ruthenium(II) arene complex developed by the Dyson group that comprises an amphiphilic 1,3,5-triaza-7-phosphaadamantane (PTA) ligand and two labile chloride ligands additional to the η^6-coordinated arene, which helps to stabilize the +2 oxidation state (Figure 1) [27]. In contrast to other phosphine ligands [28], PTA is relatively sterically undemanding and is anticipated to contribute to the increased water solubility of **RAPTA-C** relative to other Ru(II) arene complexes [29]. Similarly to cisplatin, **RAPTA-C** undergoes rapid hydrolysis of Ru–Cl at low (intracellular) chloride concentrations (4–5 mM), predominantly yielding the mono-aquated complex, $[Ru(p\text{-cymene})Cl(H_2O)(PTA)]^+$. The Ru–Cl bond remains intact at higher chloride concentrations (100 mM, i.e., in blood) and, like cisplatin, **RAPTA-C** may be considered a pro-drug in its di-chlorido form [28]. Initial studies of **RAPTA-C** indicated that damage to supercoiled pBR322 DNA was pH-dependent (occurring only below physiological pH; <7.0), so it was hypothesized that this may invoke selective targeting of the typically more acidic environment of cancer cells [27].

Unlike cisplatin, preliminary in vitro studies found **RAPTA-C** to be remarkably inactive toward TS/A adenocarcinoma and non-cancerous epithelial (HBL-100) cell lines, as well as initially having limited activity against primary tumors in vivo. However, later in vivo studies using pre-clinical models (using chicken chorioallantoic membrane and, more recently, mice) found that **RAPTA-C** was able to inhibit tumor growth by 50–75% [30,31]. In both instances, analysis of the treated tumor identified the significant anti-angiogenic properties of the complex. Both **RAPTA-C** and structurally similar **RAPTA-B** (in which the arene unit has been changed to a benzene ring) showed promise for size reduction of solid lung metastases in vivo [30]. These pre-clinical studies clearly demonstrate the requirement for careful optimization and control of experimental conditions, and for scientists to be mindful of the fact that in vitro experiments often do not accurately predict in vivo activities. Complexes, such as **RAPTA-B** or HC11, that contain cyclic, non-substituted hydrocarbons in their structure have the potential to cause liver toxicity after P450 poly-hydroxylation; hence, in vivo experiments need to closely monitor phase I metabolism of the ruthenium drugs [32]. In the case of RAPTA complexes, even at the highest dose administered to mice, negligible side effects were observed. Ruthenium was found to be excreted rapidly via the renal system and did not significantly accumulate in vital organs [28,33].

Later structural iterations of **RAPTA-C** included tethering of organic substituents on the coordinated arene. Planar anthracene groups were investigated to enhance DNA interaction by intercalation with concurrent fluorescence to map intracellular localization, with limited success [34–36]. In contrast, similar naphthalimide (a DNA intercalator) complexes were found to possess modest potencies against A2780 ovarian cancer cells (and a platinum-resistant-derived cell line, A2780Cis) (IC_{50} 2.3–9.1 µM) and displayed some selectivity over non-cancerous (HEK-293) cells [37]. Conjugation of ethacrynic acid, a glutathione transferase inhibitor, to **RAPTA-C** yielded a new ruthenium complex that was found to bind to the enzyme's H-site. Importantly, the ruthenium center is involved in the inhibition of glutathione transferase through cysteine residue binding [38]. After an extended incubation time, the ethacrynic acid moiety remained in the H-site of the enzyme but the ruthenium center had been released, which was also found to occur in vitro [39,40].

Most recently, a comparative metallomic study of the metabolism of **RAPTA-C** and cisplatin was reported. Despite **RAPTA-C** and cisplatin having similar labile chloride ligands with comparable exchange kinetics, **RAPTA-C** was found to be more inert to extracellular reactions despite being administered at a dose 40-fold higher than cisplatin. **RAPTA-C** was found to predominantly bind albumin, though extracellular metal speciation was found to be time-dependent [41]. This highlights the need to better understand speciation, and consequently drug pharmacokinetics and dynamics, to form a well-rounded pre-clinical research portfolio.

Various combination therapies involving **RAPTA-C** have also been explored in vivo. When administered alongside erlotinib (epidermal growth factor receptor (EGFR) inhibitor) and BEZ-235 (Phosphoinositide 3-kinase PI3K/mTOR inhibitor), the combination was found to synergistically inhibit tumor growth (up to 11-fold relative to single-drug experiments) [42]. An alternative combination therapy involving **RAPTA-C** and axitinib (vascular endothelial growth factor receptor (VEGFR) targeting tyrosine kinase inhibitor) led to a ca. 90% inhibition of tumor growth with a dose of only 0.4 mg/kg, despite negligible toxicity observed at doses of 100 mg/kg [43].

5. Case Study: NAMI-A

NAMI-A, originally synthesized by the groups of Alessio and Sava (Figure 2), differs from the two previously described complexes, being composed of chloride, imidazole, and dimethylsulfoxide (DMSO) in an octahedral arrangement around a Ru(III) center. Ruthenium(III) complexes are often considered to be pro-drugs, since they are typically less reactive than Ru(II) congeners and require activation via reduction to the 2+ oxidation state [44,45]. Such activation improves tumor targeting as the hypoxic cellular environment favors reduction of the metal center, thereby generating antiproliferative selectivity for cancerous cells compared to healthy cells [45,46]. Reduction of ruthenium from +3 to +2 also leads to kinetic lability; hence, the hydrolysis of chlorido ligands occurs at a faster rate, which may facilitate the reaction of ruthenium complexes with DNA [47]. In highly reactive cases, such substitution process, like hydrolysis, can effectively occur before the activation via reduction [48].

Having entered clinical trials in 2008, **NAMI-A** constitutes one success story in the development of ruthenium anticancer complexes. **NAMI-A** was first used in a phase I/II study in 32 patients suffering from an advanced form of non-small cell lung cancer (NSCLC) [49]. The metallodrug was administered intravenously in combination with gemcitabine, which is regularly used alongside cisplatin in this type of cancer [44]. Trial results indicated that such combination of **NAMI-A** and gemcitabine affected the quality of life of the patients, with side effects including gastrointestinal (GI) disturbances, neutropenia, and elevated liver enzymes [44]. Although the regimen was deemed to be "insufficiently effective for further use" [49], this problem may yet have a solution.

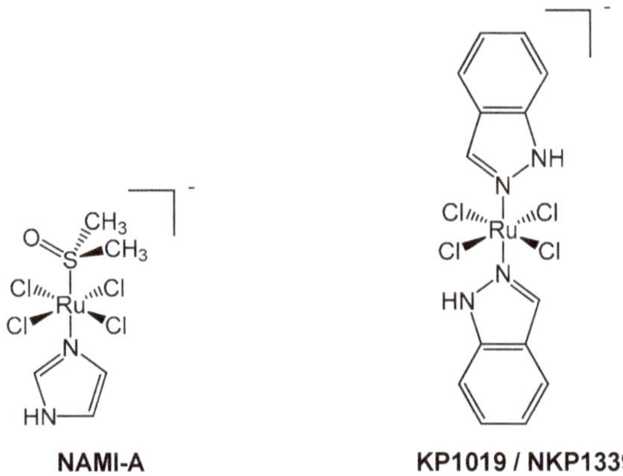

NAMI-A

KP1019 / NKP1339

Figure 2. **NAMI-A** (**left**) and **KP1019/NKP1339** (**right**) are octahedral Ru(III) pro-drugs, which are hypothesized to undergo an "activation by reduction" mechanism inside cells to form more active Ru(II) species.

NAMI-A has a synergistic ability to prevent cell invasion and hinder neo-angiogenesis [50], making it selective for metastasis rather than fully formed tumors [46]. Compared to cisplatin, **NAMI-A** has a wide variety of biological targets, most of which are extracellular rather than DNA-based [44]. One of the main mechanisms through which **NAMI-A** exerts its anti-angiogenic effects is thought to be the scavenging of nitric oxide [51]. The nitric oxide synthetase (NOS) pathway stimulates angiogenesis and endothelial cell migration and also includes VEGF-activated enzymes that catalyze the generation of NO—a signaling molecule involved in these processes [51,52]. The ruthenium center of **NAMI-A,** as well as its albumin adducts, have been proven to bind strongly to NO through displacement of the DMSO ligand [53]. This reaction is thought to be irreversible in vivo as NO release is unfavorable, even in the presence of glutathione or other reducing agents. Nitric oxide is a downstream mediator of VEGF and is implicated in endothelial cell migration [52]. NO scavengers, including **NAMI-A**, are reported to cause potent blockade of VEGF-mediated endothelial cell processes related to angiogenesis, including cell migration [51]. This may be due to **NAMI-A**'s observed effects on NO, since **NAMI-A** has no effect on the phosphorylation status of the downstream proteins, PKB (more commonly known as *Akt*) and ERK$\frac{1}{2}$ [54]. **NAMI-A** does not affect VEGF itself but instead is thought to enable cells to overcome its inductive properties, supporting the theory that **NAMI-A** prevents VEGF-mediated activity through NO scavenging [54].

Another angiogenic process effected by **NAMI-A** is inhibition of endothelial cellular proliferation [51], since NO is also a signaling molecule in the mitogen-activated protein kinase (MAPK) pathway [55]. This pathway includes a cascade of kinases that partake in cellular signaling, which coordinate various biological processes including cellular proliferation, migration, and survival [56]. After treatment with **NAMI-A**, cellular proliferation was inhibited for at least 48 h [51], an effect that could be attributed to NO's role in the MAPK pathway. Angiogenesis is essential for tumor progression, since oxygen and nutrients are needed to sustain malignant growth. Angiogenic-inhibition provides another means of suppressing metastatic development. **NAMI-A** also reduces the VEGF-dependent migration of cancerous cells [57]. Inhibition of VEGF activity hence provides a selective mechanism through which **NAMI-A** can prevent neo-angiogenesis and the formation of metastases.

NAMI-A has also been reported to act on cells via selective inhibition of the Ca^{2+}-activated potassium ion channel, *KCa3.1* [58]. In most cells, the concentration of potassium ions regulates the membrane potential, which, in turn, governs cell cycle progression [59]. However, the gene that encodes *KCa3.1* (KCNN4) is overexpressed in many cancer cells [58]. The downstream biological processes governed by *KCa3.1* differ depending on the cell type: (1) proliferation in leukemia and lymphoma cells or (2) migration in epithelial and glial cancer cells [58,60]. Such cell-type-dependent function of the *KCa* ion channel may explain why the activity of **NAMI-A** is typically limited in primary tumors, yet the complex remains potent against leukaemia and metastases. In leukemia cells, **NAMI-A**-induced inhibition of *KCa3.1* was found to induce G_2/M cell cycle arrest, leading to apoptosis [58]. Ion channels such as *KCa3.1* are considered to be indispensable in relation to cellular migration [61]; hence, **NAMI-A** is also able to inhibit cell motility. Proliferation and migration of NSCLC cells is also thought to be dictated by *KCa3.1*, as more aggressive cancers typically display up-regulation of this ion channel [62]. **NAMI-A**-mediated inhibition of *KCa3.1* effects malignant cell migration, either by diminishing or completely preventing the migration of epithelial cancer cells [61].

Another pathway thought to be involved in the mechanism of action of **NAMI-A** is the ATM/ATR (Ataxia telangiectasia mutated / RAD3-related) kinase pathway, which is activated after DNA damage and works to regulate the cellular response [63]. **NAMI-A** activates both kinases, though appears more selective for phosphorylation of ATR [54]. The downstream effects include phosphorylation of the tumor suppressor *p53* and kinases CHK1/CHK2, which are normally activated in response to genetic irregularities to coordinate cell cycle arrest [54]. **NAMI-A** also increases the expression of tumors suppressors such as *p15INK4b*, *p21*, and *p27Kip1*—CDK inhibitors that contribute to cell cycle arrest [54,64].

The anti-metastatic properties of **NAMI-A** arise from its ability to inhibit certain processes that are vital for metastatic formation and survival, including the adhesion and migration of cells. In addition to the mechanisms explained above, **NAMI-A** can inhibit these processes through protein interference, such as via α5β1 integrin [65]. Integrins are transmembrane-bound proteins that bind to other cellular components to facilitate the adhesion and migration of cells. Studies indicate that **NAMI-A** not only blocks α5β1 but significantly reduces the number of integrin receptors by modulating the expression of ITGA5 and ITGB1—the coding genes for the integrin subunits α5 and β1, respectively. This reduces downstream phosphorylation of focal adhesion kinase (FAK) and both of these processes are thought to contribute to decreased cellular adherence. Pre-treatment of HCT-116 colon cancer cells with **NAMI-A** significantly reduced adherence to Fibronectin and collagen I; an effect that was less apparent in α5β1 integrin-null cells. Notably, the observed effects on cellular adhesion and pFAK occurred in a manner that was inversely correlated with **NAMI-A** concentration, further supporting the hypothesis that lower doses of **NAMI-A** have greater efficacy [65].

MMP targeting is another anti-metastatic mechanism associated with **NAMI-A**. It inhibits the production of MMP-2 and the activity of MMP-2 and MMP-9 [66]. Within the functions of these MMPs, those of particular interest include chemotaxis and the activation of transforming growth factor (TGF β1) [67]. Since **NAMI-A** inhibits endothelial chemotaxis, this may be mediated via MMP inhibition [66]. The MMPs inhibited by **NAMI-A** and TGF-β1 partake in a feedback mechanism as TGF-β1 is also responsible for regulating MMP expression [68], which may explain the effect of **NAMI-A** on MMP-2 generation. TGF-β1 is a growth factor that selectively targets cancerous cells, inducing the epithelial–mesenchymal transition (EMT), invasion and migration; all integral processes needed for metastatic growth. **NAMI-A** counteracts this stimulus as it supposedly shares some of the pathways used by TGF-β1 but leads to opposing outcomes. Therefore, a metastatic response to TGF-β1 is prevented in tumor cells by hindering the processes mentioned above, leaving non-cancerous cell lines unaffected [69].

6. Case Study: KP1019/NKP1339

Another success story in the development of Ru(III) compounds that have reached clinical trials is **NKP1339**, reported by the Keppler group (Figure 2). The original form, **KP1019**, was modified to increase its aqueous solubility, generating the sodium salt equivalent, **NKP1339** [46]. This Ru(III) complex has some structural similarity to **NAMI-A** due to its octahedral geometry and chloride ligands [46]. **NKP1339** is a pro-drug and relatively inert compared to **NAMI-A**, as ligand loss/exchange does not occur as readily [44,47].

Both **NAMI-A** and **NKP1339** bind non-covalently with proteins in the blood, likely via hydrophobic interactions [44,46]. **NKP1339** undergoes rapid binding with albumin, so this is an important factor to consider when the metal complex is administered intravenously. The metal-protein adduct does account for the low side effect profile documented during the phase I trial of **NKP1339**, as the complex remains in its pro-drug form until it undergoes activation by reduction, after internalization by cells and release from albumin [70]. Albumin binding may also improve the selective accumulation of **NKP1339** through two possible mechanisms: the EPR effect or by binding to the glycoprotein, gp60 [44]. The former occurs due to leaky vasculature within malignant cells, resulting in a greater uptake of albumin and other substances, which are then prevented from leaving by poor lymphatic drainage. In contrast, gp60 is present along the tumor endothelium and allows receptor-mediated transcytosis of the drug-albumin adduct to occur through to the extracellular tumor matrix [44]. Adduct formation with blood proteins is more significant for **NKP1339** than **NAMI-A** since cellular uptake for the latter is limited [44], whereas uptake of **NKP1339** is considered significantly more efficient [71].

Other ruthenium complexes, including **RM175** [17], have been previously shown to interact with DNA through binding to guanine bases after reduction to their active hydrolyzed form [72], and **NKP1339** is no exception. Due to the ability of **NKP1339** to accumulate within the nucleus after activation [47], DNA may be one of the drug's intracellular targets [46]. During activation, the metal ion is reduced, and the Ru(II) species is responsible for the DNA-adduct formation. Only weak adduct signals were emitted when **NKP1339** was incubated with oligonucleotides, indicating low levels of interaction. This may be due, at least in part, to the slow ligand exchange rate of **NKP1339**, and a significant proportion of the drug remains intact [47]. In contrast with cisplatin, neither drug uptake nor genetic mutations, such as those involving *p53* and *k-ras*, are correlated with the efficacy of **NKP1339**. This *p53*-independent mechanism implies that DNA binding is not likely to be the primary mechanism of action of **NKP1339** [70], which may account for the lack of platinum-cross resistance observed with this Ru(III) complex in pre-clinical studies [73].

NKP1339 induces cell cycle arrest through mechanisms attributed to its redox activity in cancer cells [46]. Intracellular reactive oxygen species (ROS) concentrations (such as hydroxide radicals and superoxide) are already elevated in neoplastic cells compared with healthy cells [74] and this is thought to increase the vulnerability of cancer cells to further ROS fluctuations [75]. As such, redox-targeting metal complexes have attracted significant attention in recent years, and often exhibit significant selectivity for cancer cells over healthy cells [74,76–78]. **NKP1339** has been shown to increase the intracellular ROS concentration [79] and upregulate the pro-apoptotic *p38* MAPK pathway [80]. This biological cascade is normally activated in response to cellular stress, such as cytokines, DNA damage and ROS, and is implicated in cell cycle progression [81]. Activation of this pathway leads to downstream regulation of gene expression, including those that encode for cytokines, transcription factors, and cellular receptors [82]. More importantly, this pathway also regulates the G_1/S and G_2/M check points within the cell cycle [81]. By generating ROS and altering the cellular redox balance, **NKP1339** induces G_2/M cell cycle arrest [46].

Treatment of sensitive cell lines with **NKP1339** induced cell death, typically within 20–30 h [70]. Accumulation of ROS prompts the loss of the mitochondrial membrane potential by inducing membrane permeabilisation, leading to activation of the intrinsic apoptosis pathway [83,84], which is typically caspase-dependent, and involves the cleavage of the zinc finger protein, poly-(ADP-ribose)-polymerase (PARP). [70,85] However, in the case of **NKP1339**, the extent of mitochondrial membrane depolarization is limited, suggesting another mechanism of apoptosis may be at work. Caspase-8 was cleaved as part of **NKP1339**-induced apoptosis [70], a key feature of the extrinsic apoptosis pathway. Combination treatment with **NKP1339** and a caspase-8 inhibitor increased the extent of apoptosis observed within less responsive cell lines. The researchers deduced that the majority of apoptosis occurs not by the mitochondrial-intrinsic pathway, but via the extrinsic pathway [70]. Though mitochondria are a target of **NKP1339**, the effect of this interaction is limited as the majority of apoptotic activity is mediated by either death receptors on the cell surface or via other extrinsic pathways [86]. Since **NKP1339** does not enhance the expression of these death receptors or their ligands, the Ru(III) complex may act via an alternative extrinsic pathway involving endoplasmic reticulum (ER) homeostasis [70].

Since the anticancer activity of **NKP1339** is predominantly mediated through changes in redox homeostasis [46], another notable target is the endoplasmic reticulum (ER), which is also influenced by the cellular redox environment. This organelle controls the maturation, folding, and release of proteins, and initiates the unfolded protein response (UPR) as part of a feedback system, suspending the cell cycle either temporarily until the proteins are restored or, permanently, prior to apoptosis [79]. The protein *Nrf2* is able to trigger the expression of antioxidant genes to reduce the cell's exposure to oxidative stress [87]. After treatment with **NKP1339**, *Nrf2* translocates to the nucleus to commence its protective activity. This observation may be ROS-mediated and demonstrates that drug-induced hyper-oxidation is sufficiently substantial to initiate a cellular response. The redox properties of **NKP1339** lead to dysregulation of various other proteins that affect the ER. **NKP1339** increases ROS levels in vitro in colon carcinoma cell lines (HCT116 and SW480), causing upregulation of *CHOP* mRNA [79]. The targets of this transcription factor include (1) *GADD34*, implicated in cell cycle arrest; (ii) *DR5*, which promotes cell death through caspase activation and initiation of the extrinsic pathway; and (3) *Ero1α*, which leads to ER hyper-oxidation and stimulates cell death [88]. *CHOP* also reduces the expression of the anti-apoptotic protein *Bcl2*, and consequently, excess *CHOP* generation favors the induction of apoptosis [79].

Sustained ER stress may result in cellular dysfunction or death [88], so alterations in the UPR could be exploited for antiproliferative activity. In vitro treatment with **NKP1339** was shown to downregulate various ER-based proteins, including signaling molecules *PERK* and *Ire1α* and the chaperone proteins calnexin and GRP78 [70]. **NKP1339** is an effective inhibitor of the last one by preventing its activation [70,89]. These proteins all have an active role in UPR regulation: calnexin assists with protein folding, whereas the signaling molecules *PERK* and *Ire1α* are two of the main transmembrane receptors bound to the heat shock protein, GRP78 [79]. When GRP78 senses unfolded proteins, it simultaneously binds them and liberates the signaling molecules that drive the UPR [79]. GRP78 is often viewed as the "master regulator" of the UPR [89] and functions to protect the cell with the aim of restoring the ER to its original condition [90]. Downregulation or inhibition of ER proteins results in a lack of control over protein folding and accumulation of damaged proteins. Drug-induced ROS generation damages proteins, further enhancing the ER's exposure to stress [79]. Another advantage of this mechanism is that GRP78 is associated with chemo-resistance, so inhibition of this protein may lead to a better prognosis and prevent the occurrence of GRP78-mediated resistance [89]. In vitro use of **NKP1339** is also able to increase drug-sensitivity in cells that are already resistant; a promising indication that drug-resistance might be reversed.

Another reactive oxygen species, nitric oxide, can be scavenged by **NKP1339**, though its effect on NO-mediated processes is weak in comparison to **NAMI-A** [51]. Therefore, there is little evidence to support its interference with VEGF/NO-stimulated angiogenesis or cell migration [51], explaining why **NKP1339** is more effective against solid tumors than **NAMI-A**. By unsettling redox homeostasis, **NKP1339** is able to inhibit DNA synthesis, induce G_2/M cell cycle arrest and initiate both intrinsic [46] and extrinsic apoptosis [70]. Owing to its multi-targeting mechanism of action, overexpression of proteins associated with multi-drug resistance (MRP1, BCRP, LRP, and the transferrin receptor) little hinders the drug's efficacy, which also explains the lack of cross-resistance [73,85]. The phase I trial of **NKP1339** was used as a dose-escalation study to assess its use for the treatment of advanced solid tumors. The study included patient tolerability as well as pharmacodynamic and pharmacokinetic studies of the drug (Niiki Pharma Inc. and Intezyne Technologies Inc., 2017). The trial (NCT0145297) was completed in 2016 and was deemed successful [46]. **NKP1339** was also shown to be particularly effective against neuroendocrine tumors (and exhibited limited side effects in trial participants [46,73].

7. Current Developments in Ruthenium Anticancer Agents and Future Perspectives

The complexes in these four case-studies have paved the way for future research, much of which is now being completed by not only by well-established researchers but also by internationally recognized new and upcoming leaders in the field. Pushing the boundaries, new ruthenium anticancer complexes show a wider diversity of coordination spheres and a vast range of ligands that include, but are not limited to, extended intercalating aromatic units, polypyridyl rings, fluorescent derivatives, bio-active molecules, ferrocifen analogues, and carbenes, all of which exploit donor atoms such as C, N, O, S and P.

Even more interesting is observing the escalation of novel approaches regarding targeting strategies and MoA involved in the anticancer activity of Ru complexes, all of which are fueled by our increased understanding of the complexes' fates at the cellular level. Disruption of protein-protein interactions, [91] enzymatic inhibition [92,93], and redox modulation [94], as well as chromatin [95] and histone [96,97] targeting, are only a few examples of cellular events being used as means for antiproliferative activity, with in-cell catalysis taking advantage of well-established reactions in the chemistry of materials field [77,78]. Crucially, the understanding of metal anticancer complexes as efficient multi-targeting agents is starting to have a positive outlook, if not yet for funding bodies, then at least from an academic perspective. Hence, the attempts to locate a single drug target may yet develop into the search for a majoritarian cellular event in a field that seems to move at a highly fast pace.

Such advances and new trends in thought should not forget how far **NKP1339, NAMI-A, RM175,** and **RAPTA-C** have advanced the field, and how critical these complexes have been in the development of the field. As a community, we may benefit from more closely examining these four compounds to learn, amongst other things, that small structural changes generate large variations in the MoA at the cellular level [98]. Although structure activity relationships are not easily established, current and further developments of analytical and cellular techniques will provide more investigative tools and, subsequently, a better understanding of cellular behavior. The key for the new generation of anticancer complexes may well be just round the corner.

Funding: This research received no external funding.

Acknowledgments: Economic support from the School of Pharmacy and Institute of Clinical Sciences at the University of Birmingham are acknowledged.

Conflicts of Interest: The authors declare no conflict of interest.

References

1. Florea, A.M.; Büsselberg, D. Cisplatin as an Anti-Tumor Drug: Cellular Mechanisms of Activity, Drug Resistance and Induced Side Effects. *Cancers (Basel)* **2011**, *3*, 1351–1371. [CrossRef] [PubMed]

2. Pérez-Herrero, E.; Fernández-Medarde, A. Advanced targeted therapies in cancer: Drug nanocarriers, the future of chemotherapy. *Eur. J. Pharm. Biopharm.* **2015**, *93*, 52–79. [CrossRef] [PubMed]

3. Ai, Z.; Lu, Y.; Qiu, S.; Fan, Z. Overcoming cisplatin resistance of ovarian cancer cells by targeting HIF-1-regulated cancer metabolism. *Cancer Lett.* **2016**, *373*, 36–44. [CrossRef] [PubMed]

4. Shen, D.W.; Pouliot, L.M.; Hall, M.D.; Gottesman, M.M. Cisplatin resistance: A cellular self-defense mechanism resulting from multiple epigenetic and genetic changes. *Pharmacol. Rev.* **2012**, *64*, 706–721. [CrossRef] [PubMed]

5. Parker, R.J.; Eastman, A.; Bostick-Bruton, F.; Reed, E. Acquired cisplatin resistance in human ovarian cancer cells is associated with enhanced repair of cisplatin-DNA lesions and reduced drug accumulation. *J. Clin. Invest.* **1991**, *87*, 772–777. [CrossRef] [PubMed]

6. Schluga, P.; Hartinger, C.G.; Egger, A.; Reisner, E.; Galanski, M.; Jakupec, M.; Keppler, B.K. Redox behavior of tumor-inhibiting ruthenium(III) complexes and effects of physiological reductants on their binding to GMP. *Dalt. Trans.* **2006**, 1796–1802. [CrossRef] [PubMed]

7. Wiśniewska, J.; Fandzloch, M.; Łakomska, I. The reduction of ruthenium(III) complexes with triazolopyrimidine ligands by ascorbic acid and mechanistic insight into their action in anticancer therapy. *Inorg. Chim. Acta* **2019**, *484*, 305–310. [CrossRef]

8. Dasari, S.; Bernard Tchounwou, P. Cisplatin in cancer therapy: Molecular mechanisms of action. *Eur. J. Pharmacol.* **2014**, *740*, 364–378. [CrossRef] [PubMed]

9. Dougan, S.J.; Habtemariam, A.; McHale, S.E.; Parsons, S.; Sadler, P.J. Catalytic organometallic anticancer complexes. *Proc. Natl. Acad. Sci. USA* **2008**, *105*, 11628–11633. [CrossRef] [PubMed]

10. Kandioller, W.; Balsano, E.; Meier, S.M.; Jungwirth, U.; Goschl, S.; Roller, A.; Jakupec, M.A.; Berger, W.; Keppler, B.K.; Hartinger, C.G. Organometallic anticancer complexes of lapachol: Metal centre- dependent formation of reactive oxygen species and correlation with cytotoxicity. *Chem. Commun.* **2013**, *49*, 3348–3350. [CrossRef] [PubMed]

11. Zhang, P.; Sadler, P.J. Advances in the design of organometallic anticancer complexes. *J. Organomet. Chem.* **2017**, *839*, 5–14. [CrossRef]

12. Fong, J.; Kasimova, K.; Arenas, Y.; Kaspler, P.; Lazic, S.; Mandel, A.; Lilge, L. A novel class of ruthenium-based photosensitizers effectively kills in vitro cancer cells and in vivo tumors. *Photochem. Photobiol. Sci.* **2015**, *14*, 2014–2023. [CrossRef] [PubMed]

13. Morris, R.E.; Aird, R.E.; Murdoch, P.D.S.; Chen, H.; Cummings, J.; Hughes, N.D.; Parsons, S.; Parkin, A.; Boyd, G.; Jodrell, D.I.; et al. Inhibition of cancer cell growth by ruthenium(II) arene complexes. *J. Med. Chem.* **2001**, *44*, 3616–3621. [CrossRef] [PubMed]

14. Bergamo, A.; Masi, A.; Peacock, A.F.A.; Habtemariam, A.; Sadler, P.J.; Sava, G. In vivo tumour and metastasis reduction and in vitro effects on invasion assays of the ruthenium RM175 and osmium AFAP51 organometallics in the mammary cancer model. *J. Inorg. Biochem.* **2010**, *104*, 79–86. [CrossRef] [PubMed]

15. Wang, F.; Habtemariam, A.; van der Geer, E.; Fernández, R.; Melchart, M.; Deeth, R.J.; Aird, R.; Guichard, S.; Fabbiani, F.P.; Lozano-Casal, P.; et al. Controlling ligand substitution reactions of organometallic complexes: Tuning cancer cell cytotoxicity. *Proc. Natl. Acad. Sci. USA* **2005**, *102*, 18269–18274. [CrossRef] [PubMed]

16. Barry, N.P.E.; Sadler, P.J. Exploration of the medical periodic table: Towards new targets. *Chem. Commun.* **2013**, *49*, 5106–5131. [CrossRef] [PubMed]

17. Hayward, R.L.; Schornagel, Q.C.; Tente, R.; Macpherson, J.S.; Aird, R.E.; Guichard, S.; Habtemariam, A.; Sadler, P.J.; Jodrell, D.I. Investigation of the role of Bax, p21/Waf1 and p53 as determinants of cellular responses in HCT116 colorectal cancer cells exposed to the novel cytotoxic ruthenium(II) organometallic agent, RM175. *Cancer Chemother. Pharmacol.* **2005**, *55*, 577–583. [CrossRef] [PubMed]

18. Carter, R.; Westhorpe, A.; Romero, M.J.; Habtemariam, A.; Gallevo, C.R.; Bark, Y.; Menezes, N.; Sadler, P.J.; Sharma, R.A. Radiosensitisation of human colorectal cancer cells by ruthenium(II) arene anticancer complexes. *Sci. Rep.* **2016**, *6*, 1–12. [CrossRef] [PubMed]

19. Chen, H.; Parkinson, J.; Parsons, S.; Coxall, R.; Gould, R.O.; Sadler, P.J. Organometallic ruthenium(II) diamine anticancer complexes: Arene-nucleobase stacking and stereospecific hydrogen-bonding in guanine adducts. *J. Am. Chem. Soc.* **2002**, *124*, 3064–3082. [CrossRef] [PubMed]

20. Aird, R.E.; Cummings, J.; Ritchie, A.A.; Muir, M.; Morris, R.E.; Chen, H.; Sadler, P.J.; Jodrell, D.I. In vitro and in vivo activity and cross resistance profiles of novel ruthenium(II) organometallic arene complexes in human ovarian cancer. *Br. J. Cancer* **2002**, *86*, 1652–1657. [CrossRef] [PubMed]

21. Kessenbrock, K.; Plaks, V.; Werb, Z. Matrix Metalloproteinases: Regulators of the Tumor Microenvironment. *Cell* **2010**, *141*, 52–67. [CrossRef] [PubMed]

22. Nagase, H.; Visse, R.; Murphy, G. Structure and function of matrix metalloproteinases and TIMPs. *Cardiovasc. Res.* **2006**, *69*, 562–573. [CrossRef] [PubMed]

23. Fink, S.L.; Cookson, B.T. Apoptosis, Pyroptosis, and Necrosis: Mechanistic Description of Dead and Dying Eukaryotic Cells. *Infect. Immun.* **2005**, *73*, 1907–1916. [CrossRef] [PubMed]

24. Abbas, T.; Dutta, A. P21 in cancer: Intricate networks and multiple activities. *Nat. Rev. Cancer* **2009**, *9*, 400–414. [CrossRef] [PubMed]

25. Rowland, B.D.; Peeper, D.S. KLF4, p21 and context-dependent opposing forces in cancer. *Nat. Rev. Cancer* **2006**, *6*, 11–23. [CrossRef] [PubMed]

26. Jin, S.; Tong, T.; Fan, W.; Fan, F.; Antinore, M.J.; Zhu, X.; Mazzacurati, L.; Li, X.; Petrik, K.L.; Rajasekaran, B.; et al. GADD45-induced cell cycle G2-M arrest associates with altered subcellular distribution of cyclin B1 and is independent of p38 kinase activity. *Oncogene* **2002**, *21*, 8696–8701. [CrossRef] [PubMed]

27. Allardyce, C.S.; Dyson, P.J.; Ellis, D.J.; Heath, S.L. [Ru(η^6-*p*-cymene)Cl$_2$(pta)] (pta = 1,3,5-triaza-7-phosphatricyclo-[3.3.1.1]decane): A water soluble compound that exhibits pH dependent DNA binding providing selectivity for diseased cells. *Chem. Commun.* **2001**, *2*, 1396–1397. [CrossRef]

28. Murray, B.S.; Babak, M.V.; Hartinger, C.G.; Dyson, P.J. The development of RAPTA compounds for the treatment of tumors. *Coord. Chem. Rev.* **2016**, *306*, 86–114. [CrossRef]

29. Phillips, A.D.; Gonsalvi, L.; Romerosa, A.; Vizza, F.; Peruzzini, M. Coordination chemistry of 1,3,5-triaza-7-phosphaadamantane (PTA): Transition metal complexes and related catalytic, medicinal and photoluminescent applications. *Coord. Chem. Rev.* **2004**, *248*, 955–993. [CrossRef]

30. Scolaro, C.; Bergamo, A.; Brescacin, L.; Delfino, R.; Cocchietto, M.; Laurenczy, G.; Geldbach, T.J.; Sava, G.; Dyson, P.J. In vitro and in vivo evaluation of ruthenium(II)-arene PTA complexes. *J. Med. Chem.* **2005**, *48*, 4161–4171. [CrossRef] [PubMed]

31. Nowak-Sliwinska, P.; Van Beijnum, J.R.; Casini, A.; Nazarov, A.A.; Wagnières, G.; Van Den Bergh, H.; Dyson, P.J.; Griffioen, A.W. Organometallic ruthenium(II) arene compounds with antiangiogenic activity. *J. Med. Chem.* **2011**, *54*, 3895–3902. [CrossRef] [PubMed]

32. Guichard, S.M.; Else, R.; Reid, E.; Zeitlin, B.; Aird, R.; Muir, M.; Dodds, M.; Fiebig, H.; Sadler, P.J.; Jodrell, D.I. Anti-tumour activity in non-small cell lung cancer models and toxicity profiles for novel ruthenium(II) based organo-metallic compounds. *Biochem. Pharmacol.* **2006**, *71*, 408–415. [CrossRef] [PubMed]

33. Weiss, A.; Berndsen, R.H.; Dubois, M.; Müller, C.; Schibli, R.; Griffioen, A.W.; Dyson, P.J.; Nowak-Sliwinska, P. In vivo anti-tumor activity of the organometallic ruthenium(II)-arene complex [Ru(η^6-*p*-cymene)Cl$_2$(pta)] (RAPTA-C) in human ovarian and colorectal carcinomas. *Chem. Sci.* **2014**, *5*, 4742–4748. [CrossRef]

34. Nazarov, A.A.; Risse, J.; Ang, W.H.; Schmitt, F.; Zava, O.; Ruggi, A.; Groessl, M.; Scopelitti, R.; Juillerat-Jeanneret, L.; Hartinger, C.G.; et al. Anthracene-tethered ruthenium(II) arene complexes as tools to visualize the cellular localization of putative organometallic anticancer compounds. *Inorg. Chem.* **2012**, *51*, 3633–3639. [CrossRef] [PubMed]

35. Furrer, M.A.; Schmitt, F.; Wiederkehr, M.; Juillerat-Jeanneret, L.; Therrien, B. Cellular delivery of pyrenyl-arene ruthenium complexes by a water-soluble arene ruthenium metalla-cage. *Dalt. Trans.* **2012**, *41*, 7201–7211. [CrossRef] [PubMed]

36. Kilpin, K.J.; Clavel, C.M.; Edafe, F.; Dyson, P.J. Naphthalimide-tagged ruthenium–arene anticancer complexes: Combining coordination with intercalation. *Organometallics* **2012**, *31*, 7031–7039. [CrossRef]

37. Bailly, C.; Braña, M.; Waring, M.J. Sequence-Selective Intercalation of Antitumour Bis-Naphthalimides into DNA. *Eur. J. Biochem.* **2018**, *240*, 195–208. [CrossRef]

38. Ang, W.H.; Parker, L.J.; De Luca, A.; Juillerat-Jeanneret, L.; Morton, C.J.; Lo, B.M.; Parker, M.W.; Dyson, P.J. Rational design of an organometallic glutathione transferase inhibitor. *Angew. Chemi. Int. Ed.* **2009**, *48*, 3854–3857. [CrossRef] [PubMed]

39. Chakree, K.; Ovatlarnporn, C.; Dyson, P.J.; Ratanaphan, A. Altered dna binding and amplification of human breast cancer suppressor gene BRCA1 induced by a novel antitumor compound, [Ru(η^6-p-phenylethacrynate)Cl$_2$(pta)]. *Int. J. Mol. Sci.* **2012**, *13*, 13183–13202. [CrossRef] [PubMed]

40. Chatterjee, S.; Biondi, I.; Dyson, P.J.; Bhattacharyya, A. A bifunctional organometallic ruthenium drug with multiple modes of inducing apoptosis. *J. Biol. Inorg. Chem.* **2011**, *16*, 715–724. [CrossRef] [PubMed]

41. Holtkamp, H.U.; Movassaghi, S.; Morrow, S.J.; Kubanik, M.; Hartinger, C.G. Metallomic study on the metabolism of RAPTA-C and cisplatin in cell culture medium and its impact on cell accumulation. *Metallomics* **2018**, *10*, 455–462. [PubMed]

42. Weiss, A.; Ding, X.; van Beijnum, J.R.; Wong, I.; Wong, T.J.; Berndsen, R.H.; Dormond, O.; Dallinga, M.; Shen, L.; Schlingemann, R.O.; et al. Rapid optimization of drug combinations for the optimal angiostatic treatment of cancer. *Angiogenesis* **2015**, *18*, 233–244. [CrossRef] [PubMed]

43. Weiss, A.; Berndsen, R.H.; Ding, X.; Ho, C.M.; Dyson, P.J.; Van Den Bergh, H.; Griffioen, A.W.; Nowak-Sliwinska, P. A streamlined search technology for identification of synergistic drug combinations. *Sci. Rep.* **2015**, *5*, 1–11. [CrossRef] [PubMed]

44. Alessio, E. Thirty Years of the Drug Candidate NAMI-A and the Myths in the Field of Ruthenium Anticancer Compounds: A Personal Perspective. *Eur. J. Inorg. Chem.* **2017**, 1549–1560. [CrossRef]

45. Clarke, M.J. Ruthenium metallopharmaceuticals. *Coord. Chem. Rev.* **2003**, *236*, 209–233. [CrossRef]

46. Trondl, R.; Heffeter, P.; Kowol, C.R.; Jakupec, M.A.; Berger, W.; Keppler, B.K. NKP-1339, the first ruthenium-based anticancer drug on the edge to clinical application. *Chem. Sci.* **2014**, *5*, 2925–2932. [CrossRef]

47. Artner, C.; Holtkamp, H.U.; Hartinger, C.G.; Meier-Menches, S.M. Characterizing activation mechanisms and binding preferences of ruthenium metallo-prodrugs by a competitive binding assay. *J. Inorg. Biochem.* **2017**, *177*, 322–327. [CrossRef] [PubMed]

48. Mestroni, G.; Alessio, E.; Sava, G.; Pacor, S.; Coluccia, M.; Boccarelli, A. Water-soluble ruthenium(III)-dimethyl sulfoxide complexes: Chemical behaviour and pharmaceutical properties. *Met. Based. Drugs* **1993**, *1*, 41–63. [CrossRef] [PubMed]

49. Leijen, S.; Burgers, S.A.; Baas, P.; Pluim, D.; Tibben, M.; Van Werkhoven, E.; Alessio, E.; Sava, G.; Beijnen, J.H.; Schellens, J.H.M. Phase I/II study with ruthenium compound NAMI-A and gemcitabine in patients with non-small cell lung cancer after first line therapy. *Invest. New Drugs* **2015**, *33*, 201–214. [CrossRef] [PubMed]

50. Dyson, P.J.; Sava, G. Metal-based antitumour drugs in the post genomic era. *Dalt. Trans.* **2006**, 1929–1933. [CrossRef] [PubMed]

51. Morbidelli, L.; Donnini, S.; Filippi, S.; Messori, L.; Piccioli, F.; Orioli, P.; Sava, G.; Ziche, M. Antiangiogenic properties of selected ruthenium(III) complexes that are nitric oxide scavengers. *Br. J. Cancer* **2003**, *88*, 1484–1491. [CrossRef] [PubMed]

52. Feliers, D.; Chen, X.; Akis, N.; Choudhury, G.G.; Madaio, M.; Kasinath, B.S. VEGF regulation of endothelial nitric oxide synthase in glomerular endothelial cells. *Kidney Int.* **2005**, *68*, 1648–1659. [CrossRef] [PubMed]

53. Oszajca, M.; Kuliś, E.; Stochel, G.; Brindell, M. Interaction of the NAMI-A complex with nitric oxide under physiological conditions. *New J. Chem.* **2014**, *38*, 3386–3394. [CrossRef]

54. Lai, H.; Zhao, Z.; Li, L.; Zheng, W.; Chen, T. Antiangiogenic ruthenium(II) benzimidazole complexes, structure-based activation of distinct signaling pathways. *Metallomics* **2015**, *7*, 439–447. [CrossRef] [PubMed]

55. Zhang, W.; Liu, H.T. MAPK signal pathways in the regulation of cell proliferation in mammalian cells. *Cell Res.* **2002**, *12*, 9–18. [CrossRef] [PubMed]

56. Cargnello, M.; Roux, P.P. Activation and Function of the MAPKs and Their Substrates, the MAPK-Activated Protein Kinases Marie. *Microbiol. Mol. Biol. Rev.* **2011**, 50–83. [CrossRef] [PubMed]

57. Bergamo, A.; Pelillo, C.; Chambery, A.; Sava, G. Influence of components of tumour microenvironment on the response of HCT-116 colorectal cancer to the ruthenium-based drug NAMI-A. *J. Inorg. Biochem.* **2017**, *168*, 90–97. [CrossRef] [PubMed]

58. Pillozzi, S.; Gasparoli, L.; Stefanini, M.; Ristori, M.; D'Amico, M.; Alessio, E.; Scaletti, F.; Becchetti, A.; Arcangeli, A.; Messori, L. NAMI-A is highly cytotoxic toward leukaemia cell lines: Evidence of inhibition of KCa 3.1 channels. *Dalt. Trans.* **2014**, *43*, 12150–12155. [CrossRef] [PubMed]

59. Urrego, D.; Tomczak, A.P.; Zahed, F.; Stühmer, W.; Pardo, L.A. Potassium channels in cell cycle and cell proliferation. *Philos. Trans. R. Soc. B Biol. Sci.* **2014**, *369*, 1–9. [CrossRef] [PubMed]

60. Grössinger, E.M.; Weiss, L.; Zierler, S.; Rebhandl, S.; Krenn, P.W.; Hinterseer, E.; Schmölzer, J.; Asslaber, D.; Hainzl, S.; Neureiter, D.; et al. Targeting proliferation of chronic lymphocytic leukemia (CLL) cells through KCa3.1 blockade. *Leukemia* **2014**, *28*, 954–958. [CrossRef] [PubMed]

61. Schwab, A.; Fabian, A.; Hanley, P.J.; Stock, C. Role of Ion Channels and Transporters in Cell Migration. *Physiol. Rev.* **2012**, *92*, 1865–1913. [CrossRef] [PubMed]

62. Bulk, E.; Ay, A.S.; Hammadi, M.; Ouadid-Ahidouch, H.; Schelhaas, S.; Hascher, A.; Rohde, C.; Thoennissen, N.H.; Wiewrodt, R.; Schmidt, E.; et al. Epigenetic dysregulation of $K_{Ca}3.1$ channels induces poor prognosis in lung cancer. *Int. J. Cancer* **2015**, *137*, 1306–1317. [CrossRef] [PubMed]

63. Maréchal, A.; Zou, L. DNA Damage Sensing by the ATM and ATR Kinases. *Cold Spring Harb. Perspect. Biol.* **2013**, *5*, a12716. [CrossRef] [PubMed]

64. Bergamo, A.; Delfino, R.; Casarsa, C.; Sava, G. CDK1 Hyperphosphorylation Maintenance Drives the Time-course of G2-M Cell Cycle Arrest after Short Treatment with NAMI-A in Kb Cells. *Anticancer. Agents Med. Chem.* **2012**, *12*, 949–958. [CrossRef] [PubMed]

65. Pelillo, C.; Mollica, H.; Eble, J.A.; Grosche, J.; Herzog, L.; Codan, B.; Sava, G.; Bergamo, A. Inhibition of adhesion, migration and of α5β1 integrin in the HCT-116 colorectal cancer cells treated with the ruthenium drug NAMI-A. *J. Inorg. Biochem.* **2016**, *160*, 225–235. [CrossRef] [PubMed]

66. Vacca, A.; Bruno, M.; Boccarelli, A.; Coluccia, M.; Ribatti, D.; Bergamo, A.; Garbisa, S.; Sartor, L.; Sava, G. Inhibition of endothelial cell functions and of angiogenesis by the metastasis inhibitor NAMI-A. *Br. J. Cancer* **2002**, *4*, 993–998. [CrossRef] [PubMed]

67. Gialeli, G.; Theocharis, A.D.; Karamanos, N.K. Roles of MMP in cancer progression and their pharmacological targeting. *FEBS J.* **2011**, *278*, 16–27. [CrossRef] [PubMed]

68. Gomes, L.R.; Terra, L.F.; Wailemann, R.A.M.; Labriola, L.; Sogayar, M.C. TGF-β1 modulates the homeostasis between MMPs and MMP inhibitors through p38 MAPK and ERK1/2 in highly invasive breast cancer cells. *BMC Cancer* **2012**, *12*, 1–15. [CrossRef] [PubMed]

69. Sava, L.B.A.M.G.; Bergamo, A. Effects of the ruthenium-based drug NAMI-A on the roles played by TGF-β1 in the metastatic process. *J. Biol. Inorg. Chem.* **2015**, *20*, 1163–1173.

70. Schoenhacker-Alte, B.; Mohr, T.; Pirker, C.; Kryeziu, K.; Kuhn, P.S.; Buck, A.; Hofmann, T.; Gerner, C.; Hermann, G.; Koellensperger, G.; et al. Sensitivity towards the GRP78 inhibitor KP1339/IT-139 is characterized by apoptosis induction via caspase 8 upon disruption of ER homeostasis. *Cancer Lett.* **2017**, *404*, 79–88. [CrossRef] [PubMed]

71. Kapitza, S.; Pongratz, M.; Jakupec, M.A.; Heffeter, P.; Berger, W.; Lackinger, L.; Keppler, B.K.; Marian, B. Heterocyclic complexes of ruthenium(III) induce apoptosis in colorectal carcinoma cells. *J. Cancer Res. Clin. Oncol.* **2005**, *131*, 101–110. [CrossRef] [PubMed]

72. Chen, H.; Parkinson, J.; Morris, R.E.; Sadler, P.J. Highly selective binding of organometallic ruthenium ethylenediamine complexes to nucleic acids: Novel recognition mechanisms. *J. Am. Chem. Soc.* **2003**, *125*, 173–186. [CrossRef] [PubMed]

73. Dickson, N.R.; Jones, S.F.; Burris, H.A.; Ramanathan, R.K.; Weiss, G.J.; Infante, J.R.; Bendell, J.C.; McCulloch, W.; Von Hoff, D.D. A phase I dose-escalation study of NKP-1339 in patients with advanced solid tumors refractory to treatment. *J. Clin. Oncol.* **2011**, *29*, 2607. [CrossRef]

74. Liu, J.; Wang, Z. Increased oxidative stress as a selective anticancer therapy. *Oxid. Med. Cell. Longev.* **2015**, *2015*. [CrossRef] [PubMed]

75. Diehn, M.; Cho, R.W.; Lobo, N.; Kalisky, T.; Dorie, M.J.; Kulp, A.N.; Qian, D.; Lam, J.S.; Ailles, L.E.; Wong, M.; et al. Association of reactive oxygen species levels and radioresistance in cancer stem cells. *Nature* **2009**, *458*, 780–783. [CrossRef] [PubMed]

76. Gorrini, C.; Harris, I.S.; Mak, T.W. Modulation of oxidative stress as an anticancer strategy. *Nat. Rev. Drug Discov.* **2013**, *12*, 931–947. [CrossRef] [PubMed]

77. Coverdale, J.P.C.; Romero-Canelón, I.; Sanchez-Cano, C.; Clarkson, G.J.; Habtemariam, A.; Wills, M.; Sadler, P.J. Asymmetric transfer hydrogenation by synthetic catalysts in cancer cells. *Nat. Chem.* **2018**, *10*, 347–354. [CrossRef] [PubMed]

78. Soldevila-Barreda, J.J.; Romero-Canelón, I.; Habtemariam, A.; Sadler, P.J. Transfer hydrogenation catalysis in cells as a new approach to anticancer drug design. *Nat. Commun.* **2015**, *6*, 6582. [CrossRef] [PubMed]

79. Flocke, L.S.; Trondl, R.; Jakupec, M.A.; Keppler, B.K. Molecular mode of action of NKP-1339—A clinically investigated ruthenium-based drug—Involves ER- and ROS-related effects in colon carcinoma cell lines. *Invest. New Drugs* **2016**, *34*, 261–268. [CrossRef] [PubMed]

80. Heffeter, P.; Atil, B.; Kryeziu, K.; Groza, D.; Koellensperger, G.; Körner, W.; Jungwirth, U.; Mohr, T.; Keppler, B.K.; Berger, W. The ruthenium compound KP1339 potentiates the anticancer activity of sorafenib in vitro and in vivo. *Eur. J. Cancer* **2013**, *49*, 3366–3375. [CrossRef] [PubMed]

81. Thornton, T.M.; Rincon, M. Non-classical p38 map kinase functions: Cell cycle checkpoints and survival. *Int. J. Biol. Sci.* **2009**, *5*, 44–52. [CrossRef] [PubMed]

82. Zarubin, T.; Han, J. Activation and signaling of the p38 MAP kinase pathway. *Cell Res.* **2005**, *15*, 11–18. [CrossRef] [PubMed]

83. Kowaltowski, A.J.; Castilho, R.F.; Vercesi, A.E. Mitochondrial permeability transition and oxidative stress. *FEBS Lett.* **2001**, *495*, 12–15. [CrossRef]

84. Indran, I.R.; Tufo, G.; Pervaiz, S.; Brenner, C. Recent advances in apoptosis, mitochondria and drug resistance in cancer cells. *Biochim. Biophys. Acta* **2011**, *1807*, 735–745. [CrossRef] [PubMed]

85. Hartinger, C.G.; Zorbas-Seifried, S.; Jakupec, M.A.; Kynast, B.; Zorbas, H.; Keppler, B.K. From bench to bedside—Preclinical and early clinical development of the anticancer agent indazolium *trans*-[tetrachlorobis(1*H*-indazole)ruthenate(III)] (KP1019 or FFC14A). *J. Inorg. Biochem.* **2006**, *100*, 891–904. [CrossRef] [PubMed]

86. Kalkavan, H.; Green, D.R. MOMP, cell suicide as a BCL-2 family business. *Cell Death Differ.* **2018**, *25*, 46–55. [CrossRef] [PubMed]

87. Ma, Q. Role of Nrf2 in Oxidative Stress and Toxicity. *Annu. Rev. Pharmacol. Toxicol.* **2013**, *53*, 401–426. [CrossRef] [PubMed]

88. Sano, R.; Reed, J.C. ER stress-induced cell death mechanisms. *Biochim. Biophys. Acta, Mol. Cell. Res.* **2013**, *1833*, 3460–3470. [CrossRef] [PubMed]

89. Gifford, J.B.; Huang, W.; Zeleniak, A.E.; Hindoyan, A.; Wu, H.; Donahue, T.R.; Hill, R. Expression of GRP78, Master Regulator of the Unfolded Protein Response, Increases Chemoresistance in Pancreatic Ductal Adenocarcinoma. *Mol. Cancer Ther.* **2016**, *15*, 1043–1052. [CrossRef] [PubMed]

90. Fiskus, W.; Saba, N.; Shen, M.; Ghias, M.; Liu, J.; Gupta, S.D.; Chauhan, L.; Rao, R.; Gunewardena, S.; Schorno, K.; et al. Auranofin induces lethal oxidative and endoplasmic reticulum stress and exerts potent preclinical activity against chronic lymphocytic leukemia. *Cancer Res.* **2014**, *74*, 2520–2532. [CrossRef] [PubMed]

91. Liu, L.-J.; Wang, W.; Huang, S.-Y.; Hong, Y.; Li, G.; Lin, S.; Tian, J.; Cai, Z.; Wang, H.-M.D.; Ma, D.-L.; et al. Inhibition of the Ras/Raf interaction and repression of renal cancer xenografts in vivo by an enantiomeric iridium(III) metal-based compound. *Chem. Sci.* **2017**, *8*, 4756–4763. [CrossRef] [PubMed]

92. Ang, W.H.; Casini, A.; Sava, G.; Dyson, P.J. Organometallic ruthenium-based antitumor compounds with novel modes of action. *J. Organomet. Chem.* **2011**, *696*, 989–998. [CrossRef]

93. Mitrović, A.; Kljun, J.; Sosič, I.; Gobec, S.; Turel, I.; Kos, J. Clioquinol–ruthenium complex impairs tumour cell invasion by inhibiting cathepsin B activity. *Dalt. Trans.* **2016**, *45*, 16913–16921. [CrossRef] [PubMed]

94. Jungwirth, U.; Kowol, C.R.; Keppler, B.K.; Hartinger, C.G.; Berger, W.; Heffeter, P. Anticancer activity of metal complexes: Involvement of redox processes. *Antioxidants Redox Signal.* **2011**, *15*, 1085–1127. [CrossRef] [PubMed]

95. Palermo, G.; Magistrato, A.; Riedel, T.; von Erlach, T.; Davey, C.A.; Dyson, P.J.; Rothlisberger, U. Fighting Cancer with Transition Metal Complexes: From Naked DNA to Protein and Chromatin Targeting Strategies. *ChemMedChem* **2016**, *11*, 1199–1210. [CrossRef] [PubMed]

96. Adhireksan, Z.; Davey, G.E.; Campomanes, P.; Groessl, M.; Clavel, C.M.; Yu, H.; Nazarov, A.A.; Yeo, C.H.F.; Ang, W.H.; Dröge, P.; et al. Ligand substitutions between ruthenium–cymene compounds can control protein versus DNA targeting and anticancer activity. *Nat. Commun.* **2014**, *5*, 3462. [CrossRef]

97. Wu, B.; Ong, M.S.; Groessl, M.; Adhireksan, Z.; Hartinger, C.G.; Dyson, P.J.; Davey, C.A. A ruthenium antimetastasis agent forms specific histone protein adducts in the nucleosome core. *Chem. Eur. J.* **2011**, *17*, 3562–3566. [CrossRef] [PubMed]

98. Meier-Menches, S.M.; Gerner, C.; Berger, W.; Hartinger, C.G.; Keppler, B.K. Structure-activity relationships for ruthenium and osmium anticancer agents-towards clinical development. *Chem. Soc. Rev.* **2018**, *47*, 909–928. [CrossRef]

 *inorganics*

Communication

Functionalizing NaGdF$_4$:Yb,Er Upconverting Nanoparticles with Bone-Targeting Phosphonate Ligands: Imaging and In Vivo Biodistribution

Silvia Alonso-de Castro [1,2], Emmanuel Ruggiero [1,†], Aitor Lekuona Fernández [1], Unai Cossío [1], Zuriñe Baz [1], Dorleta Otaegui [1], Vanessa Gómez-Vallejo [1], Daniel Padro [1], Jordi Llop [1,*] and Luca Salassa [2,3,*]

[1] CIC biomaGUNE, Paseo de Miramón 182, 20014 Donostia, Spain; silviaalonsodc@gmail.com (S.A.-d.C.); emmanuel.ruggiero@basf.com (E.R.); alekuona@cicbiomagune.es (A.L.F.); ucossio@cicbiomagune.es (U.C.); zbaz@cicbiomagune.es (Z.B.); dotaegui@cicbiomagune.es (D.O.); vgomez@cicbiomagune.es (V.G.-V.); dpadro@cicbiomagune.es (D.P.)
[2] Donostia International Physics Center (DIPC), Manuel Lardizabal Ibilbidea 4, 20018 Donostia, Spain
[3] Ikerbasque, Basque Foundation for Science, 48011 Bilbao, Spain
*   Correspondence: jllop@cicbiomagune.es (J.L.); lsalassa@dipc.org (L.S.)
†   Current address: BASF SE, Department of Material Physics & Analytics, Carl-Bosch-Strasse 38, 67056 Ludwigshafen, Germany.

Received: 11 December 2018; Accepted: 19 April 2019; Published: 30 April 2019

Abstract: Lanthanide-doped upconverting nanoparticles (UCNPs) transform near infrared light (NIR) into higher-energy UV and visible light by multiphotonic processes. Owing to such unique feature, UCNPs have found application in optical imaging and have been investigated for the NIR light activation of prodrugs, including transition metal complexes of interest in photochemotherapy. Besides, UCNPs also function as magnetic resonance imaging (MRI) contrast agents and positron emission tomography (PET) probes when labelled with radionuclides such as ^{18}F. In this contribution, we report on a new series of phosphonate-functionalized NaGdF$_4$:Yb,Er UCNPs that show affinity for hydroxyapatite (inorganic constituent of bones), and we discuss their potential as bone targeting multimodal (MRI/PET) imaging agents. In vivo biodistribution studies of ^{18}F-labelled NaGdF$_4$:Yb,Er UCNPs in rats indicate that surface functionalization with phosphonates favours the accumulation of nanoparticles in bones over time. PET results reveal leakage of ^{18}F$^-$ for phosphonate-functionalized NaGdF$_4$:Yb,Er and control nanomaterials. However, Gd was detected in the femur for phosphonate-capped UCNPs by ex vivo analysis using ICP-MS, corresponding to 6–7% of the injected dose.

Keywords: upconverting nanoparticles; phosphonates; bones; imaging; MRI; PET

---

1. Introduction

In recent years, our group developed a number of unconventional approaches for the photoactivation of metal-based anticancer prodrug complexes [1–3], including the use of lanthanide-doped upconverting nanoparticles (UCNPs) as near infrared (NIR) light triggers [4–6]. In this context, we and others reported on the capacity of UCNPs to prompt the photochemistry of several transition metal complexes such as Ru, Pt, Mn and Fe [7]. However, UCNPs have much more to offer besides their photophysics [8] since they can simultaneously act as multimodal imaging agents and be employed in theranostics. Gd-containing UCNPs are magnetic resonance imaging (MRI) contrast agents, while labelling with radionuclides such as ^{18}F and ^{153}Sm yields positron emission and single photon emission tomography (PET and SPECT) probes, respectively [9,10].

In this work, we explored the potential of phosphonate-functionalized $NaGdF_4$:Yb,Er UCNPs to behave as MRI and PET imaging tools for bones, motivated by the long-term prospect of designing theranostic nanoconstructs for the photoactivated delivery of metal-based drugs. Actually, recent results highlight the possibility of adopting photodynamic therapy and related strategies for the treatment of spinal metastases [11] and to ablate tumors within vertebral bone [12]. Therefore, phosphonate ligands were selected for functionalization of UCNPs because of their capacity of accumulating in bones. The use of bisphosphonates for targeting bones has been recently explored in combination of superparamagnetic iron oxide nanoparticles (SPIONs) [13]. Several phosphonates, e.g., alendronate, etidronate, are used in the treatment of osteoporosis and investigated for their use against bone cancers [14]. Furthermore, a recent study by Li et al. [15] demonstrated that phosphonates have high affinity for the surface of UCNPs and show the ability of stabilizing colloidal dispersions of UCNPs in acidic conditions (e.g., in lysosomes), thus lowering their pro-inflammatory effects both in vitro and in vivo. Lastly, we recently demonstrated the functionalization and UCNP loading of photoactivatable Ru complexes modified with a phosphonate arm [4].

Herein, we investigated a set of 10 nm $NaGdF_4$:Yb,Er NPs functionalized with three phosphonic acids, namely etidronic acid, alendronic acid and nitrilo(trimethylphosphonic acid) (3P), as MRI and PET agents and evaluated their biodistribution and in vivo accumulation capacity in bones.

2. Results and Discussion

2.1. Synthesis and Surface Decoration of UCNPs $NaGdF_4$:Yb,Er

Oleate-capped $NaGdF_4$:Yb,Er nanoparticles were synthesized by thermal decomposition following a procedure previously reported by us and others [6,16]. The obtained UCNPs had homogeneous distribution in size and shape, displaying an average diameter of 9.5 ± 1.1 nm according to TEM measurements (Figures 1a and S1). The relatively small size of these UCNPs is advantageous for MRI because of their high surface area, increasing the amount of Gd^{3+} ions accessible to water molecules and hence the relaxivity per Gd atom [17–19]. Conversely, small UCNPs such as the ones prepared in this work have modest upconversion properties due to solvent deactivation processes (Figure S2) [20].

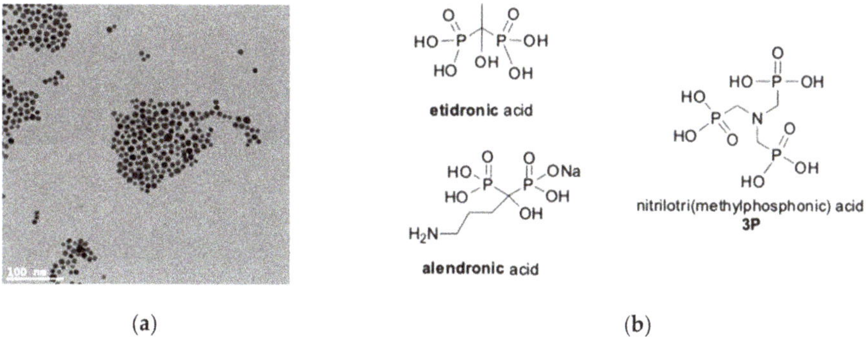

(a)   (b)

Figure 1. (a) TEM image of Gd-UCNPs capped with oleic acid; (b) chemical structure of the phosphonate capping ligands.

Next, the surface of the oleate-capped nanoparticles was decorated with phosponates to achieve aqueous dispersibility and to assess the capacity of these ligands to drive accumulation of UCNPs into bones. Surface functionalization was performed by ligand exchange [21]. Accordingly to the different ligand solubility, oleic acid was exchanged for etidronic acid, alendronic acid or nitrilo(trimethylphosphonic acid) (3P) in a $CHCl_3$:H_2O mixture (Figure 1b), affording water dispersible nanoparticles where the negatively charged phosphonates are electrostatically tethered to the positive surface of the Gd-UCNPs.

All phosphonate-decorated UCNPs were characterized with FT-IR (Figure S3) and ICP-MS (Table S1) to confirm ligand functionalization and determine the quantity of Gd^{3+} per gram of functionalized nanoparticle.

2.2. MRI Characterization and Interaction with Hydroxyapatite

UCNPs serve as contrast agents for T1-weighted MR imaging. T1 relaxation times in UCNPs depend on the accessibility of Gd^{3+} ions on the surface of the nanomaterials and can be strongly affected by different surface coatings [17]. For this reason, we investigated the longitudinal r_1 relaxivity of the three phosphonate-functionalized $NaGdF_4$:Yb,Er UCNPs and how this parameter responded to different media, including human serum as a mimic of blood components.

Relaxivity values for the three UCNPs were determined at 1.5 T (Figures S4–S9 and Table S2) and are summarized in Table 1. In aqueous solution, UCNPs coated with etidronate have the highest r_1 (3.47 mM^{-1} s^{-1}) compared to the alendronate and 3P analogues ($r_1 < 0.70$ mM^{-1} s^{-1}). In the case of etidronate UCNPs, a significant increase in r_1 was observed upon changing the medium, while this was not the case for the other phosphonates.

Table 1. Longitudinal r_1 relaxivity values determined at 1.5 T (37 °C) for the three phosphonate-functionalized UCNPs in different media.

	r_1 (mM^{-1} s^{-1})		
	H_2O	MES Buffer	Human Serum
Gd-UCNP-etidronate	3.47 ± 0.02	8.47 ± 0.02	9.73 ± 0.02
Gd-UCNP-alendronate	0.59 ± 0.05	0.66 ± 0.03	0.27 ± 0.03
Gd-UCNP-3P	0.70 ± 0.01	0.69 ± 0.07	0.76 ± 0.05

In MES buffer, the r_1 value of etidronate-capped UCNPs is five units higher than in water. In human serum r_1 reached a value of 9.73 ± 0.02 mM^{-1} s^{-1}, that is twice the relaxivity of Gadobutrol (gadovist) in water (3.30 mM^{-1} s^{-1}) and in human blood plasma (4.70 mM^{-1} s^{-1}) [22,23].

We also evaluated the capacity of the etidronate UCNPs to provide contrast by acquiring T1 weighted MRI images at 7 T in H_2O (Figure S10) and in human serum (Figure 2) using a phantom. As shown in Figure 2, good contrast levels were obtained for etidronate UCNPs already at Gd concentrations of 0.25 mM.

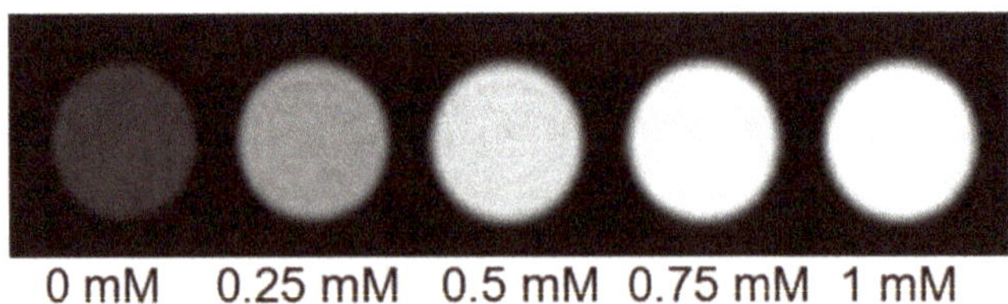

Figure 2. 7T MRI phantom (37 °C) of etidronate-capped UCNPs in human serum.

The different behaviour of etidronate UCNPs in terms of relaxivity compared to the other phosphonate analogues can be rationalized in terms of colloidal stability. As shown by results obtained from DLS (dynamic light scattering, Figure S11), etidronate favours the dispersion of UCNPs, particularly in MES buffer, while nanoparticles capped with alendronate or 3P tend to aggregate and eventually precipitate in solution (Table S3), ultimately displaying a reduced number of Gd atoms exposed to solvent and hence lower r_1 values.

Hydroxyapatite (HA) is a recognized analogue of bone mineral [14] and was employed as model to evaluate the affinity for bones of our phosphonate-functionalized UCNPs. To this aim, we performed relaxivity r_1 measurements on solutions containing different concentrations of phosphonate-capped

UCNPs that were incubated in the presence of 12.5 mg/mL HA (insoluble). In all cases, binding of UCNPs to HA resulted in lower r_1 values compared to samples without HA. For example, etidronate-UCNPs showed remarkable affinity with HA already after 2 h of incubation. Figure 3a clearly demonstrated that UCNPs incubated with HA do not induce any T1 shortening effect in the medium until a concentration of 1 mM Gd is reached. At higher concentrations, the T1 shortening effects of etidronate-UCNPs are restored due to the saturation of HA surface. In order to quantify the affinity of UCNPs for HA, we compared relaxivity values for 2 mM Gd solutions of UCNPs in the presence and absence of the mineral. For etidronate nanoparticles a 75% affinity was found in human serum (Figure 3b). Similar behaviour was obtained for alendronate and 3P UCNPs upon 24 h of incubation with HA in human serum (Figure S12). However, the capacity of these nanomaterials to interact with HA was significantly lower (Figure 3b). For comparison, the same type of measurements was performed using UCNPs functionalized with EDTA (ethylenediaminetetraacetic acid) which decorates the nanoparticles through carboxylate groups instead of phosphonates. EDTA UCNPs have an r_1 value in water of 3.9 mM^{-1} s^{-1} that is not modified in the presence of HA (Figure 3b), confirming that affinity of our UCNPs for this mineral is associated to phosphonate groups.

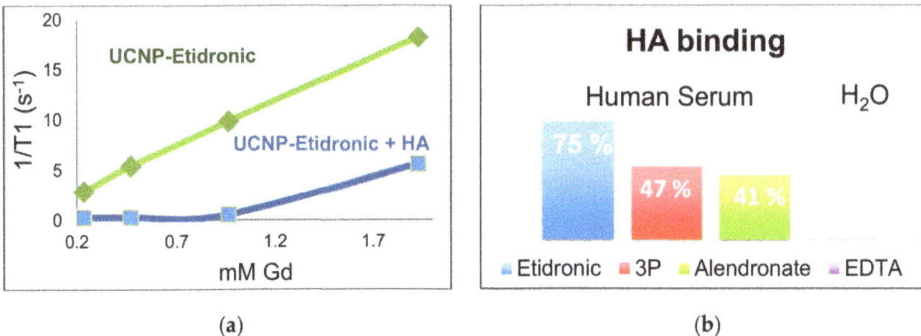

Figure 3. (**a**) 1/T1 profile of etidronate UCNPs at 1.5 T (37 °C) before and after addition of 12.5 mg/mL hydroxyapatite (HA) in human serum; (**b**) HA binding of UCNPs capped with etidronate, alendronate, 3P and EDTA in human serum and H$_2$O.

2.3. Biodistribution Study by Radiolabeling UCNPs

Nuclear imaging, and in particular PET, is routinely employed in the clinics to study the whole-body distribution of radiolabeled species with high sensitivity [24]. ^{18}F-labelling of UCNPs has demonstrated to be an effective strategy to monitor the biodistribution of these nanomaterials in vivo [25–27], often in combination with other imaging modalities [28]. In this study, the observation of the UCNPs in vivo adhesion to the bones using MRI was discarded due to the intrinsic low sensitivity of the technique added to the specific characteristics of the bones: low water content, very short T2 relaxation times due to its rigid nature and susceptibility effects. Adopting the procedure reported by Li and coworkers [25], we radiolabelled phosphonate UCNPs by exposing the nanoparticles to [^{18}F]F$^-$ ions with the aim of visualizing via PET their biodistribution in healthy rodents and gain information about their eventual accumulation in bones. As previously described [25], this straightforward synthetic method afforded high incorporation ratios, with non-decay corrected radiochemical yields of 36%, 43% and 43% for citrate, 3P and alendronate-UCNPs, respectively, resulting in estimated specific activity values of 105 MBq/mg for citrate-UCNPs and of 126 MBq/mg for 3P and alendronate-UCNPs. Only etidronate functionalised UCNPs resulted in low radiochemical yields (<10%), and therefore were no further investigated in vivo. Nanoparticles decorated with citrate were used as control since this carboxylate ligand did not show any specific affinity to HA (as confirmed by MRI, data not shown). The stability (Figure S13) and radiochemical purity of alendronate, 3P and citrate UCNPs were confirmed by radio-TLC at different time points. The radiochemical purity of the labelled particles was >95% at $t - 5$ h of incubation in physiologic saline solution (see Figure S14 for representative example).

Biodistribution and clearance of the ^{18}F-labeled UCNPs was obtained by PET imaging (Figure 4) after intravenous injection in the tail of 100 µL of a saline solution of the alendronate, 3P and citrate UCNPs (0.5 mg/mL). Direct inspection of the images (see Figure 4) clearly showed initial accumulation in the lungs and the liver, probably due to sequestration of the NPs by the mononuclear phagocyte system (MPS) which might be a consequence of aggregation of the NPs under physiologic conditions. At 3 h and 6 h after nanoparticle injection, ^{18}F radioactivity was also clearly identified in the bones and joints, independently of the ligand coating. These initial results were confirmed by image quantification (Figure S15). All NPs, irrespective of the surface decoration, exhibited initial accumulation in the lungs, with values close to 4% of injected dose per cm^3 of tissue (%ID/cc). Initial localization was also observed in the liver and kidneys, which was paralleled by high accumulation in the urine (results not shown) suggesting partial elimination of the NPs via urinary excretion or detachment of the label.

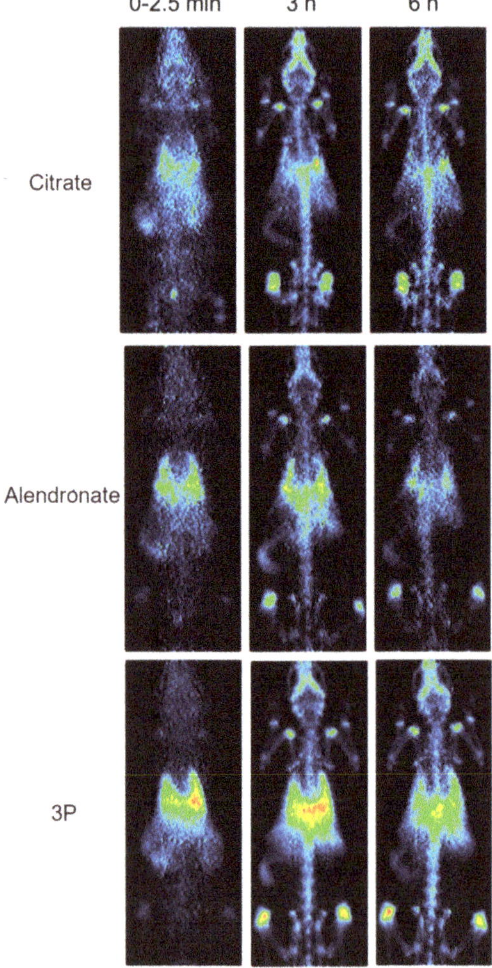

Figure 4. Biodistribution of ^{18}F-labelled phosphonate and control UCNPs at different time points after intravenous injection (0.5 mg/mL).

At early times, citrate nanoparticles (controls) displayed a pronounced accumulation in bladder and kidney compared to the phosphonate-functionalized UCNPs, suggesting that they were excreted more rapidly.

A progressive accumulation of the radioactivity in the bones was observed, which reached values close to 2%ID/cc in the case of citrate UCNPs. This result, confirmed by dissection/γ-counting (Figure 5a), was unexpected considering that the citrate had no affinity for HA, and suggested that $^{18}F^-$ ions could leak partially from the nanoparticles, hence masking the real uptake of the labelled particles in bone tissue. For this reason, we further investigated the accumulation of NPs in bone by ex vivo analysis using ICP-MS. With that aim, the amount of Gd in the femurs was quantified (Figure 5b). The samples to be analysed by ICP-MS were obtained after digestion of the femurs harvested from the animals with HNO_3. In the case of the citrate control (Figure 5b), no sign of Gd was detected by ICP-MS indicating that the totality of the $[^{18}F]F^-$ signal in the femur could be associated to free $[^{18}F]F^-$. Conversely, Gd was detected by ICP-MS in the femur for phosphonate-capped UCNPs, corresponding to 6–7% (Figure 5b) of the injected dose per gram of bone. Although these results confirmed the leakage of $^{18}F^-$ ions for alendronate, 3P and citrate UCNPs, they also indicated that phosphonate ligands were capable of promoting accumulation of UCNPs in bones to some extent, while the control nanomaterial did not. Although further studies are required, we speculate that phosphonate-capped UCNPs can escape the endothelium and reach the skeleton via a mechanism known as nanomaterial induced endothelial leakiness. In this process, nanomaterials enter and enlarge the nanosized gaps between microvascular capillary endothelial cells favouring their leakage to other tissue sites [29,30].

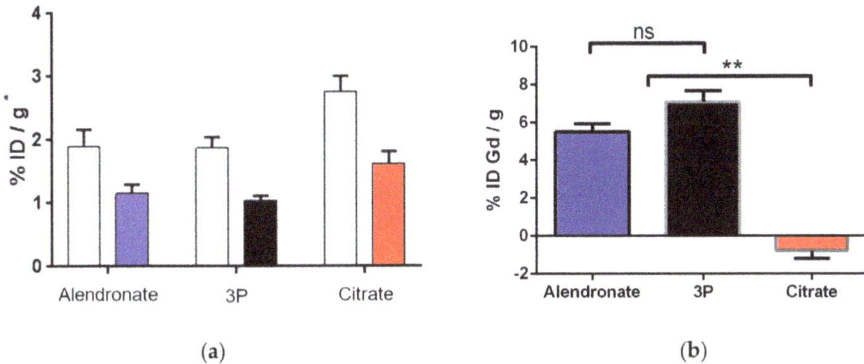

(a)   (b)

Figure 5. (**a**) ^{18}F detection in bones in vivo (color) and ex vivo (white), where in vivo units are % ID/mL (where ID is the injected dose), and corresponds to all skeleton, while ex vivo is only femur; (**b**) Gd detection in femur bones by ICP-MS. ICP-MS data is presented as mean ± SEM of two independent measurements. * = statistically significant difference ($p < 0.05$); ** = statistically significant difference ($p < 0.01$), ns = non-significant by one-way ANOVA followed by Tukey's test.

3. Materials and Methods

3.1. Materials

All chemical reagents, yttrium(III) acetate hydrate, ytterbium(III) acetate tetrahydrate, erbium(III) chloride hexahydrate, 1-octadecene (technical grade), oleic acid (technical grade), sodium hydroxide, ammonium fluoride, etidronic acid, alendronic acid, nitrilo(trimethylphosphonic acid) and solvents were purchased from Sigma-Aldrich (St. Louis, MO, USA).

3.2. Synthesis of UCNPs NaGdF₄:Yb,Er

The $NaGdF_4$:Yb,Er (rare earth element ratio 78/20/2 mol %) synthesis was carried out following a slightly modified literature procedure [6]. In a typical synthesis (total rare earth amount, 3 mmol), gadolinium (III) chloride hydrate (618 mg, 2.34 mmol), ytterbium(III) acetate tetrahydrate (253.8 mg, 0.60 mmol) and erbium (III) chloride hexahydrate (23 mg, 0.06 mmol) were dissolved in 1-octadecene (25 mL) and oleic acid (15 mL) in a 100 mL three-neck round-bottom flask with coil condenser. The suspension was heated up to 120 °C with a slow temperature ramp of 3.2 °C/min under stirring and vacuum. Once the reaction mixture reached such temperature, it was kept in these conditions for 30 min in order to form a clear solution and eliminate residual water and oxygen. The system was then allowed to cool to 50 °C under a flow of nitrogen gas.

A solution of sodium hydroxide (300 mg, 7.5 mmol) and ammonium fluoride (444 mg, 12.0 mmol) in methanol (8 mL) was added to the reaction flask dropwise in 10 min. The subsequent cloudy solution was stirred for 30 min at 50 °C and for 30 min at 70 °C under nitrogen to evaporate the entire amount of methanol from the solution. Successively, the system was heated up to 300 °C with a fast temperature ramp of 13.5 °C/min under stirring and nitrogen and maintained in such conditions for 1 h. The reaction mixture progressively changed from turbid to slightly yellow and transparent.

Next, the flask was left cooling to room temperature, and nanoparticles were purified by centrifugation (4500 rpm at 20 °C for 15 min) to remove any excess of reagents and solvents. The white pellet was washed once with of ethanol (30 mL) and once with THF/ethanol (5/30 mL) and recollected by centrifugation. Upconverting nanoparticles were dried at room temperature overnight. Typically, ca. 700 mg of $NaGdF_4$:Yb,Er nanoparticles are obtained employing the reaction conditions described.

3.3. Functionalization of UCNPs/Ligand Exchange

Typically, 30 mg of UCNPs were dissolved in 2 mL of chloroform and 100 mg of organic ligand were dissolved in 10 mL of MES buffer (20 mM, pH 6). The solution mixture was left stirring overnight and the aqueous phase (supernatant) lyophilized. The solid obtained was then washed three times with 3 mL of ethanol (or a mixture of H_2O:EtOH 1:3 in the case of etidronate) and left drying at room temperature for two days.

3.4. Relaxivity Measurements and HA Binding

All relaxivity results were obtained in triplicates by measuring at 37 °C 300 µL of aqueous solution containing different concentrations of UCNPs on a 1.5 T MiniSpec MQ60 (Bruker, Madrid, Spain. The affinity of UCNPs for HA was evaluated by adding a fixed quantity of HA (insoluble) in every sample and were directly measured on the 1.5 T MiniSpec at two time points (2 and 24 h).

3.5. Labeling of UCNPs with ^{18}F

UCNPs with four different capping ligands (citrate, 3P, alendronate and etidronate) were labelled with fluorine-18 (positron emitter with $T_{1/2}$ = 109.7 min). In brief, $[^{18}F]F^-$ (ca. 440 MBq) dissolved in 150 µL of ^{18}O-enriched water (used for the production of ^{18}F by proton irradiation) was added to the suspension of UCNPs (1.5 mg in 10 µL of distilled water). The reaction mixture was incubated for 10 min at room temperature and centrifuged to remove the unreacted $[^{18}F]F^-$ from $[^{18}F]$UCNPs. The $[^{18}F]$UCNPs were washed three times with distilled water by centrifugation (5 min, 10,000 rpm). Finally, the pellet was suspended in 150 µL of physiological saline solution (NaCl 9.0 g per liter). The amount of radioactivity in the final suspension was determined using a dose calibrator (CRC-25R PET dose calibrator, Capintec Inc., Ramsey, NJ, USA) and the non-decay corrected radiochemical yield was calculated as the ratio between the final and the starting amounts of radioactivity. Estimated specific activity values were calculated as the ratio between the final amount of radioactivity and the starting mass amount of UCNPs.

The stability of the NPs in physiologic saline solution was determined. With that aim, NPs were incubated at 37 °C and fractions were withdrawn at pre-selected time points, diluted with purified water (1:20) and further analysed by radio-Thin Layer Chromatography (radio-TLC, silica and 9:1 saline/ethanol as eluent). Under experimental conditions, ^{18}F-labeled UCNPs remained at the seeding spot while free $[^{18}F]F^-$ eluted with the solvent front.

3.6. Gd^{3+} Determination from Ex Vivo Femur Bones

After the last imaging session (6.5 h post-administration, see Section 3.7.5) the animals were sacrificed and the bones harvested in order to conduct ex vivo analysis and determine the amount of Gd^{3+} by ICP-MS. We followed a protocol reported elsewhere [31]. Briefly, the femurs ($n = 2$ per UCNP type) treated with citrate, 3P, alendronate UCNPs were dissolved in 3 mL of HNO_3 overnight. An aliquot of this solution was then diluted 50 times to use it as ICP sample. A non-treated femur was also digested in HNO_3 overnight, to build a matrix for the calibration curve of Gd^{3+}.

3.7. Instrumentation

3.7.1. Transmission Electron Microscopy (TEM)

TEM was performed on a JEOL JEM-1400 PLUS-HC microscope (Peabody, MA, USA) operating at 120 kV. The TEM samples were prepared by dropping 3 μL sample solutions (0.1–1 mg/mL in THF or in water) onto a 400-mesh carbon coated copper grid (3 mm in diameter) followed by the evaporation of the solvent under vacuum. The nanoparticle sizes were estimated from over 150 nanoparticles.

3.7.2. Fourier Transform Infrared (FTIR)

FTIR spectra of $NaGdF_4$:Yb,Er UCNPs coated with oleic acid, citrate, EDTA, 3P, alendronate and etidronate were recorded on a Nicolet FTIR 6700 spectrometer (Thermo Fisher, Waltham, MA, USA) as KBr pellet.

3.7.3. Dynamic Light Scattering

The hydrodynamic radius of the different functionalized $NaGdF_4$:Yb,Er UCNPs was obtained measuring an aqueous solution containing the nanoparticles (0.5 mg/mL) with 173° scattering angle at 25 °C using a NanoSizer Malvern Nano-Zs (Malvern Panalytical, Malvern, UK).

3.7.4. MRI

T1 and T2 experiments were performed on a 1.5 T MiniSpec TD-NMR (Bruker) at 37 °C. Also, they were performed on a 7 Tesla Bruker Biospec 70/30 USR MRI system (Bruker Biospin GmbH, Ettlingen, Germany) at 37 °C, interfaced to an AVANCE III console. The BGA12-S imaging gradient (maximum gradient strength 400 mT/m switchable within 80 us and a 40 mm inner diameter quadrature volume resonator were used. R1 values were determined using a series of Spin echo measurements with increasing repetition time (TR) of 110.0, 130.0, 160.0, 190.0, 220.0, 280.0, 400.0, 600.0, 850.0, 1110.0, 1250.0, 1500.0, 1800.0, 2100.0, 2500.0, 3000.0, 4000.0 and 5000.0 ms, with an echo time (TE) of 10 ms. R2 values were measured using a multislice multiecho (MSME) spin echo sequence with a TR of 20,000 ms and TE ranging from 50 to 1600 ms, with an echo interval of 50 ms. All images were acquired with one average, 256 × 256 points and a resolution of 125 μm in plane, with a slice thickness of 2.0 mm. The images were fitted into Levenberg-Margardt method to calculate T1 and T2 values using Bruker's Paravision 5.1 software.

3.7.5. PET

Healthy, 10–12 weeks aged Sprague-Dawley rats (Harlan, Udine, Italy) were used to investigate the biodistribution of the 18F-labelled NPs. Animals were maintained and handled in accordance with the Guidelines for Accommodation and Care of Animals (European Convention for the Protection of Vertebrate Animals Used for Experimental and Other Scientific Purposes) and internal guidelines, and experimental procedures were approved by the Ethical Committee of CIC biomaGUNE and local authorities (AE-biomaGUNE-0216, 26 June 2016). Rats were acclimated to the housing facility at 22–24 °C and 40–60% of humidity under light/dark conditions for at least five days prior to experiments.

For imaging sessions, animals were anesthetized with 5% isoflurane in pure oxygen. Anaesthesia was maintained afterwards with 2–3% isoflurane in pure oxygen. ^{18}F-labelled NPs suspended in physiologic saline solution and prepared as previously described were appropriately diluted to a concentration of ca. 370 MBq/mL and 100 µL (containing ca. 37 MBq) were intravenously administered via one of the lateral tail veins.

After administration of the NPs, PET images were dynamically acquired on an eXplore Vista-CT camera (GE Healthcare, Las Rozas, Madrid, Spain) in four bed positions to cover the whole body of the animal (frames: 4×30 s, 4×1 min, 3×2 min, 4×4 min; total acquisition time = 28 min). Static imaging sessions (20 min each) were repeated at $t = 3$ and 6 h after administration of the labelled NPs. After each PET acquisition, a CT scan was performed for later attenuation correction during image reconstruction. Random and scatter corrections were also applied to the reconstructed images (2DOSEM iterative algorithm, four iterations). PET-CT images were co-registered and analysed using PMOD image processing tool. Volumes of interest (VOIs) were drawn in major organs on the CT images, translated to the PET images and the concentration of radioactivity in each organ was determined as a function of time. Injected dose and organ mass normalizations were finally applied to data to achieve percentage of injected dose per gram of tissue (%ID/g).

3.7.6. ICP-MS

After diluting the digested samples to a suitable concentration, Gd was determined by iCAP-Q ICPMS (Thermo Scientific, Bremen, Germany) equipped with an autosampler ASX-520 (Cetac Technologies Inc., Omaha, NE, USA). The instrumental operating conditions for the determination of the elements are summarized in Table 2. The quantification was based on at least five point external calibrations and Qtegra™ v2.6 (Thermo Scientific) software was used for the analysis.

Prior to analyses, instrument settings were optimized by infusion of TUNE B iCAP Q solution (Thermo Scientific).

The analyses were carried out in KED mode using He as collision gas in order to reduce possible polyatomic interferences. All the samples were measured in triplicate using Iridium 193 as internal standard. Blank samples containing 2% HNO_3/0.5% HCl were infused before the calibration curve and wash samples (2% HNO_3/0.5% HCl) were also infused and measured after calibration and between samples. Calibration samples were prepared at 25, 10, 5, 1, 0.1 µg/L in 2% HNO_3/0.5% HCl from a certified stock solution from Inorganic Ventures (Lakewood, NJ, USA).

Table 2. ICP-MS operating conditions.

Parameter	Condition
Nebulizer gas flow	1.075 L min^{-1}
Spray chamber	2.70 °C
Extraction Lens	−111.3 V
CCT focus lens	0.6 V
Plasma power	1550 w
Cooling gas flow	14 L min^{-1}
Auxiliary gas flow	0.8 L min^{-1}
Colision gas flow	4.733 L min^{-1}
Pole bias	−18 V
CCT bias	−21 V
Wash time	30 s
Uptake time	50 s

4. Conclusions

Our study shows that phosphonates are suitable ligands for functionalizing the surface of UCNPs in pursues of multimodal imaging agents. They confer promising MRI properties to these nanomaterials in term of r_1 relaxivity and allow efficient labelling with [18]F$^-$ ions. These simple ligands favour the affinity of UCNPs for HA. The integration of in vivo PET imaging and ex vivo ICP-MS analysis demonstrated that indeed alendronate and 3P UCNPs have the capacity to accumulate in bones, although ^{18}F-labeling is not an appropriate strategy to monitor in vivo such accumulation, as leakage of the radiolabel can mask the accumulation of UCNPs in bones.

Supplementary Materials: The following are available online at http://www.mdpi.com/2304-6740/7/5/60/s1. Table S1: Gd^{3+} determination per gram of functionalized UCNPs by ICP-MS; Table S2: Relaxivity r_1 and r_2 measurements and r_1/r_2 ratios for phosphonate-functionalized UCNPs obtained at 37 °C and 1.5 T; Table S3: Dynamic light scattering data for UCNPs in water and MES buffer (pH 6). Figure S1: TEM images (n = 150) of $NaGdF_4$:Yb,Er capped with oleic acid and dissolved in THF; Figure S2: Emission spectra of $NaYF_4$ Yb:Er UCNPs capped with oleic acid irradiated at 980 nm with 4.0 W; Figure S3: FT-IR spectra of $NaGdF_4$:Yb,Er UCNPs capped with different ligand coatings and of their corresponding free ligand; Figures S4–S9: Relaxivity values for the three UCNPs in different solvents; Figure S10: T1 weighted MRI image of etidronate UCNPs (7 T, 37 °C) in H_2O at different concentrations; Figure S11: Dymanic Light Scatering of UCNPs coated with etidronate, alendronate, 3P and citrate measured in H_2O and MES buffer (pH 6); Figure S12: Relaxivity r_1 values for 3P and alendronate UCNPs in the presence and in the absence of hydroxyapatite HA (12.5 mg/mL); Figure S13: TEM images of 3P and alendronate UCNPs in aqueous solution measured at different time points (t = 0 and 1 day); Figure S14: Diluted solution of ^{18}F-labeled UCNPs (1:20) analysed by radio-TLC at t = 5 h of incubation in physiologic saline solution; Figure S15: Concentration of radioactivity in different organs at different time points for the different labeled NPs, as determined from PET imaging.

Author Contributions: Conceptualization, S.A.-d.C., E.R., J.L. and L.S.; methodology, S.A.-d.C., E.R., A.L.F., U.C., Z.B., D.O., V.G.-V., D.P.; formal analysis, S.A.-d.C., U.C., Z.B., D.O., V.G.-V., D.P.; investigation, S.A.-d.C., E.R., A.L.F., U.C., Z.B., D.O.; data curation, V.G.-V., D.P.; writing—original draft preparation, S.A.-d.C., J.L. and L.S.; writing—review and editing, S.A.-d.C., J.L. and L.S.; supervision, L.S. and J.L.; funding acquisition, L.S.

Funding: This research was funded by the Spanish Ministry of Economy and Competitiveness (grant CTQ2012-39315, CTQ2016-80844-R and BES-2013-065642) and by the MC CIG fellowship UCnanomat4iPACT (grant no. 321791).

Conflicts of Interest: The authors declare no conflict of interest.

References

1. Alonso-de Castro, S.; Ruggiero, E.; Ruiz-de-Angulo, A.; Rezabal, E.; Mareque-Rivas, J.C.; Lopez, X.; López-Gallego, F.; Salassa, L. Riboflavin as a bioorthogonal photocatalyst for the activation of a Pt^{IV} prodrug. *Chem. Sci.* **2017**, *8*, 4619–4625. [CrossRef]

2. Alonso-de Castro, S.; Cortajarena, A.L.; López-Gallego, F.; Salassa, L. Bioorthogonal Catalytic Activation of Platinum and Ruthenium Anticancer Complexes by FAD and Flavoproteins. *Angew. Chem. Int. Ed.* **2018**, *57*, 3143–3147. [CrossRef] [PubMed]

3. Alonso-de Castro, S.; Terenzi, A.; Hager, S.; Englinger, B.; Faraone, A.; Martínez, J.C.; Galanski, M.; Keppler, B.K.; Berger, W.; Salassa, L. Biological activity of Pt^{IV} prodrugs triggered by riboflavin-mediated bioorthogonal photocatalysis. *Sci. Rep.* **2018**, *8*, 17198. [CrossRef] [PubMed]

4. Ruggiero, E.; Garino, C.; Mareque-Rivas, J.C.; Habtemariam, A.; Salassa, L. Upconverting Nanoparticles Prompt Remote Near-Infrared Photoactivation of Ru(II)–Arene Complexes. *Chem. Eur. J.* **2016**, *22*, 2801–2811. [CrossRef]

5. Ruggiero, E.; Hernández-Gil, J.; Mareque-Rivas, J.C.; Salassa, L. Near infrared activation of an anticancer Pt^{IV} complex by Tm-doped upconversion nanoparticles. *Chem. Commun.* **2015**, *51*, 2091–2094. [CrossRef]

6. Ruggiero, E.; Habtemariam, A.; Yate, L.; Mareque-Rivas, J.C.; Salassa, L. Near infrared photolysis of a Ru polypyridyl complex by upconverting nanoparticles. *Chem. Commun.* **2014**, *50*, 1715–1718. [CrossRef]

7. Ruggiero, E.; Alonso-De Castro, S.; Habtemariam, A.; Salassa, L. Upconverting nanoparticles for the near infrared photoactivation of transition metal complexes: New opportunities and challenges in medicinal inorganic photochemistry. *Dalton Trans.* **2016**, *45*, 13012–13020. [CrossRef]

8. Hemmer, E.; Acosta-Mora, P.; Méndez-Ramos, J.; Fischer, S. Optical nanoprobes for biomedical applications: Shining a light on upconverting and near-infrared emitting nanoparticles for imaging, thermal sensing, and photodynamic therapy. *J. Mater. Chem. B* **2017**, *5*, 4365–4392. [CrossRef]

9. Zhou, J.; Liu, Z.; Li, F. Upconversion nanophosphors for small-animal imaging. *Chem. Soc. Rev.* **2012**, *41*, 1323–1349. [CrossRef] [PubMed]

10. Cao, T.; Yang, Y.; Sun, Y.; Wu, Y.; Gao, Y.; Feng, W.; Li, F. Biodistribution of sub-10nm PEG-modified radioactive/upconversion nanoparticles. *Biomaterials* **2013**, *34*, 7127–7134. [CrossRef] [PubMed]

11. Wise-Milestone, L.; Akens, M.K.; Lo, V.C.K.; Yee, A.J.; Wilson, B.C.; Whyne, C.M. Local treatment of mixed osteolytic/osteoblastic spinal metastases: Is photodynamic therapy effective? *Breast Cancer Res. Treat.* **2012**, *133*, 899–908. [CrossRef]

12. Lo, V.C.K.; Akens, M.K.; Wise-Milestone, L.; Yee, A.J.M.; Wilson, B.C.; Whyne, C.M. The benefits of photodynamic therapy on vertebral bone are maintained and enhanced by combination treatment with bisphosphonates and radiation therapy. *J. Orthop. Res.* **2013**, *31*, 1398–1405. [CrossRef]

13. Panahifar, A.; Mahmoudi, M.; Doschak, M.R. Synthesis and in Vitro Evaluation of Bone-Seeking Superparamagnetic Iron Oxide Nanoparticles as Contrast Agents for Imaging Bone Metabolic Activity. *ACS Appl. Mater. Interfaces* **2013**, *5*, 5219–5226. [CrossRef] [PubMed]

14. Zhang, S.; Gangal, G.; Uludağ, H. 'Magic bullets' for bone diseases: Progress in rational design of bone-seeking medicinal agents. *Chem. Soc. Rev.* **2007**, *36*, 507–531. [CrossRef]

15. Li, R.; Ji, Z.; Dong, J.; Chang, C.H.; Wang, X.; Sun, B.; Wang, M.; Liao, Y.-P.; Zink, J.I.; Nel, A.E.; et al. Enhancing the Imaging and Biosafety of Upconversion Nanoparticles through Phosphonate Coating. *ACS Nano* **2015**, *9*, 3293–3306. [CrossRef] [PubMed]

16. Boyer, J.-C.; Carling, C.-J.; Chua, S.Y.; Wilson, D.; Johnsen, B.; Baillie, D.; Branda, N.R. Photomodulation of Fluorescent Upconverting Nanoparticle Markers in Live Organisms by Using Molecular Switches. *Chem. Eur. J.* **2012**, *18*, 3122–3126. [CrossRef] [PubMed]

17. Johnson, N.J.J.; Oakden, W.; Stanisz, G.J.; Scott Prosser, R.; van Veggel, F.C.J.M. Size-Tunable, Ultrasmall $NaGdF_4$ Nanoparticles: Insights into Their T_1 MRI Contrast Enhancement. *Chem. Mater.* **2011**, *23*, 3714–3722. [CrossRef]

18. Wang, Z.; Carniato, F.; Xie, Y.; Huang, Y.; Li, Y.; He, S.; Zang, N.; Rinehart, J.D.; Botta, M.; Gianneschi, N.C. High Relaxivity Gadolinium-Polydopamine Nanoparticles. *Small* **2017**, *13*, 1701830. [CrossRef]

19. Jacques, V.; Dumas, S.; Sun, W.-C.; Troughton, J.S.; Greenfield, M.T.; Caravan, P. High-relaxivity magnetic resonance imaging contrast agents. Part 2. Optimization of inner- and second-sphere relaxivity. *Investig. Radiol.* **2010**, *45*, 613–624. [CrossRef]

20. Wen, S.; Zhou, J.; Zheng, K.; Bednarkiewicz, A.; Liu, X.; Jin, D. Advances in highly doped upconversion nanoparticles. *Nat. Commun.* **2018**, *9*, 2415. [CrossRef]

21. Sedlmeier, A.; Gorris, H.H. Surface modification and characterization of photon-upconverting nanoparticles for bioanalytical applications. *Chem. Soc. Rev.* **2015**, *44*, 1526–1560. [CrossRef] [PubMed]

22. Rohrer, M.; Bauer, H.; Mintorovitch, J. Comparison of Magnetic Properties of MRI Contrast Media Solutions at Different Magnetic Field Strengths. *Investig. Radiol.* **2005**, *40*, 715–724. [CrossRef]

23. Pintaske, J.; Martirosian, P.; Graf, H.; Erb, G.; Lodemann, K.-P.; Claussen, C.D.; Schick, F. Relaxivity of Gadopentetate Dimeglumine (Magnevist), Gadobutrol (Gadovist), and Gadobenate Dimeglumine (MultiHance) in Human Blood Plasma at 0.2, 1.5, and 3 Tesla. *Investig. Radiol.* **2006**, *41*, 213–221. [CrossRef]

24. Koziorowski, J.; Stanciu, A.; Gomez-Vallejo, V.; Llop, J. Radiolabeled Nanoparticles for Cancer Diagnosis and Therapy. *Anticancer Agents Med. Chem.* **2017**, *17*, 333–354. [CrossRef]

25. Zhou, J.; Yu, M.; Sun, Y.; Zhang, X.; Zhu, X.; Wu, Z.; Wu, D.; Li, F. Fluorine-18-labeled $Gd^{3+}/Yb^{3+}/Er^{3+}$ co-doped $NaYF_4$ nanophosphors for multimodality PET/MR/UCL imaging. *Biomaterials* **2011**, *32*, 1148–1156. [CrossRef] [PubMed]

26. Sun, Y.; Yu, M.; Liang, S.; Zhang, Y.; Li, C.; Mou, T.; Yang, W.; Zhang, X.; Li, B.; Huang, C.; et al. Fluorine-18 labeled rare-earth nanoparticles for positron emission tomography (PET) imaging of sentinel lymph node. *Biomaterials* **2011**, *32*, 2999–3007. [CrossRef]

27. Liu, Q.; Sun, Y.; Li, C.; Zhou, J.; Li, C.; Yang, T.; Zhang, X.; Yi, T.; Wu, D.; Li, F. 18F-labeled magnetic-upconversion nanophosphors via rare-earth cation-assisted ligand assembly. *ACS Nano* **2011**, *5*, 3146–3157. [CrossRef]

28. Li, Z.; Zhang, Y.; Han, G. Lanthanide-Doped Upconversion Nanoparticles for Imaging-Guided Drug Delivery and Therapy. In *Advances in Nanotheranostics I: Design and Fabrication of Theranosic Nanoparticles*; Dai, Z., Ed.; Springer Series in Biomaterials Science and Engineering; Springer: Berlin/Heidelberg, Germany, 2016; pp. 139–164. ISBN 978-3-662-48544-6.

29. Wang, J.; Zhang, L.; Peng, F.; Shi, X.; Leong, D.T. Targeting Endothelial Cell Junctions with Negatively Charged Gold Nanoparticles. *Chem. Mater.* **2018**, *30*, 3759–3767. [CrossRef]
30. Tan, Y.L.; Leong, D.T.; Ho, H.K. Inorganic Nanomaterials as Highly Efficient Inhibitors of Cellular Hepatic Fibrosis. *ACS Appl. Mater. Interfaces* **2018**, *10*, 31938–31946. [CrossRef]
31. Darrah, T.H.; Prutsman-Pfeiffer, J.J.; Poreda, R.J.; Campbell, M.E.; Hauschka, P.V.; Hannigan, R.E. Incorporation of excess gadolinium into human bone from medical contrast agents. *Metallomics* **2009**, *1*, 479–488. [CrossRef] [PubMed]

MDPI

St. Alban-Anlage 66

4052 Basel

Switzerland

Tel. +41 61 683 77 34

Fax +41 61 302 89 18

www.mdpi.com

Inorganics Editorial Office

E-mail: inorganics@mdpi.com

www.mdpi.com/journal/inorganics